Biodiversity, Distribution and Conservation of Plants and Fungi; Effects of Global Warming and Environmental Stress

Biodiversity, Distribution and Conservation of Plants and Fungi; Effects of Global Warming and Environmental Stress

Editors

Anush Kosakyan
Rodica Catana
Alona Yu. Biketova

MDPI • Basel • Beijing • Wuhan • Barcelona • Belgrade • Manchester • Tokyo • Cluj • Tianjin

Editors

Anush Kosakyan
Biology Centre, Institute of
Parasitology
Czech Academy of Sciences
Ceske Budejovice
Czech Republic

Rodica Catana
Developmental Biology
Department
Institute of Biology Bucharest
of Romanian Academy
Bucharest
Romania

Alona Yu. Biketova
Jodrell Laboratory
Royal Botanic Gardens, Kew
Richmond
United Kingdom

Editorial Office
MDPI
St. Alban-Anlage 66
4052 Basel, Switzerland

This is a reprint of articles from the Special Issue published online in the open access journal *Journal of Fungi* (ISSN 2309-608X) (available at: www.mdpi.com/journal/jof/special_issues/plants_fungi_joint).

For citation purposes, cite each article independently as indicated on the article page online and as indicated below:

LastName, A.A.; LastName, B.B.; LastName, C.C. Article Title. *Journal Name* **Year**, *Volume Number*, Page Range.

ISBN 978-3-0365-4406-9 (Hbk)
ISBN 978-3-0365-4405-2 (PDF)

Cover image courtesy of Alona Yu. Biketova

Contents

About the Editors

Anush Kosakyan

Anush Kosakyan's research interests span three different (yet very interconnected) topics: mycology, protistology and microbial parasitology (myxozoan parasites). She gained her first PhD from University of Haifa, Israel, focusing on the taxonomy and systematics of *Agaricomycota*, and her second PhD from University of Neuchatel, Switzerland, focusing on testate amoeba diversity, evolution and ecology. Currently, she is a postdoctoral researcher at Institute of Parasitology at the Biology Center of Czech Academy of Sciences (Czech Republic), where she studies gene expression and host–parasite interactions in microbial eukaryotes (*Myxozoa*).

Rodica Catana

Catana Rodica Daniela is a senior researcher at the Institute of Biology Bucharest of Romanian Academy. Between 2004 and 2005, she enrolled in the post-university course "Horticulture Genetics and Biotechnology", at Mediterranean Agronomic Institute of Chania, Crete. In 2009, she obtained a Ph.D. in the field of biology from the Romanian Academy. She has taken part in numerous research projects in the field of plant conservation, plant micropropagation, in vitro techniques, secondary metabolites, pollution, and urban lakes. She has been the author and co-author of more than 45 peer-reviewed publications and 2 book chapters. She acts as a reviewer for different journals.

Alona Yu. Biketova

Alona Yu. Biketova's research interests cover mycology, molecular and synthetic biology. She gained her PhD from the University of Haifa (Israel) focusing on the taxonomy, distribution, molecular phylogeny of *Boletales* in Israel, with special attention paid to the former *Boletus* sect. *Luridi* as well as xerocomoid genera *Hortiboletus* and *Xerocomellus*. Her current occupation is Fungal DNA Curator in Royal Botanic Gardens, Kew (UK). Most of her ongoing research projects are focused on the systematics, diversity, phylogeny, evolution, and species conservation of *Agaricomycotina*, especially the family *Boletaceae*.

Preface to "Biodiversity, Distribution and Conservation of Plants and Fungi; Effects of Global Warming and Environmental Stress"

The estimation of global biodiversity and its conservation is an old, but still unresolved, concern in biology. On the one hand, the number of described species is constantly increasing, especially with the accumulation of modern morphological and molecular data; on the other hand, the existence of many species is threatened due to environmental and anthropogenic pressures.

Plants and fungi are essential components of the ecosystem and are widely used by humanity. However, fungi are much understudied in many aspects in comparison with animals and plants. They are also poorly represented in national and international red lists; i.e., there are only 550 species of fungi in the IUCN Red List, while Plantae have more than 58,300 and Animalia—more than 83,600 species. This is due to a significant lack of knowledge in fungal biodiversity and distribution, especially in tropical areas. There are no comprehensive studies on how global climate changes affect fungi and their interactions with plants.

The goal of this Special Issue is to highlight the research concerning subjects related to biodiversity and the conservation of fungi and plants, plant–fungal interactions, the effects of global warming, pollution, parasites, and other abiotic factors affecting their conservation, as well as phylogenetic, genomic, and transcriptomic studies with the direction of biodiversity estimation and gene expression studies related to environmental stressors.

Anush Kosakyan, Rodica Catana, and Alona Yu. Biketova
Editors

Journal of
Fungi

Editorial

Biodiversity, Distribution, and Conservation of Plants and Fungi: Effects of Global Warming and Environmental Stress

Alona Yu. Biketova [1,2,*], Rodica Catana [3] and Anush Kosakyan [4]

1   Jodrell Laboratory, Royal Botanic Gardens, Kew, Richmond TW9 3DS, UK
2   Mycological Society of Israel, P.O. Box 164, Pardesiya 42815, Israel
3   Institute of Biology Bucharest of Romanian Academy, 296 Splaiul Independentei, 060031 Bucharest, Romania; rodica.catana@ibiol.ro
4   Institute of Parasitology, Biology Centre, Czech Academy of Sciences, 37005 Ceske Budejovice, Czech Republic; anna.kosakyan@gmail.com
*   Correspondence: alyona.biketova@gmail.com

Citation: Biketova, A.Yu.; Catana, R.; Kosakyan, A. Biodiversity, Distribution, and Conservation of Plants and Fungi: Effects of Global Warming and Environmental Stress. *J. Fungi* 2022, *8*, 441. https://doi.org/10.3390/jof8050441

Received: 7 April 2022
Accepted: 20 April 2022
Published: 24 April 2022

Publisher's Note: MDPI stays neutral with regard to jurisdictional claims in published maps and institutional affiliations.

The estimation of global biodiversity and its conservation is an old, but still unresolved, concern in biology. On the one hand, the number of described species is constantly increasing, especially with the accumulation of modern morphological and molecular data; on the other hand, the existence of many species is threatened due to environmental and anthropogenic pressures.

Most of the articles in this Special Issue are devoted to the diversity, taxonomy and molecular phylogeny of fungi [1–6]. Ten new species and three new taxonomic combinations are described for science [1–3,5,6]. It is noteworthy that half of these manuscripts are devoted to three genera of the family *Boletaceae*—one of the most remarkable and most vulnerable groups of fungi to the destruction of ecosystems. Novel comprehensive phylogenetic and taxonomic analyses of *Leccinum, Hemileccinum, Exsudoporus, Amoenoboletus*, and allied genera, together with descriptions of eight new species and two taxonomic combinations, and the typification of *Exsudoporus floridanus* indicate that there are many unresolved issues, even related to such a relatively well-studied group of macroscopic fungi as *Boletaceae* [2,3,6]. Mešić et al. (2021) described a new agaricoid species *Inocybe brijunica*, growing in the Mediterranean Biogeographical Region, one of the most prominent global climate change hot spots [1]. Two other articles presented investigations of the biodiversity of the *Ascomycota* genera *Calonectria* and *Wickerhamomyces* [4,5].

Another group of manuscripts was dedicated to the ecological, physiological, and applied aspects of mycobiota [7–11]. Jabborova et al. (2021) studied the interactions between biochar and arbuscular mycorrhizal fungi (AMF) and spinach. It was shown that these fungi can promote plant growth, improve soil properties, and maintain microbial activity [7]. A review by Boorboori and Zhang (2022) provided comprehensive up-to-date information on the use of AMF in the phytoremediation of arsenic, cadmium, lead, and chromium [8]. Another study was devoted to the composition of major, trace, and rare-earth elements in 15 different species of wild edible mushrooms. The data obtained did not indicate a significant exposure to anthropogenic influences, regardless of the sampling location. While the contents of major elements seem to be influenced by species–specific affinities, this is not true for trace elements, whose contents probably reflect the geochemical characteristics of the sampling site [9].

Adamo et al. (2021) conducted a metabarcoding analysis of ectomycorrhizal fungi in five different Mediterranean pine forests (*Pinus nigra, P. halepensis, P. sylvestris*, and two mixed) and concluded that fungal communities did not differ in phylogenetic composition, structure, or phylogenetic diversity among tree hosts [10]. Mihai et al. (2022) offered bio-friendly solutions to reduce the waste of coconut coir, pine sawdust, and paper (as some of the main pollutants in Ecuador) by using them as suitable growth substrates for the edible fungus *Pleurotus ostreatus*. The results showed that all waste products represent desirable substrates for fungal growth, with an emphasis on coconut coir waste, whose usage increased the desirable characteristics of the fungi [11].

The editors of the Special Issue express their gratitude to all the authors who contributed to this issue as well as to MDPI's staff for their valuable support and prompt decisions.

Author Contributions: Conceptualization, A.Yu.B., R.C. and A.K.; writing—original draft preparation, A.Yu.B. and R.C.; writing—review and editing, R.C., A.K. and A.Yu.B. All authors have read and agreed to the published version of the manuscript.

Funding: This research received no external funding.

Conflicts of Interest: The authors declare no conflict of interest.

References

1. Mešić, A.; Haelewaters, D.; Tkalčec, Z.; Liu, J.; Kušan, I.; Aime, M.C.; Pošta, A. *Inocybe brijunica* sp. nov., a New Ectomycorrhizal Fungus from Mediterranean Croatia Revealed by Morphology and Multilocus Phylogenetic Analysis. *J. Fungi* **2021**, *7*, 199. [CrossRef]
2. Meng, X.; Wang, G.-S.; Wu, G.; Wang, P.-M.; Yang, Z.L.; Li, Y.-C. The Genus *Leccinum* (*Boletaceae*, *Boletales*) from China Based on Morphological and Molecular Data. *J. Fungi* **2021**, *7*, 732. [CrossRef] [PubMed]
3. Li, M.-X.; Wu, G.; Yang, Z.L. Four New Species of *Hemileccinum* (*Xerocomoideae*, *Boletaceae*) from Southwestern China. *J. Fungi* **2021**, *7*, 823. [CrossRef] [PubMed]
4. Liu, L.; Wu, W.; Chen, S. Species Diversity and Distribution Characteristics of *Calonectria* in Five Soil Layers in a Eucalyptus Plantation. *J. Fungi* **2021**, *7*, 857. [CrossRef]
5. Nundaeng, S.; Suwannarach, N.; Limtong, S.; Khuna, S.; Kumla, J.; Lumyong, S. An Updated Global Species Diversity and Phylogeny in the Genus *Wickerhamomyces* with Addition of Two New Species from Thailand. *J. Fungi* **2021**, *7*, 957. [CrossRef] [PubMed]
6. Biketova, A.Y.; Gelardi, M.; Smith, M.E.; Simonini, G.; Healy, R.A.; Taneyama, Y.; Vasquez, G.; Kovács, Á.; Nagy, L.G.; Wasser, S.P.; et al. Reappraisal of the Genus *Exsudoporus* (*Boletaceae*) Worldwide Based on Multi-Gene Phylogeny, Morphology and Biogeography, and Insights on *Amoenoboletus*. *J. Fungi* **2022**, *8*, 101. [CrossRef]
7. Jabborova, D.; Annapurna, K.; Paul, S.; Kumar, S.; Saad, H.A.; Desouky, S.; Ibrahim, M.F.M.; Elkelish, A. Beneficial Features of Biochar and Arbuscular Mycorrhiza for Improving Spinach Plant Growth, Root Morphological Traits, Physiological Properties, and Soil Enzymatic Activities. *J. Fungi* **2021**, *7*, 571. [CrossRef]
8. Boorboori, M.R.; Zhang, H.-Y. Arbuscular Mycorrhizal Fungi Are an Influential Factor in Improving the Phytoremediation of Arsenic, Cadmium, Lead, and Chromium. *J. Fungi* **2022**, *8*, 176. [CrossRef] [PubMed]
9. Ivanić, M.; Furdek Turk, M.; Tkalčec, Z.; Fiket, Ž.; Mešić, A. Distribution and Origin of Major, Trace and Rare Earth Elements in Wild Edible Mushrooms: Urban vs. Forest Areas. *J. Fungi* **2021**, *7*, 1068. [CrossRef] [PubMed]
10. Adamo, I.; Castaño, C.; Bonet, J.A.; Colinas, C.; Martínez de Aragón, J.; Alday, J.G. Lack of Phylogenetic Differences in Ectomycorrhizal Fungi among Distinct Mediterranean Pine Forest Habitats. *J. Fungi* **2021**, *7*, 793. [CrossRef] [PubMed]
11. Mihai, R.A.; Melo Heras, E.J.; Florescu, L.I.; Catana, R.D. The Edible Gray Oyster Fungi *Pleurotus ostreatus* (Jacq. ex Fr.) P. Kumm a Potent Waste Consumer, a Biofriendly Species with Antioxidant Activity Depending on the Growth Substrate. *J. Fungi* **2022**, *8*, 274. [CrossRef] [PubMed]

Journal of
Fungi

Review

Arbuscular Mycorrhizal Fungi Are an Influential Factor in Improving the Phytoremediation of Arsenic, Cadmium, Lead, and Chromium

Mohammad Reza Boorboori and Hai-Yang Zhang *

College of Environment and Surveying and Mapping Engineering, Suzhou University, Suzhou 234000, China; m.boorboori@yahoo.com
* Correspondence: szxyhczp@ahszu.edu.cn

Abstract: The increasing expansion of mines, factories, and agricultural lands has caused many changes and pollution in soils and water of several parts of the world. In recent years, metal(loid)s are one of the most dangerous environmental pollutants, which directly and indirectly enters the food cycle of humans and animals, resulting in irreparable damage to their health and even causing their death. One of the most important missions of ecologists and environmental scientists is to find suitable solutions to reduce metal(loid)s pollution and prevent their spread and penetration in soil and groundwater. In recent years, phytoremediation was considered a cheap and effective solution to reducing metal(loid)s pollution in soil and water. Additionally, the effect of soil microorganisms on increasing phytoremediation was given special attention; therefore, this study attempted to investigate the role of arbuscular mycorrhizal fungus in the phytoremediation system and in reducing contamination by some metal(loid)s in order to put a straightforward path in front of other researchers.

Keywords: mycorrhizal fungi; metal(loid)s pollution; soil; plants

Citation: Boorboori, M.R.; Zhang, H.-Y. Arbuscular Mycorrhizal Fungi Are an Influential Factor in Improving the Phytoremediation of Arsenic, Cadmium, Lead, and Chromium. *J. Fungi* 2022, 8, 176. https://doi.org/10.3390/jof8020176

Academic Editors: Anush Kosakyan, Rodica Catana and Alona Biketova

Received: 19 January 2022
Accepted: 10 February 2022
Published: 12 February 2022

Publisher's Note: MDPI stays neutral with regard to jurisdictional claims in published maps and institutional affiliations.

1. Introduction

1.1. Metal(loid)s Pollution and Phytoremediation

Soil is the basis of life and the most valuable ecosystem globally, which plays a crucial role in producing food and filtering air and water, and is a good platform for building our homes and cities [1,2]. Due to the rapid urbanization and industrialization of the world community, agricultural land is declining day by day while the demand for food is increasing [3]. Since chemical and physical disorders easily damage soils, thousands of years of human activity have left many contaminated soils worldwide [1,2]. One of the most critical soil pollutions is metal(loid)s pollution, which has become a global problem and poses a severe threat to human health and the environment [4]. Various activities such as mining, use of sewage sludge in agricultural lands, excessive use of pesticides and fertilizers containing metal(loid)s, smelting of metals, and irrational emission of waste in industrial activities increase the concentration of metal(loid)s in soil and surface water, and also affect concentrations underground [5,6]. Although metal(loid)s of lithogenic origin are found naturally, and in different concentrations in the soil [7], arsenic, chromium, lead, cadmium, copper, and zinc are among the most common metal(loid)s pollutants commonly found near industrial, mining, and agricultural sites [8,9].

The constant increase of metal(loid)s in the soil, and their non-degradability causes changes in biogeochemical cycles (including the impact on soil microorganisms and enzymatic activities) and ultimately causes imbalances in ecosystems [10,11]. Since metal(loid)s react chemically in different environments, this increases their mobility and bioavailability in the soil [12]; therefore, their uptake by plants is increased and thus causes the accumulation of metal(loid)s in plant seeds [13]. The accumulation of metal(loid)s in plants and contaminated drinking water causes them to enter the food chain of humans and animals and causes severe problems for their health [13,14] (Table 1). Therefore, there is an urgent need for the permanent removal of metal(loid)s or at least their long-term immobilization in soils and water in the present era [15].

Table 1. Range of some metal(loid)s in terrestrial plants and regulatory standards for them in food and drinking water in different countries.

Metal(loid)s	Content Measured in Different Plants (μg/g DW)	WHO * (mg/kg)	Canada (in Row Herbal Materials) (mg/kg)	China (Herbal Material) (mg/kg)	India	
					Food (mg/kg)	Water (mg/L)
Arsenic	0.02–7	Nil	5	2	1.1	0.05
Cadmium	0.1–2.4	0.3	0.3	1	1.5	0.01
Lead	1–13	10	10	10	2.5	0.1
Chromium	0.2–1	Nil	2	Nil	20	0.05

Nomenclature is as proposed by Gjorgieva Ackova D [16]. * World Health Organization.

Increasing soil contamination with metal(loid)s has prompted humans to think of ways to clear and restore these marginal, toxic, and contaminated soils to their reuse cycle [3]. Various techniques that are developed for this purpose include drilling contaminants, adding chemical regenerators, pumping contaminants, and physical stabilization by adding non-toxic materials [3,17]. Other methods used to remediate metal(loid)s contaminated soils include improving the soil physical and chemical traits by using mineral tailings, such as steel slag, zeolite, limestone, and fly ash [18]. It should be noted that steel slag is an alkali product that is a remnant of the steel industry and is composed of compounds of calcium, silicon, phosphorus, etc. [19], and due to its cheapness and availability, it can be used as a modifier of soils contaminated with metal(loid)s or fertilizers in agriculture [18,19].

Many methods of remediating soils contaminated with metal(loid)s are expensive and, at the same time, reduce the ability of soil to be reused in food production [20]. Recently, carbon-based materials to remediate contaminated soils have received more attention because it is more environmentally friendly and cheaper [21]. Biochar is one of these carbon-based materials produced by charring animal and plant biomass at a temperature of 300 to 600 °C under anaerobic conditions [22]. Biochar has alkaline properties, is stable, has a large surface area, high aromatic molecular structure, and can bind cations and prevent the migration and bioavailability of metal(loid)s in soil [23,24]. The addition of biochar to agricultural soils improves soil physical and chemical properties such as cation exchange capacity, soil pH, mineral retention, and adsorption capacity and positively affects plants growth [24]. Various studies showed that adding biochar to soils contaminated with metal(loid)s improves plant biomass and antioxidant enzymes and thus reduces the uptake of metal(loid)s by plants [25,26].

As mentioned, when soil ecosystems are contaminated with metal(loid)s, intervention is necessary to improve soil conditions. It is usually possible to stabilize metal(loid)s in the soil in the short term, but in the long run, it is costly [27]. Phytoremediation is one of the best ways to stabilize metal(loid)s in technosols in the long run [28], and it is the usage of plants to remove or move soil contaminants, in which plants are used to absorb or immobilize metal(loid)s in contaminated soils [29]. Phytoremediation is a cheap, effective, on site, and in general, eco-friendly process that does not require advanced engineering work and has raised great hopes for clearing soil ecosystems of metal(loid)s contaminants [30,31]. This technique can be conducted on vast scales and prevents metal(loid)s from penetrating groundwater aquifers [32,33].

According to this method, plants with high accumulation power in cooperation with soil microorganisms reduce the accumulation of metal(loid)s in soils [32]; on the other hand, this technique increases soil organic matter and microbial activity as well as reduces erosion processes and ultimately improves ecosystems [34]. The classification of phytoremediation involves the destruction, immobilization, inhibition, extraction of metal(loid)s, or a combination of these processes [35]. Most research on phytoremediation is related to plant extraction and stabilization. In the process of plant extraction, plants remove metal(loid)s from the soils by concentrating them in their shoots, while in plant stabilization, metal(loid)s become immobile in the roots of the plants [36].

This technique uses plants that are resistant to metal(loid)s that can accumulate high levels of metal(loid)s in their roots and shoots [37]. A plant's tolerance to heavy metals depends on various factors such as the secretion of chelating agents in the rhizosphere, increased proline concentration, separation of metals, increased antioxidant enzymes activity, and the storage of metal(loid)s in the intracellular parts of the roots [38,39]. The resistance of plant species to metal(loid)s varies considerably between different plant families and genera, but in general vascular plants are slightly more tolerant than metal(loid)s [40]. Plants with high resistance to metal(loid)s are the best option for phytoremediation if they have an extensive root system and the ability to produce suitable aerial biomass (such as sunflower) [4,41]. On the other hand, nitrogen-fixing species that can survive in degraded environments are suitable examples of phytoremediation due to adding more organic matter and nutrients to the soil [42]. However, tree species have more potential to regenerate contaminated sites due to their ability to accumulate more metal(loid)s in plant tissues [43]. In aquatic ecosystems, wetland plants such as macrophytes are suitable tools for phytoremediation and reducing metal(loid)s pollution due to fast growth and high biomass production [44].

Plant stabilization is one of the essential methods in phytoremediation, which due to the accumulation of metal(loid)s in the roots of plants, prevents metal(loid)s spread and reduces their bioavailability in the soil [45]. Numerous studies showed that the cell walls of plants are the primary site of metal(loid)s immobilization [46]. Polysaccharides and proteins are the main components of the cell wall of plants, and the stabilization of metal(loid)s in the cell walls of plants can reduce metabolic damage of the cell [47] The polysaccharide portion of the plant cell wall is composed of cellulose, hemicellulose, and pectin, and plants can adjust their polysaccharide content to cope with the stress of metal(loid)s [46,47]. Peroxidases are the major proteins in cell walls involved in plant responses to stresses and cell wall dynamics [48]. Additionally, the growth of plant roots adds secretions such as low molecular weight organic acids to the soil, which cause the weathering of soil minerals and the release of metals in the soil [49]. With the growth of plant roots, pores are created in the soil that provides pathways for rapid intercurrent movement and facilitates leaching of soil solution and preferential migration [50].

One of the positive points of phytoremediation is the restoration of the structure of the soil microbial community, which is a significant advantage over other soil remediation methods [51], and the success of phytoremediation is highly dependent on the presence of soil microorganisms [52]. Soil microorganisms improve plant growth and bioremediate contaminated soils by separating or degrading metal(loid)s [53]. Other factors that increase the efficiency of phytoremediation include the addition of cattle manure or earthworms and other microorganisms to contaminated soils [54–56].

1.2. Arbuscular Mycorrhizal Fungi (AMF)

AMF is one of the most common probiotic microorganisms, known as the most widespread fungi that coexist with plant roots [35,57]. About 240 different species of AMF belong to the Glomeromycota subfamily, which are found in almost all natural and agricultural ecosystems and coexist with most plants [58,59], to the extent that they are reported to coexist with 90% of terrestrial plants [60]. In addition, it was reported that various AMF, such as *Funneliformis mosseae*, *Rhizophagus irregularis*, and *Claroideoglomus claroideum* exist in wetland habitats [61], and various studies have shown that AMF can colonize the roots of plants such as rice [62,63].

Different studies showed that AMF biodiversity varies significantly in different environments, and their presence in different ecosystems results from different factors, including soil type, host plant, agricultural practices, and environmental conditions [64]. It should be noted that host plants also have a significant impact on AMF growth and efficiency [65]. AMF hyphae divide, grow, and form an extensive, dense network of hyphae in the soil [66]; in addition, AMF hyphae can drastically alter the structure and chemical properties of aggregates by releasing compounds such as proteins and polysaccharides [67]. In addition to creating stable aggregates resistant to wind and water erosion by binding soil particles together, AMF can also affect soil microbial activity and communities [68,69].

According to findings, AMF creates a direct relationship between plant roots and the substrate [70] and gives closer access to plant roots to absorb nutrients and water, among other soil microorganisms, thus significantly improving plant growth [71]. Research has also shown that the effects of AMF vary depending on the type of soil substrate and plant species [72]; however, AMF causes the establishment and survival of host plants in various environments such as saline soils, alkaline soils, agricultural lands, mine tailings, soils contaminated with metal(loid)s, etc. [73]. As mentioned, AMF improves plant nutrition and plant–water relationships by creating efficient coexistence with plants and increasing the uptake level and volume of available soil for the root system, thereby increasing plant tolerance to environmental stresses such as root pathogen damage [10,74,75].

AMF grows inside the roots of plants and allows them to get their necessary nutrients through networks of extra-radical mycelium (ERM) that spread in the soil [76]. AMF can dissolve, activate, and ultimately absorb nutrients such as phosphorus, nitrogen, and zinc [76,77], and in return, it receives carbohydrates and lipids from plants [9]. Researchers have shown that AMF has the most significant effect on plants' phosphorus uptake [76], and it seems that up to 100% of the phosphorus required by plants is obtained through AMF phosphorus carriers [78]. AMF coexistence with plants causes phosphorus to be absorbed from a distance through fungal phosphate transporters such as GiPT [79,80] and effectively reaches plant roots through phosphate transporters (such as MtPT4) [81]. AMF also improves plant nutrition by increasing phosphate uptake from the soil because AMF increases the area of nutrient uptake through its hyphae [57], thus directly affecting plant growth [82], but on the other hand, increasing phosphorus uptake by the plant reduces this element in the soil solution [83]. Previous studies have shown that AMF symbiosis with plants has a positive effect on S uptake through regulating the expression of sulfate transporter genes (MtSULTR 1.1, MtSULTR2.1, MtSULTR 1.2, MtSULTR2.2, MtSULTR3.1, MtSULTR4.1, etc.) [81,84]; on the other hand, this coexistence directly affects S uptake through ERM activities [84].

As stated, AMF symbiosis with plants under stress conditions makes the plant more resistant to stress. Different studies have shown that exposed AMF spores to different stresses reduce ROS production with increasing antioxidant activities such as superoxide dismutases (SOD), glutathione (GSH), Vitamin B6, Vitamin C, and E [85,86]. AMF also activates plants antioxidant mechanisms such as glutathione peroxidase (GPx), ascorbate peroxidase (APX), catalyzes (CAT), and SOD [85]. In addition, AMF mycelium releases a particular glycoprotein called glomalin-related soil protein (GRSP), which helps improve soil conditions by forming complexes with heavy metals, helping to accumulate soil particles, increasing organic carbon content, and resulting in carbon sequestration [35,38,87]. Zhang et al. showed that AMF significantly increased the number of aggregates more prominent than 2 mm in contaminated soils [88]; a similar study also showed AMF-inoculation in Calcaric Regosol under drought stress, and proper irrigation increased the percentage of macro-particles larger than 5 mm [89].

Plants are the basis of phytoremediation, but various soil microorganisms such as AMF can significantly improve phytoremediation efficiency in different ecosystems [45,90], and this has shown that AMF naturally survives on high levels of metal(loid)s and helps plants withstand these contaminants [58]. The effect of AMF on soil improvement and phytoremediation depends on AMF species, metal(loid)s concentration, plant–metal tolerance, and metal(loid)s bioavailability [91]. AMF can also help the phytoremediation system by increasing immobilization, conversion, detoxification, and extraction of heavy metals [92]. Therefore, the use of plants that are more resistant to metal(loid)s and have more ability to accumulate more metal(loid)s, if they coexist with a suitable species of AMF (Glomeraceae family), have a significant effect on reducing pollution of heavy metal contaminated environments [85,93].

The growth of hyphae in environments contaminated with metal(loid)s is reduced [94], but extensive growth of AMF-mycelium by affecting the surface properties of aggregates increases the level of heavy metal uptake in phytoremediation systems, increasing the storage of metal(loid)s in soil and roots and thus reducing the transfer of metal(loid)s to aerial parts [95,96]. As mentioned, AMF has various solutions to reduce the bioavailability and biological absorption of metal(loid)s in the soil, including GRSP and antioxidant activities [86,97]. Phosphate, sulfhydryl, and other compounds in mycelium can prevent the transfer of metal(loid)s from the roots to the soil by converting their ions into forms such as oxalic acid extract, which have poor biological activity [98].

AMF-inoculation increases the resistance of host plants to metal(loid)s [99], which various factors can cause. The coexistence of host plants with AMF increases the absorption of water and nutrients such as nitrogen and phosphorus by plants and thus improves their growth [100]. Additionally, AMF, with the help of ERM, causes the deposition of polyphosphate complexes and the preservation of metal(loid)s in the roots of host plants, increasing the adaptation of plants to metal stresses [101]. The detoxification ability of AMF in plants largely depends on how AMF affects the host plant, the ability of AMF to precipitate metal(loid)s in the plant, the type of toxic metal(loid), and the availability of the toxic metal(loid) [25,98,99].

1.3. Arsenic (As)

Arsenic is one of the most toxic elements in nature, seriously endangering plants, animals, and even humans [102], As is generally found in all crustal rocks but can be released due to natural factors or human activities in the environment [102,103]. Among the natural factors that cause the release of As in nature are volcanic activity and weathering of rocks [104], but the human factors that cause the release of As in the environment are mining, fossil fuels, tannery, pesticides, herbicides, and chemical fertilizers [102–104]. The most important cause of arsenic poisoning is contaminated groundwater found in Bangladesh, Italy, Argentina, Hungary, India, China, Mexico, Chile, and the United States [105,106].

Gupta et al. reported that As affects growth and productivity due to the morphological, biochemical, and physiological changes it causes in plants [104]. As in plants reduces transpiration rate and leaf water potential, chlorophyll content (Chl), nutrient uptake, CO_2 stabilization rate, photosystem II activity, photosynthesis rate, heat loss capacity, carbon splitting, and sugar metabolism [107–112]; also, one of the most dangerous biochemical effects of As is the production of intracellular reactive oxygen species (ROS), which causes irreversible damage to DNA, lipids, carbohydrates, and proteins [107]. Symptoms of As in plants include reduced germination, biomass, leaf area, number of leaves, and yield [113]. The reaction of different plant species to As is different; for example, the yield of potatoes (*Solanum tuberosum* L.) decreases at a concentration of 300 mg/kg As in soil [114], while in the case of soybeans (*Glycine max* L.), it is 35 mg/kg As in soil [115]. However, this concentration for rice (*Oryza sativa* L.) is equal to 25 mg/kg As in soil [116], which can significantly reduce rice yield from 7–9 tons to 2–3 tons per hectare [117] (Table 1). Arsenic is a class 1 human carcinogen [118] that is not only transmitted to the human body through food and drinking water but also its prolonged inhalation causes poisoning in humans [105], which can cause skin ulcers, cardiovascular problems, lung and bladder diseases, cancer, and eventually death [105,119]. Therefore, finding effective ways to remove more arsenic from contaminated environments is essential.

Arsenic exists in inorganic and organic forms in nature, and its organic species are more toxic and mobile [120]. The most common types of As that are uptaken by plants are arsenate, arsenite, MMA, and DMA [113] (Table 2). Although different forms of arsenic are present in the environment simultaneously, plants receive it from the soil with a special preferential system (arsenite, arsenate, dimethylarsinate, and methylarsonate) [113].

It should be noted that As(V) is predominant in oxidizing media, and As(III) is more prevalent in reduced environments [120]. The most common way As enters plants is through roots, but the distribution of As in plant organs is very different so that the highest amount of As accumulates in plant roots and then in leaves, shoots, pods, and seeds [121]. Various studies showed that As(V) enters plants through phosphate transporters (Pht1 and Pht4) and As(III) through silicon transporters (OsLsi1 and OsLsi2) [122,123]. Arsenic enters plant cells mainly in the form of As(III) and As(V), but eventually As(V) is also catalyzed and converted to As(III) by arsenate reductase enzyme and is pumped through special cells transmitters or stored in vacuoles [124,125].

Table 2. The list of various As species in nature.

Arsenic Compounds	Acronyms	Chemical Formula
Arsenate	As (V)	$As(O^-)_3$
Arsenite	As (III)	$O{=}As(O^-)_3$
Methylarsonate	MMA	$CH_3AsO(O^-)_2$
Dimethylarsinate	DMA	$(CH_3)_2AsO(O^-)$
Trimethylarsin oxide	TMAO	$(CH_3)_3AsO$
Tetramethylarsonium ion	TETRA	$(CH_3)_4As^+$
Arsenobetain	AB	$(CH_3)_3As^+CH_2COO^-$
Trimethylarsoniopropionate	TMAP	$(CH_3)_3As^+CH_2CH_2COO^-$
Arsenocholine	AC	$(CH_3)_3As^+CH_2CH_2O^-$
Dimethylarsinoylacetate	DMAA	$(CH_3)_2(O)As^+CH_2COO^-$
Dimethylarsinoylpropionate	DMAP	$(CH_3)_2(O)As^+CH_2CH_2COO^-$

Nomenclature is as proposed by Boorboori et al. [126].

Since AMF is naturally present in As-contaminated environments [127], it can be an important component of increasing the efficiency of phytoremediation methods [128]. Research has also shown that arsenic can damage AMF and inhibit the primary stages of its development cycle [129], which can ultimately reduce mycorrhizal colonization [85]. Various studies showed that the prevalence of different species of AMF varies in arsenic stress environments [130], for example, Gonzalez-Chavez et al. found that *Glomeraceae* and *Acaulosporaceae* are predominant species in contaminated environments in Brazil [131]; however, *Glomeraceae* and *Glomus* are the predominant family and genus in arsenic-contaminated ecosystems [132].

Spagnoletti et al. reported that mycorrhiza plants are more tolerant of As toxicity, which may be due to various factors [85], which we will address below. AMF-inoculation increases plant biomass and dilute As in plants, and as a result, it increases the plant's resistance to As toxicity [85,100]. Studies showed that AMF-inoculation increases the uptake of nutrients, including phosphorus (P), nitrogen (N), and magnesium (Mg), by plants [104,133]. It can also result in increased photosynthetic pigment concentration, higher Hill reaction activity, an optimum chlorophyll a/b ratio, and higher photosystem II activity [104]. Increasing the absorption of CO_2 and improving the metabolism of essential carbohydrates are other positive roles of AMF in plants [104]. All of the above factors increase the resistance of mycorrhiza plants in As-contaminated environments.

On the other hand, the effect of AMF on the uptake, displacement, and speciation of As is well identified [134]. Research showed that As is absorbed by AMF hyphae (through RiPT and GiPT gene) [135,136], AMF reduces As(V) to As(III) (through RiarsC gene) [136], and finally, As is released through RiArsB, ATPase pump, and GiArsA into the soil [135,136]. It was shown that AMF can evaporate and methylate inorganic As through RiMT-11 into a wide range of organic As [134,137]; also, AMF causes more DMA release, especially when high concentrations of As(V) are present in the environment [134]. Numerous reports showed that coexistence with AMF has increased As evaporation and methylation as well as increased the As(III) to As(V) ratio in various crops, including rice and alfalfa [134,138,139].

Another AMF tool to counter As toxicity is GRSP secretion in the soil [140], and given that GRSP has a high amount of iron, it can produce AsIII-FeIII and ultimately immobilize As in the soil through bio-adsorption [85,140]. Spagnoletti et al. showed that with increasing As concentration in soil, GRSP content in the soil also increases [141] (Figure 1). Other benefits of AMF include helping to increase the activity of plant antioxidants in As-contaminated environments [85]. In this case, Spagnoletti et al. reported an increase in SOD, CAT, GPX, and GSH activities in AMF-inoculated soybeans in arsenic-contaminated soils [85,135]. In addition, the researchers found that AMF regulated the uptake of As and some other elements into plant roots by affecting the protein synthesis of channels related to As uptake and other elements such as P [105,135,142]. Christophersen et al. found that AMF reduced the expression of MtPht1 and MtPht2 genes, which are involved in the uptake of heavy metals into the root membrane of *Medicago truncatula*, but instead improved the expression of the MtPT4 gene, which carries P [142]; Li et al. Additionally, reported similar results for alfalfa [139].

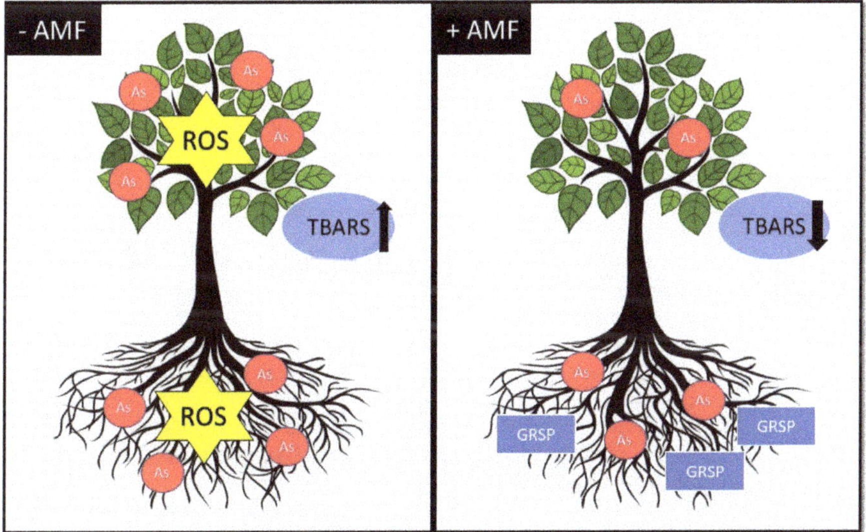

Figure 1. An overview of the role of AMF in increasing plant tolerance to arsenic contamination. Nomenclature is as proposed by Spagnoletti et al. [85]. AMF: arbuscular mycorrhizal fungi; As: arsenic; ROS: reactive oxygen species; GRSP: glomalin-related soil protein; TBARS: thiobarbituric acid-reactive species.

So far, only a handful of arsenic hyperaccumulator plants are identified, some of which are listed below: Pteris vittata, Pteris ryukyensis, Peris biaurita, Pteris capadogramatica, Pteris longifolia, Pteris fauriei, Pteris umbrosa, Pteris capadograma, Pteris quadriaurita, Pteris cretica, Pteris oshimensis, and Pteris aspericanlis [132]. Therefore, recognizing more plant species that are highly compatible with As and also suitable AMF species with which plants coexist more significantly impacts the phytoremediation of As-contaminated environments [45].

1.4. Cadmium (Cd)

Cd is an unnecessary element for animals and plants and is naturally present in low concentrations in soil and rocks [143,144]. This element is stable and does not decompose, so it is one of the most common contaminants in agricultural lands, especially in China. If the concentration of Cd in the soil exceeds 0.5 mg kg^{-1}, it will cause damage to plants and animals [144–146] (Table 1). In nature, the increase in Cd pollution is mainly due to human activities, including mining, metal smelting processes, Cd-rich phosphate fertilizers, industrial effluents, and fuel production [147,148]; the International Agency for Research on Cancer categorized it as a group 1 carcinogen [149]. Due to the high solubility of Cd, heavy rains, field irrigation, fine soil particles, and preferential flow cause the leaching of Cd from the surface layers of the soil to the subsoil, which is a factor in increasing Cd pollution in groundwater [150–152].

Cd is transported to plant tissues due to its common pathway with essential elements such as potassium (K) and calcium (Ca) [153]. High accumulation of it in plant tissues is toxic and reduces water and nutrient uptake, reducing plant growth, chlorosis, and eventually, the plant dies [154]. Plants have several important strategies for increasing their tolerance to Cd contamination, which can generally be divided into four categories: reduction of cell membrane transmission, Cd attachment to the cell wall, chelation, and compartmentalization [82]. Therefore, finding solutions to improve plant performance and increase their resistance to cadmium can be a good way to absorb, stabilize, accumulate, and reduce the toxicity of this element in contaminated environments and thus prevent its penetration into groundwater or the food chain [99,155]. One of these strategies is the coexistence of plant roots with soil microorganisms, especially AMF, which reduces Cd's toxicity, bioavailability, and environmental migration in the soil [49].

Although some studies have suggested that high concentrations of Cd inhibit mycorrhizal colonization, an important part of the research showed that different Cd concentrations do not significantly affect AMF colonization [4,156]. Additionally, different researches have shown that different species of AMF have different effects on the uptake of Cd by plants or its stabilization in soil [4,157,158]. For example, Hassan et al. showed that *Rhizophagus irregularis* in contaminated environments compared with *Funneliformis mosseae* caused more uptake of Cd by the sunflower and, as a result, increased its accumulation in the shoots of this plant. In contrast, *Funneliformis mosseae* inoculation increased the deposition of Cd in the soil [4]. Jiang et al. also reported that *Glomus versiforme* significantly reduced Cd accumulation in shoots and roots, while *Rhizophagus intraradices* increased Cd concentrations in roots and decreased cadmium concentrations in shoots [159].

AMF increases the resistance of host plants to Cd toxicity by improving photosynthesis, antioxidant enzymes, water and nutrition absorption, and growth [88]. Various studies showed that AMF inoculation increased the activity of SOD, CAT, APX, peroxidases (POD), and the total soluble protein content of plants tissues grown in Cd-contaminated environments while decreasing malondialdehyde (MDA) content in plants under similar conditions [160,161]. In addition, AMF helps increase the growth and biomass of the host plant by helping to increase the root length of plants and improve photosynthesis conditions, resulting in greater resistance of plants in Cd-contaminated environments [49,162]. He et al. also showed that AMF hyphae are prevented from N and P leaching in the soil, making them more accessible to plants [49] (Figure 2). Better absorption of nutrients and improved water absorption are other factors to increase the resistance of plants to Cd [160], which were already mentioned.

In addition, AMF hyphae are easily intertwined on the soil particle surfaces and reduce their bioavailability and migration by adsorption and stabilizing Cd ions [96,163]. In addition, GRSP, which is the result of the degradation of AMF mycelium, can form a complex with Cd and significantly increase its uptake by soil particles [97,164]. On the other hand, AMF increases the formation of coarse-grained soil, which, with other factors mentioned above, reduced the Cd in porous water, thus reducing Cd leaching to the depth of soil and groundwater [49,105] (Figure 2).

The researchers showed that inoculation of some species of AMF in different plants increases the uptake of cadmium by the roots of plants, in which case, depending on the AMF species, increases the accumulation of cadmium in the roots of plants or its transfer to the shoot [4,49], which in any case causes reduced Cd contamination in soil and improves phytoremediation in contaminated environments [166] (Figure 2). For example, Audet and Charest showed that *Rhizophagus irregularis* inoculation allows the sunflower plant to be used as a Cd accumulator [166].

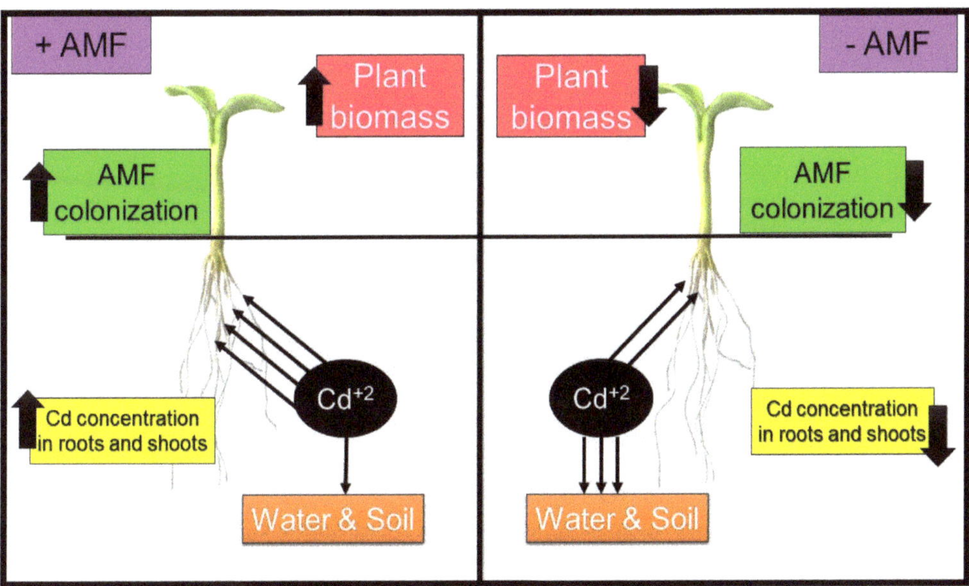

Figure 2. The role of AMF in cadmium contamination in the plant, water, and soil. Nomenclature is as proposed by Gunathilakae et al. [165].

Numerous studies showed that inoculation of AMF with the use of biochar and steel slag can firstly increase the resistance of plants to cadmium contamination and secondly increase the rate of stabilization and accumulation of cadmium in the soil and roots of plants [5,25]. In a study on the corn plant, Hu et al. found that inoculation of three different species of AMF, including *Glomus versiforme*, *Funneliformis mosseae*, and *Rhizophagus intraradices* with the application of steel slag in Cd-contaminated environments not only increased plant growth and decreased Cd uptake into plant organs, but also caused increased pH and total glomalin content in the soil [5]. Liu et al. also observed that AMF-inoculation and biochar application in maize simultaneously increased plant growth, increased antioxidant activity such as SOD, CAT and POD, and decreased Cd concentration in different plant organs. This combination treatment also caused increased soil pH and cadmium stabilization and finally decreased cadmium bioavailability in soil [25].

1.5. Lead (Pb)

Pb is one of the most toxic heavy metals in nature, and due to its high stability and lack of decomposition, it pollutes nature and accumulates in plants and living organisms [167]. This heavy metal is one of the most common pollutants of the present age, which is very dangerous due to the soil and climate pollution. The removal of this pollution helps maintain the health of humans, other creatures, and nature [145,168]. Pb is naturally distributed in the Earth's crust, and due to human activities, has become the most widespread toxic metal in nature [169]. Among the human factors that cause environmental pollution to Pb are mining, smelting metals, and industrial effluents [147]. Other factors that lead to the spread of Pb pollution in the environment include the performance of small companies in some countries, including Pakistan, which repair obsolete lead-acid batteries and, because they are not able to comply with strict environmental regulations, discharge wastewater directly into the soil, waterways, and the surrounding environment, which causes severe environmental pollution with Pb. This danger is exacerbated when Pb-contaminated water is used to irrigate fields [138].

Increasing the concentration of Pb in the environment causes adverse effects such as the inhibition of seed germination, slow plant growth and stunting, reduced metabolism and photosynthesis, accumulation of Pb in plant tissues, chlorosis, and eventually plant death [170,171]. Therefore, the increase of Pb in the environment is a serious threat to food security [172]. On the other hand, prolonged exposure of children to Pb contamination adversely affects the development of their nervous system and is likely to lead to mental retardation [173]. Other side effects of Pb in humans include anaemia and kidney disease, to the extent that research showed that human kidney cells are destroyed by high concentrations of Pb [174,175]. Therefore, phytoremediation can reduce Pb's adverse effects on human, animal, and plant communities.

Salazar et al. showed that AMF has suitable Pb accumulation mechanisms in their spores and mycelium, but the amount of Pb uptake may depend on AMF species and host plants, and in general, Glomeraceae is the most important diverse species of AMF present in Pb-contaminated soils [176]. AMF, on the other hand, helps prevent the transfer of Pb to the shoot by helping to increase the accumulation of Pb in plant roots [70]. Another important role of AMF excretes total glomalin-associated soil protein (TGSP), which increases the retention of Pb in soil, reduces the bioavailability of it, and thus reduces the toxicity of Pb to plants [35,98,177].

The researchers also showed that AMF increases plant resistance in lead-contaminated environments by increasing plant nutritional efficiency and improving the antioxidant defence system [35,98]. Chen et al. showed that *Populus euphratica* inoculation with *F. mosseae* in Pb-contaminated environments significantly increased SOD and CAT activities [178]; however, Spagnoletti et al. also reported that *Cichorium intybus* inoculation by *R. irregularis* increased SOD and CAT activity in Pb infected environments [85]. Research also showed that the inoculation of plants with AMF by increasing the content of polysaccharides in the hemicellulose and pectin of the cell wall, and increasing peroxidase activities in the cell walls and thus increasing Pb fixation in the root cell wall of host plants, makes them more resistant to Pb toxicity [105]. Zhang et al. showed that the expression of MtPrx05 and MtPrx10 genes related to cell wall polysaccharide cross-linking was increased by AMF-inoculation in Pb-contaminated media [105] (Figure 3).

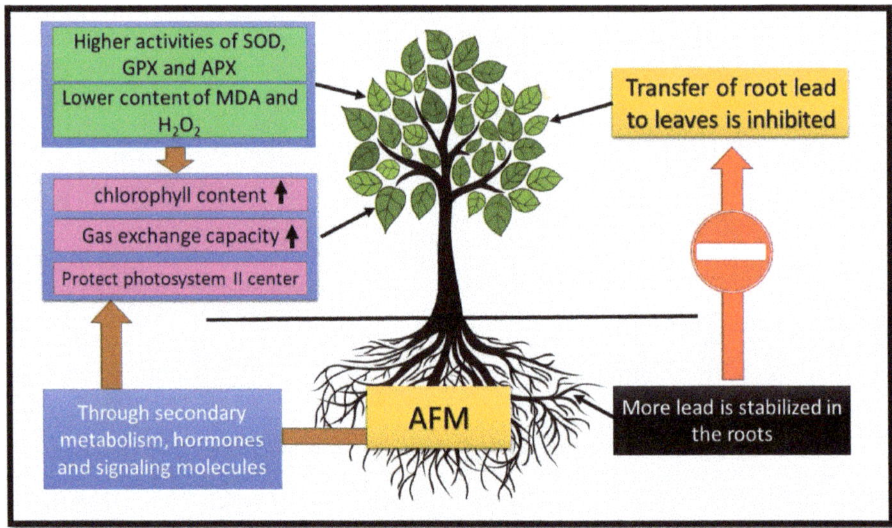

Figure 3. AMF inoculation role in plant tolerance to Pb. Nomenclature is as proposed by Yang et al. [91].

In recent years, scientists have noted the positive effect of inoculating plants with AMF and adding earthworms, biochar, cow manure, lignin, and steel slag to increase uptake and stabilize Pb from contaminated environments [5,24,179,180]. They found that combining AMF with other factors improves phytoremediation in Pb-contaminated environments through increasing the uptake of nitrogen, phosphorus, potassium, and iron, reducing Pb transfer from root to shoot, increasing root colonization, improving soil pH, further plant growth, and increasing soil glomalin content [5,24,179,180].

1.6. Chromium (Cr)

In addition to being one of the most abundant elements in the Earth's crust, Cr is also one of the most dangerous heavy metals [181]. The abundance of Cr in the soil indicates an environmental problem that is most likely originated from human activities, including chemicals used in agriculture, paint and leather industries, alloy production, and stainless steel [182,183]. It can be said that India's tanning industry alone imports between 2000 and 32,000 tons of Cr into the environment annually [88]. Cr is very dangerous because it does not decompose chemically or biologically, and its concentration in living organisms increases as it moves along the food cycle, turning it into a dangerous environmental contaminant in soil, water, and air [182].

Cr usually exists in two stable forms, trivalent [Cr(III)] and hexavalent [Cr(VI)], both of which can be exchanged through the precipitation/dissolution, oxidation/reduction, and adsorption/desorption processes [184]. Cr(III) is the most abundant and stable form of Cr [181], which is non-toxic and is usually immobile and insoluble in water [185]. Cr(III) is susceptible to adsorption on the soil surface or deposition in chromium hydroxide form in slightly acidic or alkaline environments [60], which plants can not easily absorb [183]. Cr(VI), on the other hand, is a class A carcinogen substance that can kill living cells [183]. This substance, which is completely soluble in water in the pH range, is highly mobile and is usually present in neutral to alkaline soils, mainly in the form of a chromate anion (CrO^{2-}_4) or relatively sparing chromate salts such as $PbCrO_4$, $BaCrO_4$, and $CaCrO_4$ [60,183]. It should be noted that Cr(III) is considered an essential human substance that can interfere with cholesterol, glucose metabolism, and increased insulin secretion [181], but Cr(VI) is highly toxic by inhalation and, in high concentrations, cause adverse effects such as renal failure, hemolysis and liver failure [183,185]; therefore, the permissible dose in water is 8 micrograms per litre for Cr(III) and 1 microgram per litre for Cr(VI) [186].

Cr is the most toxic pollutant that negatively affects plants' performance and metabolic activity [187]. Cr(VI), as a strong oxidizing with redox potential between 1.33 to 1.38, causes rapid production of reactive oxygen species (ROS) such as superoxide and hydroxyl radicals [188], and its negative effects on plants include changes in membrane structure and root damage, carbon uptake, antioxidant defence activity, nutrient uptake, DNA damage, ion transport imbalance, reduced photosynthesis and growth, and eventually plant death [39,186,189,190]. Plants have different mechanisms for combating the toxicity of Cr, and the most important of which is the chemical reduction of Cr(VI) to Cr(III), which can be carried out enzymatically and non-enzymatically [191]. The addition of salts containing Fe(III), animal manure, or organic acids to the culture medium helps the plants in this direction [9]. Therefore, plants that accumulate Cr, such as *Prosopis laevigata*, *Spartina argentinensis*, and *Amaranthus dubius*, can convert Cr(VI) to Cr(III) and prevent chromium erosion and leaching in the soil [32,92]. Due to the low cost, phytoremediation can be a good strategy for improving and cleaning Cr-contaminated environments [9].

According to studies, factors such as AMF play an important role in regulating Cr uptake and detoxification of plants [192], but the effect of this factor depends on the type of plant, the type of fungus and soil conditions [182]. Research showed that AMF absorbs more chromium through its various structures (such as hyphae, ERM, and spores), and complexing Cr with histidine or phosphate prevents them from being transferred to plants [101,182]. AMF can also use various strategies such as helping to absorb nutrients (P and N) to increase plant resistance to Cr stress and prevent severe damage or death due to Cr toxicity [182]. Another important role of AMF is to stabilize Cr in plant roots through external and internal radical mycelium and ultimately help reduce its displacement to plant stems [101]. In addition, BGlomalin-secretion by AMF can immobilize Cr [164]. Due to their similar chemical structure to Cr(VI) and P, these potentially compete during the adsorption process by plants, and since AMF increases P uptake by plant roots, they can play an important role in reducing the uptake of Cr(VI) [193]. Gil-Cardoza et al. also found that AMF could help detoxify Cr(VI) by reducing it to Cr(III) through ERM [193].

As mentioned, plant coexistence with AMF may help detoxify Cr by improving plant mineral nutrition and producing important metabolites such as sulfur compounds [81]. Since sulfur metabolites can combine with Cr through thiol groups to reduce the toxicity of Cr in microorganisms and plants [194], increasing the uptake of S by plants can provide the conditions for reducing the toxicity of Cr [195]. Among the S metabolites are glutathione (GSH), phytochelatins (PCs), and cysteine (Cys), which can act as non-enzymatic antioxidants in the elimination of ROS induced Cr [81,196]. Wu et al. also showed that AMF inoculation increases the expression of sulfate transporter genes with high affinity (MtSULTR1.1 and MtSULTR1.2) in plant roots and thus increases S uptake, which ultimately increases Cr(VI) detoxification [81,101].

2. Conclusions

As observed, AMF-inoculation in various forms increases phytoremediation efficiency in environments contaminated with arsenic, cadmium, lead, and chromium. AMF increases the accumulation of these metal(loid)s in the soil and roots of plants, prevents them from washing deeper into the soil and penetration into groundwater, and increases the resistance of plants to the high toxicity of these metal(loid)s. Increasing awareness of ways to improve the performance of AMF in phytoremediation, especially in the case of lead, about which there is limited information, can introduce phytoremediation as one of the most practical and cheapest ways to improve contaminated sites in many parts of the world. It should be noted that knowing more about plants that accumulate metal(loid)s and AMF species that coexist better with plants will help in this way.

Author Contributions: M.R.B. wrote the manuscript. M.R.B. and H.-Y.Z. edited the manuscript. All authors have read and agreed to the published version of the manuscript.

Funding: This work was supported by the Anhui Province Large-scale Online Open Course (MOOC) Demonstration Project (2018mooc428) and School-level quality engineering project of Suzhou University: College of Geographic Information and Energy Agriculture Modern Industry (szxy2021cyxy06).

Institutional Review Board Statement: Not applicable.

Informed Consent Statement: Not applicable.

Conflicts of Interest: The authors declare that they have no competing interest.

References

1. McLaughlin, M.J.; Hamon, R.; McLaren, R.; Speir, T.; Rogers, S. A bioavailability-based rationale for controlling metal and metalloid contamination of agricultural land in Australia and New Zealand. *Soil Res.* **2000**, *38*, 1037–1086. [CrossRef]
2. Swartjes, F.A. *Dealing with Contaminated Sites: From Theory towards Practical Application*; Springer Science & Business Media: Berlin/Heidelberg, Germany, 2011.
3. Liu, L.; Li, W.; Song, W.; Guo, M. Remediation techniques for heavy metal-contaminated soils: Principles and applicability. *Sci. Total Environ.* **2018**, *633*, 206–219. [CrossRef] [PubMed]
4. Hassan, S.E.; Hijri, M.; St-Arnaud, M. Effect of arbuscular mycorrhizal fungi on trace metal uptake by sunflower plants grown on cadmium contaminated soil. *New Biotechnol.* **2013**, *30*, 780–787. [CrossRef] [PubMed]
5. Hu, Z.-H.; Zhuo, F.; Jing, S.-H.; Li, X.; Yan, T.-X.; Lei, L.-L.; Lu, R.-R.; Zhang, X.-F.; Jing, Y.-X. Combined application of arbuscular mycorrhizal fungi and steel slag improves plant growth and reduces Cd, Pb accumulation in Zea mays. *Int. J. Phytoremediat.* **2019**, *21*, 857–865. [CrossRef] [PubMed]
6. Cui, S.; Zhou, Q.-X.; Wei, S.-H.; Zhang, W.; Cao, L.; Ren, L.-P. Effects of exogenous chelators on phytoavailability and toxicity of Pb in Zinnia elegans Jacq. *J. Hazard. Mater.* **2007**, *146*, 341–346. [CrossRef]
7. Kabata-Pendias, A. *Trace Elements in Soils and Plants*, 4th ed.; CRC Press: Boca Raton, FL, USA, 2011.

8.   Ali, S.; Abbas, Z.; Rizwan, M.; Zaheer, I.E.; Yavaş, İ.; Ünay, A.; Abdel-Daim, M.M.; Bin-Jumah, M.; Hasanuzzaman, M.; Kalderis, D. Application of floating aquatic plants in phytoremediation of heavy metals polluted water: A review. *Sustainability* **2020**, *12*, 1927. [CrossRef]
9.   Lourdes, G.-C.M.; Stéphane, D.; Maryline, C.-S. Impact of increasing chromium (VI) concentrations on growth, phosphorus and chromium uptake of maize plants associated to the mycorrhizal fungus Rhizophagus irregularis MUCL 41833. *Heliyon* **2021**, *7*, e05891. [CrossRef]
10. Ogar, A.; Sobczyk, Ł.; Turnau, K. Effect of combined microbes on plant tolerance to Zn–Pb contaminations. *Environ. Sci. Pollut. Res.* **2015**, *22*, 19142–19156. [CrossRef]
11. Li, Y.; Becquer, T.; Quantin, C.; Benedetti, M.; Lavelle, P.; Dai, J. Effects of heavy metals on microbial biomass and activity in subtropical paddy soil contaminated by acid mine drainage. *Acta Ecol. Sin.* **2004**, *24*, 2430–2436.
12. Sharpley, A.N.; Weld, J.L.; Beegle, D.B.; Kleinman, P.J.; Gburek, W.; Moore, P.; Mullins, G. Development of phosphorus indices for nutrient management planning strategies in the United States. *J. Soil Water Conserv.* **2003**, *58*, 137–152.
13. Adrees, M.; Ali, S.; Rizwan, M.; Ibrahim, M.; Abbas, F.; Farid, M.; Zia-ur-Rehman, M.; Irshad, M.K.; Bharwana, S.A. The effect of excess copper on growth and physiology of important food crops: A review. *Environ. Sci. Pollut. Res.* **2015**, *22*, 8148–8162. [CrossRef] [PubMed]
14. Krishnamoorthy, R.; Kim, C.-G.; Subramanian, P.; Kim, K.-Y.; Selvakumar, G.; Sa, T.-M. Arbuscular mycorrhizal fungi community structure, abundance and species richness changes in soil by different levels of heavy metal and metalloid concentration. *PLoS ONE* **2015**, *10*, e0128784. [CrossRef] [PubMed]
15. Jadia, C.D.; Fulekar, M. Phytoremediation of heavy metals: Recent techniques. *Afr. J. Biotechnol.* **2009**, *8*, 921–928.
16. Gjorgieva Ackova, D. Heavy metals and their general toxicity on plants. *Plant Sci. Today* **2018**, *5*, 15–19. [CrossRef]
17. Chang, Q.; Diao, F.-W.; Wang, Q.-F.; Pan, L.; Dang, Z.-H.; Guo, W. Effects of arbuscular mycorrhizal symbiosis on growth, nutrient and metal uptake by maize seedlings (*Zea mays* L.) grown in soils spiked with Lanthanum and Cadmium. *Environ. Pollut.* **2018**, *241*, 607–615. [CrossRef]
18. Basta, N.; McGowen, S. Evaluation of chemical immobilization treatments for reducing heavy metal transport in a smelter-contaminated soil. *Environ. Pollut.* **2004**, *127*, 73–82. [CrossRef]
19. Navarro, C.; Díaz, M.; Villa-García, M.A. Physico-chemical characterization of steel slag. Study of its behavior under simulated environmental conditions. *Environ. Sci. Technol.* **2010**, *44*, 5383–5388. [CrossRef]
20. Khan, Z.; Doty, S. Endophyte-assisted phytoremediation. *Plant Biol.* **2011**, *12*, 97–105.
21. Xu, P.; Chen, M.; Zeng, G.; Huang, D.; Lai, C.; Wang, Z.; Yan, M.; Huang, Z.; Gong, X.; Song, B. Effects of multi-walled carbon nanotubes on metal transformation and natural organic matters in riverine sediment. *J. Hazard. Mater.* **2019**, *374*, 459–468. [CrossRef]
22. Qian, L.; Zhang, W.; Yan, J.; Han, L.; Gao, W.; Liu, R.; Chen, M. Effective removal of heavy metal by biochar colloids under different pyrolysis temperatures. *Bioresour. Technol.* **2016**, *206*, 217–224. [CrossRef]
23. Kumar, A.; Tsechansky, L.; Lew, B.; Raveh, E.; Frenkel, O.; Graber, E.R. Biochar alleviates phytotoxicity in Ficus elastica grown in Zn-contaminated soil. *Sci. Total Environ.* **2018**, *618*, 188–198. [CrossRef] [PubMed]
24. Zhuo, F.; Zhang, X.-F.; Lei, L.-L.; Yan, T.-X.; Lu, R.-R.; Hu, Z.-H.; Jing, Y.-X. The effect of arbuscular mycorrhizal fungi and biochar on the growth and Cd/Pb accumulation in Zea mays. *Int. J. Phytoremediat.* **2020**, *22*, 1009–1018. [CrossRef] [PubMed]
25. Liu, L.; Li, J.; Yue, F.; Yan, X.; Wang, F.; Bloszies, S.; Wang, Y. Effects of arbuscular mycorrhizal inoculation and biochar amendment on maize growth, cadmium uptake and soil cadmium speciation in Cd-contaminated soil. *Chemosphere* **2018**, *194*, 495–503. [CrossRef]
26. Vejvodová, K.; Száková, J.; García-Sánchez, M.; Praus, L.; Romera, I.G.; Tlustoš, P. Effect of dry olive residue–based biochar and arbuscular mycorrhizal fungi inoculation on the nutrient status and trace element contents in wheat grown in the As-, Cd-, Pb-, and Zn-contaminated soils. *J. Soil Sci. Plant Nutr.* **2020**, *20*, 1067–1079. [CrossRef]
27. Mahar, A.; Ping, W.; Ronghua, L.; Zhang, Z. Immobilization of lead and cadmium in contaminated soil using amendments: A review. *Pedosphere* **2015**, *25*, 555–568. [CrossRef]
28. Deng, L.; Li, Z.; Wang, J.; Liu, H.; Li, N.; Wu, L.; Hu, P.; Luo, Y.; Christie, P. Long-term field phytoextraction of zinc/cadmium contaminated soil by *Sedum plumbizincicola* under different agronomic strategies. *Int. J. Phytoremediat.* **2016**, *18*, 134–140. [CrossRef]
29. Dos Santos, J.V.; Varón-López, M.; Soares, C.R.F.S.; Leal, P.L.; Siqueira, J.O.; de Souza Moreira, F.M. Biological attributes of rehabilitated soils contaminated with heavy metals. *Environ. Sci. Pollut. Res.* **2016**, *23*, 6735–6748. [CrossRef]
30. Vamerali, T.; Bandiera, M.; Mosca, G. Field crops for phytoremediation of metal-contaminated land. A review. *Environ. Chem. Lett.* **2010**, *8*, 1–17. [CrossRef]
31. Malaviya, P.; Singh, A.; Anderson, T.A. Aquatic phytoremediation strategies for chromium removal. *Rev. Environ. Sci. Bio/Technol.* **2020**, *19*, 897–944. [CrossRef]
32. Ali, H.; Khan, E.; Sajad, M.A. Phytoremediation of heavy metals—Concepts and applications. *Chemosphere* **2013**, *91*, 869–881. [CrossRef]
33. Yu, X.; Kang, X.; Li, Y.; Cui, Y.; Tu, W.; Shen, T.; Yan, M.; Gu, Y.; Zou, L.; Ma, M. Rhizobia population was favoured during in situ phytoremediation of vanadium-titanium magnetite mine tailings dam using *Pongamia pinnata*. *Environ. Pollut.* **2019**, *255*, 113167. [CrossRef] [PubMed]
34. Khan, A.G.; Kuek, C.; Chaudhry, T.; Khoo, C.S.; Hayes, W.J. Role of plants, mycorrhizae and phytochelators in heavy metal contaminated land remediation. *Chemosphere* **2000**, *41*, 197–207. [CrossRef]

35. Wang, B.; Wang, Q.; Liu, W.; Liu, X.; Hou, J.; Teng, Y.; Luo, Y.; Christie, P. Biosurfactant-producing microorganism *Pseudomonas* sp. SB assists the phytoremediation of DDT-contaminated soil by two grass species. *Chemosphere* **2017**, *182*, 137–142. [CrossRef] [PubMed]

36. Pajević, S.; Borišev, M.; Nikolić, N.; Arsenov, D.D.; Orlović, S.; Županski, M. Phytoextraction of heavy metals by fast-growing trees: A review. *Phytoremediation* **2016**, *2016*, 29–64.

37. Souri, Z.; Karimi, N.; Sandalio, L.M. Arsenic hyperaccumulation strategies: An overview. *Front. Cell Dev. Biol.* **2017**, *5*, 67. [CrossRef]

38. Pedroso, D.d.F.; Barbosa, M.V.; dos Santos, J.V.; Pinto, F.A.; Siqueira, J.O.; Carneiro, M. Arbuscular mycorrhizal fungi favor the initial growth of *Acacia mangium, Sorghum bicolor,* and *Urochloa brizantha* in soil contaminated with Zn, Cu, Pb, and Cd. *Bull. Environ. Contam. Toxicol.* **2018**, *101*, 386–391. [CrossRef]

39. Patra, D.K.; Pradhan, C.; Patra, H.K. Chromium bioaccumulation, oxidative stress metabolism and oil content in lemon grass *Cymbopogon flexuosus* (Nees ex Steud.) W. Watson grown in chromium rich over burden soil of Sukinda chromite mine, India. *Chemosphere* **2019**, *218*, 1082–1088. [CrossRef]

40. Rosa, C.; Sierra, M.; Radetski, C. Use of plant tests in the evaluation of textile effluent toxicity. *Ecotoxicol. Environ. Res.* **1999**, *2*, 56–61.

41. Gomes, M.; Carvalho, M.; Carvalho, G.; Marques, T.; Garcia, Q.; Guilherme, L.; Soares, A. Phosphorus improves arsenic phytoremediation by *Anadenanthera peregrina* by alleviating induced oxidative stress. *Int. J. Phytoremediat.* **2013**, *15*, 633–646. [CrossRef]

42. Chaer, G.M.; Resende, A.S.; Campello, E.F.C.; de Faria, S.M.; Boddey, R.M. Nitrogen-fixing legume tree species for the reclamation of severely degraded lands in Brazil. *Tree Physiol.* **2011**, *31*, 139–149. [CrossRef]

43. Marques, T.C.L.L.d.S.; Moreira, F.M.d.S.; Siqueira, J.O. Growth and metal concentration of seedlings of woody species in a heavy metal contaminated soil. *Pesqui. Agropecu. Bras.* **2000**, *35*, 121–132. [CrossRef]

44. Vymazal, J.; Kröpfelová, L.; Švehla, J.; Chrastný, V.; Štíchová, J. Trace elements in Phragmites australis growing in constructed wetlands for treatment of municipal wastewater. *Ecol. Eng.* **2009**, *35*, 303–309. [CrossRef]

45. Schneider, J.; Bundschuh, J.; de Melo Rangel, W.; Guilherme, L.R.G. Potential of different AM fungi (native from As-contaminated and uncontaminated soils) for supporting *Leucaena leucocephala* growth in As-contaminated soil. *Environ. Pollut.* **2017**, *224*, 125–135. [CrossRef] [PubMed]

46. Xu, Q.; Min, H.; Cai, S.; Fu, Y.; Sha, S.; Xie, K.; Du, K. Subcellular distribution and toxicity of cadmium in *Potamogeton crispus* L. *Chemosphere* **2012**, *89*, 114–120. [CrossRef]

47. Zhu, X.F.; Lei, G.J.; Jiang, T.; Liu, Y.; Li, G.X.; Zheng, S.J. Cell wall polysaccharides are involved in P-deficiency-induced Cd exclusion in *Arabidopsis thaliana*. *Planta* **2012**, *236*, 989–997. [CrossRef]

48. Ralph, J.; Bunzel, M.; Marita, J.M.; Hatfield, R.D.; Lu, F.; Kim, H.; Schatz, P.F.; Grabber, J.H.; Steinhart, H. Peroxidase-dependent cross-linking reactions of p-hydroxycinnamates in plant cell walls. *Phytochem. Rev.* **2004**, *3*, 79–96. [CrossRef]

49. He, Y.-M.; Yang, R.; Lei, G.; Li, B.; Jiang, M.; Yan, K.; Zu, Y.-Q.; Zhan, F.-D.; Li, Y. Arbuscular mycorrhizal fungi reduce cadmium leaching from polluted soils under simulated heavy rainfall. *Environ. Pollut.* **2020**, *263*, 114406. [CrossRef]

50. Jiang, X.J.; Liu, W.; Chen, C.; Liu, J.; Yuan, Z.-Q.; Jin, B.; Yu, X. Effects of three morphometric features of roots on soil water flow behavior in three sites in China. *Geoderma* **2018**, *320*, 161–171. [CrossRef]

51. Krämer, U. Phytoremediation: Novel approaches to cleaning up polluted soils. *Curr. Opin. Biotechnol.* **2005**, *16*, 133–141. [CrossRef]

52. Vickers, N.J. Animal communication: When i'm calling you, will you answer too? *Curr. Biol.* **2017**, *27*, R713–R715. [CrossRef]

53. Quintella, C.M.; Mata, A.M.; Lima, L.C. Overview of bioremediation with technology assessment and emphasis on fungal bioremediation of oil contaminated soils. *J. Environ. Manag.* **2019**, *241*, 156–166. [CrossRef] [PubMed]

54. Nguyen, B.T.; Trinh, N.N.; Le, C.M.T.; Nguyen, T.T.; Tran, T.V.; Thai, B.V.; Le, T.V. The interactive effects of biochar and cow manure on rice growth and selected properties of salt-affected soil. *Arch. Agron. Soil Sci.* **2018**, *64*, 1744–1758. [CrossRef]

55. Rajkumar, M.; Sandhya, S.; Prasad, M.; Freitas, H. Perspectives of plant-associated microbes in heavy metal phytoremediation. *Biotechnol. Adv.* **2012**, *30*, 1562–1574. [CrossRef] [PubMed]

56. Gupta, P.; Rani, R.; Chandra, A.; Varjani, S.J.; Kumar, V. Effectiveness of plant growth-promoting Rhizobacteria in phytoremediation of chromium stressed soils. In *Waste Bioremediation*; Springer: Singapore, 2018; pp. 301–312.

57. Smith, S.E.; Read, D.J. *Mycorrhizal Symbiosis*; Academic Press: Cambridge, UK, 2010.

58. Parvin, S.; Van Geel, M.; Yeasmin, T.; Lievens, B.; Honnay, O. Variation in arbuscular mycorrhizal fungal communities associated with lowland rice (*Oryza sativa*) along a gradient of soil salinity and arsenic contamination in Bangladesh. *Sci. Total Environ.* **2019**, *686*, 546–554. [CrossRef]

59. Nelson, L.W.; Cheeke, T.E.; Cifizzari, K. An inquiry-based lab activity to investigate potential effects of arbuscular mycorrhizal fungi on seed germination. *Am. Biol. Teach.* **2021**, *83*, 537–541. [CrossRef]

60. Gil-Cardeza, M.L.; Müller, D.; Amaya-Martin, S.M.; Viassolo, R.; Gómez, E. Differential responses to high soil chromium of two arbuscular mycorrhizal fungi communities isolated from Cr-polluted and non-polluted rhizospheres of *Ricinus communis*. *Sci. Total Environ.* **2018**, *625*, 1113–1121. [CrossRef]

61. Hu, S.; Hu, B.; Chen, Z.; Vosátka, M.; Vymazal, J. Arbuscular mycorrhizal fungi modulate the chromium distribution and bioavailability in semi-aquatic habitats. *Chem. Eng. J.* **2021**, *420*, 129925. [CrossRef]

62. Vallino, M.; Fiorilli, V.; Bonfante, P. Rice flooding negatively impacts root branching and arbuscular mycorrhizal colonization, but not fungal viability. *Plant Cell Environ.* **2013**, *37*, 557–572. [CrossRef]

63. Lumini, E.; Vallino, M.; Alguacil, M.M.; Romani, M.; Bianciotto, V. Different farming and water regimes in Italian rice fields affect arbuscular mycorrhizal fungal soil communities. *Ecol. Appl.* **2011**, *21*, 1696–1707. [CrossRef]
64. Miransari, M. Arbuscular mycorrhizal fungi and heavy metal tolerance in plants. In *Arbuscular Mycorrhizas and Stress Tolerance of Plants*; Springer: Singapore, 2017; pp. 147–161.
65. Smith, S.; Read, D. *Mycorrhizal Symbiosis*, 3rd ed.; Academic Press: Cambridge, UK, 2008.
66. Simard, S.W.; Beiler, K.J.; Bingham, M.A.; Deslippe, J.R.; Philip, L.J.; Teste, F.P. Mycorrhizal networks: Mechanisms, ecology and modelling. *Fungal Biol. Rev.* **2012**, *26*, 39–60. [CrossRef]
67. Rillig, M.C.; Mummey, D.L. mMycorrhizas and soil structure. *New Phytol.* **2006**, *171*, 41–53. [CrossRef] [PubMed]
68. Rillig, M.C.; Steinberg, P.D. Glomalin production by an arbuscular mycorrhizal fungus: A mechanism of habitat modification? *Soil Biol. Biochem.* **2002**, *34*, 1371–1374. [CrossRef]
69. Xu, Z.; Wu, Y.; Xiao, Z.; Ban, Y.; Belvett, N. Positive effects of *Funneliformis mosseae* inoculation on reed seedlings under water and TiO$_2$ nanoparticles stresses. *World J. Microbiol. Biotechnol.* **2019**, *35*, 1–13. [CrossRef]
70. Gu, H.-H.; Zhou, Z.; Gao, Y.-Q.; Yuan, X.-T.; Ai, Y.-J.; Zhang, J.-Y.; Zuo, W.-Z.; Taylor, A.A.; Nan, S.-Q.; Li, F.-P. The influences of arbuscular mycorrhizal fungus on phytostabilization of lead/zinc tailings using four plant species. *Int. J. Phytoremediat.* **2017**, *19*, 739–745. [CrossRef]
71. Solís-Domínguez, F.A.; Valentín-Vargas, A.; Chorover, J.; Maier, R.M. Effect of arbuscular mycorrhizal fungi on plant biomass and the rhizosphere microbial community structure of mesquite grown in acidic lead/zinc mine tailings. *Sci. Total Environ.* **2011**, *409*, 1009–1016. [CrossRef]
72. Carrasco, L.; Azcón, R.; Kohler, J.; Roldán, A.; Caravaca, F. Comparative effects of native filamentous and arbuscular mycorrhizal fungi in the establishment of an autochthonous, leguminous shrub growing in a metal-contaminated soil. *Sci. Total Environ.* **2011**, *409*, 1205–1209. [CrossRef]
73. Akhtar, O.; Kehri, H.K.; Zoomi, I. *Arbuscular mycorrhiza* and *Aspergillus terreus* inoculation along with compost amendment enhance the phytoremediation of Cr-rich technosol by *Solanum lycopersicum* under field conditions. *Ecotoxicol. Environ. Saf.* **2020**, *201*, 110869. [CrossRef]
74. Ismail, Y.; Hijri, M. Arbuscular mycorrhisation with *Glomus irregulare* induces expression of potato PR homologues genes in response to infection by *Fusarium sambucinum*. *Funct. Plant Biol.* **2012**, *39*, 236–245. [CrossRef]
75. Tan, S.-Y.; Jiang, Q.-Y.; Zhuo, F.; Liu, H.; Wang, Y.-T.; Li, S.-S.; Ye, Z.-H; Jing, Y.-X. Effect of inoculation with *Glomus versiforme* on cadmium accumulation, antioxidant activities and phytochelatins of *Solanum photeinocarpum*. *PLoS ONE* **2015**, *10*, e0132347. [CrossRef]
76. Parniske, M. Arbuscular mycorrhiza: The mother of plant root endosymbioses. *Nat. Rev. Genet.* **2008**, *6*, 763–775. [CrossRef]
77. Dong, Y.; Zhu, Y.-G.; Smith, F.A.; Wang, Y.; Chen, B. Arbuscular mycorrhiza enhanced arsenic resistance of both white clover (*Trifolium repens* Linn.) and ryegrass (*Lolium perenne* L.) plants in an arsenic-contaminated soil. *Environ. Pollut.* **2008**, *155*, 174–181. [CrossRef] [PubMed]
78. Smith, F.A.; Smith, S.E. How harmonious are arbuscular mycorrhizal symbioses? Inconsistent concepts reflect different mindsets as well as results. *New Phytol.* **2015**, *205*, 1381–1384. [CrossRef] [PubMed]
79. Karandashov, V.; Bucher, M. Symbiotic phosphate transport in arbuscular mycorrhizas. *Trends Plant Sci.* **2005**, *10*, 22–29. [CrossRef] [PubMed]
80. Campos-Soriano, L.; García-Garrido, J.M.; Segundo, B.S. Activation of basal defense mechanisms of rice plants by *Glomus intraradices* does not affect the arbuscular mycorrhizal symbiosis. *New Phytol.* **2010**, *188*, 597–614. [CrossRef]
81. Wu, S.; Hu, Y.; Zhang, X.; Sun, Y.; Wu, Z.; Li, T.; Lv, J.; Li, J.; Zhang, J.; Zheng, L. Chromium detoxification in arbuscular mycorrhizal symbiosis mediated by sulfur uptake and metabolism. *Environ. Exp. Bot.* **2018**, *147*, 43–52. [CrossRef]
82. Gai, J.; Fan, J.; Zhang, S.; Mi, N.; Christie, P.; Li, X.; Feng, G. Direct effects of soil cadmium on the growth and activity of arbuscular mycorrhizal fungi. *Rhizosphere* **2018**, *7*, 43–48. [CrossRef]
83. Martínez-García, L.B.; De Deyn, G.B.; Pugnaire, F.I.; Kothamasi, D.; van der Heijden, M.G. Symbiotic soil fungi enhance ecosystem resilience to climate change. *Glob. Chang. Biol.* **2017**, *23*, 5228–5236. [CrossRef]
84. Gigolashvili, T.; Kopriva, S. Transporters in plant sulfur metabolism. *Front. Plant Sci.* **2014**, *5*, 442. [CrossRef]
85. Spagnoletti, F.N.; Lavado, R.S.; Giacometti, R. Interaction of plants and arbuscular mycorrhizal fungi in responses to arsenic stress: A collaborative tale useful to manage contaminated soils. In *Mechanisms of Arsenic Toxicity and Tolerance in Plants*; Springer: Singapore, 2018; pp. 239–255.
86. González-Guerrero, M.; Benabdellah, K.; Valderas, A.; Azcón-Aguilar, C.; Ferrol, N. GintABC1 encodes a putative ABC transporter of the MRP subfamily induced by Cu, Cd, and oxidative stress in *Glomus intraradices*. *Mycorrhiza* **2010**, *20*, 137–146. [CrossRef]
87. Zhu, R.; Zheng, Z.; Li, T.; He, S.; Zhang, X.; Wang, Y.; Liu, T. Effect of tea plantation age on the distribution of glomalin-related soil protein in soil water-stable aggregates in southwestern China. *Environ. Sci. Pollut. Res.* **2018**, *26*, 1973–1982. [CrossRef]
88. Zhang, Z.; Mallik, A.; Zhang, J.; Huang, Y.; Zhou, L. Effects of arbuscular mycorrhizal fungi on inoculated seedling growth and rhizosphere soil aggregates. *Soil Tillage Res.* **2019**, *194*, 104340. [CrossRef]
89. Ji, L.; Tan, W.; Chen, X. Arbuscular mycorrhizal mycelial networks and glomalin-related soil protein increase soil aggregation in Calcaric Regosol under well-watered and drought stress conditions. *Soil Tillage Res.* **2019**, *185*, 1–8. [CrossRef]
90. Ker, K.; Charest, C. Nickel remediation by AM-colonized sunflower. *Mycorrhiza* **2010**, *20*, 399–406. [CrossRef] [PubMed]

91. Yang, Y.; Song, Y.; Scheller, H.V.; Ghosh, A.; Ban, Y.; Chen, H.; Tang, M. Community structure of arbuscular mycorrhizal fungi associated with *Robinia pseudoacacia* in uncontaminated and heavy metal contaminated soils. *Soil Biol. Biochem.* **2015**, *86*, 146–158. [CrossRef]
92. Wu, S.; Zhang, X.; Huang, L.; Chen, B. Arbuscular mycorrhiza and plant chromium tolerance. *Soil Ecol. Lett.* **2019**, *1*, 94–104. [CrossRef]
93. Miransari, M. Hyperaccumulators, arbuscular mycorrhizal fungi and stress of heavy metals. *Biotechnol. Adv.* **2011**, *29*, 645–653. [CrossRef]
94. Shalaby, A.M. Responses of arbuscular mycorrhizal fungal spores isolated from heavy metal-polluted and unpolluted soil to Zn, Cd, Pb and their interactions in vitro. *Pak. J. Biol. Sci.* **2003**, *6*, 1416–1422. [CrossRef]
95. Kullu, B.; Patra, D.K.; Acharya, S.; Pradhan, C.; Patra, H.K. AM fungi mediated bioaccumulation of hexavalent chromium in *Brachiaria mutica*-a mycorrhizal phytoremediation approach. *Chemosphere* **2020**, *258*, 127337. [CrossRef]
96. Janoušková, M.; Pavlíková, D. Cadmium immobilization in the rhizosphere of arbuscular mycorrhizal plants by the fungal extraradical mycelium. *Plant Soil* **2010**, *332*, 511–520. [CrossRef]
97. Wang, Q.; Mei, D.; Chen, J.; Lin, Y.; Liu, J.; Lu, H.; Yan, C. Sequestration of heavy metal by glomalin-related soil protein: Implication for water quality improvement in mangrove wetlands. *Water Res.* **2019**, *148*, 142–152. [CrossRef]
98. Wang, F.Y.; Wang, L.; Shi, Z.Y.; Li, Y.J.; Song, Z.M. Effects of AM inoculation and organic amendment, alone or in combination, on growth, P nutrition, and heavy-metal uptake of tobacco in Pb-Cd-contaminated soil. *J. Plant Growth Regul.* **2012**, *31*, 549–559. [CrossRef]
99. Chen, B.; Nayuki, K.; Kuga, Y.; Zhang, X.; Wu, S.; Ohtomo, R. Uptake and intraradical immobilization of cadmium by arbuscular mycorrhizal fungi as revealed by a stable isotope tracer and synchrotron radiation μX-ray fluorescence analysis. *Microbes Environ.* **2018**, *33*, 257–263. [CrossRef] [PubMed]
100. Chen, B.; Xiao, X.; Zhu, Y.-G.; Smith, F.A.; Xie, Z.M.; Smith, S.E. The arbuscular mycorrhizal fungus *Glomus mosseae* gives contradictory effects on phosphorus and arsenic acquisition by *Medicago sativa* Linn. *Sci. Total Environ.* **2007**, *379*, 226–234. [CrossRef] [PubMed]
101. Wu, S.; Zhang, X.; Sun, Y.; Wu, Z.; Li, T.; Hu, Y.; Lv, J.; Li, G.; Zhang, Z.; Zhang, J. Chromium immobilization by extra-and intraradical fungal structures of arbuscular mycorrhizal symbioses. *J. Hazard. Mater.* **2016**, *316*, 34–42. [CrossRef]
102. Danh, L.T.; Truong, P.; Mammucari, R.; Foster, N. A critical review of the arsenic uptake mechanisms and phytoremediation potential of Pteris vittata. *Int. J. Phytoremediat.* **2014**, *16*, 429–453. [CrossRef]
103. Otones, V.; Álvarez-Ayuso, E.; García-Sánchez, A.; Santa Regina, I.; Murciego, A. Arsenic distribution in soils and plants of an arsenic impacted former mining area. *Environ. Pollut.* **2011**, *159*, 2637–2647. [CrossRef]
104. Gupta, S.; Thokchom, S.D.; Kapoor, R. Arbuscular mycorrhiza improves photosynthesis and restores alteration in sugar metabolism in *Triticum aestivum* L. grown in arsenic contaminated soil. *Front. Plant Sci.* **2021**, *12*, 334. [CrossRef]
105. Zhang, X.; Hu, W.; Xie, X.; Wu, Y.; Liang, F.; Tang, M. Arbuscular mycorrhizal fungi promote lead immobilization by increasing the polysaccharide content within pectin and inducing cell wall peroxidase activity. *Chemosphere* **2021**, *267*, 128924. [CrossRef]
106. Pigna, M.; Cozzolino, V.; Violante, A.; Meharg, A.A. Influence of phosphate on the arsenic uptake by wheat (*Triticum durum* L.) irrigated with arsenic solutions at three different concentrations. *Water Air Soil Pollut.* **2009**, *197*, 371–380. [CrossRef]
107. Chandrakar, V.; Naithani, S.C.; Keshavkant, S. Arsenic-induced metabolic disturbances and their mitigation mechanisms in crop plants: A review. *Biologia* **2016**, *71*, 367–377. [CrossRef]
108. Stoeva, N.; Berova, M.; Zlatev, Z. Physiological response of maize to arsenic contamination. *Biol. Plant.* **2003**, *46*, 449–452. [CrossRef]
109. Emamverdian, A.; Ding, Y.; Mokhberdoran, F.; Xie, Y. Heavy metal stress and some mechanisms of plant defense response. *Sci. World J.* **2015**, *2015*, 756120. [CrossRef] [PubMed]
110. Majumder, B.; Das, S.; Biswas, S.; Mazumdar, A.; Biswas, A.K. Differential responses of photosynthetic parameters and its influence on carbohydrate metabolism in some contrasting rice (*Oryza sativa* L.) genotypes under arsenate stress. *Ecotoxicology* **2020**, *29*, 912–931. [CrossRef] [PubMed]
111. Garg, N.; Singla, P. Arsenic toxicity in crop plants: Physiological effects and tolerance mechanisms. *Environ. Chem. Lett.* **2011**, *9*, 303–321. [CrossRef]
112. Debona, D.; Rodrigues, F.A.; Datnoff, L.E. Silicon's role in abiotic and biotic plant stresses. *Ann. Rev. Phytopathol.* **2017**, *55*, 85–107. [CrossRef]
113. Finnegan, P.; Chen, W. Arsenic toxicity: The effects on plant metabolism. *Front. Physiol.* **2012**, *3*, 182. [CrossRef]
114. Codling, E.; Chaney, R.; Green, C. Accumulation of lead and arsenic by potato grown on lead–arsenate-contaminated orchard soils. *Commun. Soil Sci. Plant Anal.* **2016**, *47*, 799–807. [CrossRef]
115. Bustingorri, C.; Balestrasse, K.B.; Lavado, R.S. Effects of high arsenic and fluoride soil concentrations on soybean plants. *Phyton* **2015**, *84*, 407–416.
116. Das, H.; Mitra, A.K.; Sengupta, P.; Hossain, A.; Islam, F.; Rabbani, G. Arsenic concentrations in rice, vegetables, and fish in Bangladesh: A preliminary study. *Environ. Int.* **2004**, *30*, 383–387. [CrossRef]
117. Panaullah, G.M.; Alam, T.; Hossain, M.B.; Loeppert, R.H.; Lauren, J.G.; Meisner, C.A.; Ahmed, Z.U.; Duxbury, J.M. Arsenic toxicity to rice (*Oryza sativa* L.) in Bangladesh. *Plant Soil* **2009**, *317*, 31–39. [CrossRef]
118. Valko, M.; Rhodes, C.; Moncol, J.; Izakovic, M.; Mazur, M. Free radicals, metals and antioxidants in oxidative stress-induced cancer. *Chem. Biol. Interact.* **2006**, *160*, 1–40. [CrossRef] [PubMed]

119. Gamboa-Loira, B.; Cebrian, M.E.; Franco-Marina, F.; Lopez-Carrillo, L. Arsenic metabolism and cancer risk: A meta-analysis. *Environ. Res.* **2017**, *156*, 551–558. [CrossRef] [PubMed]
120. Khalid, S.; Shahid, M.; Niazi, N.K.; Rafiq, M.; Bakhat, H.F.; Imran, M.; Abbas, T.; Bibi, I.; Dumat, C. Arsenic behaviour in soil-plant system: Biogeochemical reactions and chemical speciation influences. In *Enhancing Cleanup of Environmental Pollutants*; Springer: Cham, Switzerland, 2017; pp. 97–140.
121. Lee, J.-T.; Yu, W.-. C Evaluation of legume growth in arsenic-polluted acidic soils with various pH values. *J. Water Sustain.* **2012**, *2*, 1.
122. Ullrich-Eberius, C.; Sanz, A.; Novacky, A. Evaluation of arsenate-and vanadate-associated changes of electrical membrane potential and phosphate transport in *Lemna gibba* G1. *J. Exp. Bot.* **1989**, *40*, 119–128. [CrossRef]
123. Pommerrenig, B.; Diehn, T.A.; Bienert, G.P. Metalloido-porins: Essentiality of Nodulin 26-like intrinsic proteins in metalloid transport. *Plant Sci.* **2015**, *238*, 212–227. [CrossRef] [PubMed]
124. Zhao, F.J.; Ma, J.F.; Meharg, A.; McGrath, S. Arsenic uptake and metabolism in plants. *New Phytol.* **2009**, *181*, 777–794. [CrossRef] [PubMed]
125. Bleeker, P.M.; Hakvoort, H.W.; Bliek, M.; Souer, E.; Schat, H. Enhanced arsenate reduction by a CDC25-like tyrosine phosphatase explains increased phytochelatin accumulation in arsenate-tolerant *Holcus lanatus*. *Plant J.* **2006**, *45*, 917–929. [CrossRef]
126. Boorboori, M.R.; Gao, Y.; Wang, H.; Fang, C. Usage of Si, P, Se, and Ca decrease arsenic concentration/toxicity in rice, a review. *Appl. Sci.* **2021**, *11*, 8090. [CrossRef]
127. Wu, F.; Ye, Z.; Wong, M.H. Intraspecific differences of arbuscular mycorrhizal fungi in their impacts on arsenic accumulation by *Pteris vittata* L. *Chemosphere* **2009**, *76*, 1258–1264. [CrossRef]
128. Xu, J.-Y.; Han, Y.-H.; Chen, Y.; Zhu, L.-J.; Ma, L.Q. Arsenic transformation and plant growth promotion characteristics of As-resistant endophytic bacteria from As-hyperaccumulator *Pteris vittata*. *Chemosphere* **2016**, *144*, 1233–1240. [CrossRef]
129. Spagnoletti, F.; Lavado, R.S. The arbuscular mycorrhiza Rhizophagus intraradices reduces the negative effects of arsenic on soybean plants. *Agronomy* **2015**, *5*, 188–199. [CrossRef]
130. Schneider, J.; Stürmer, S.L.; Guilherme, L.R.G.; de Souza Moreira, F.M.; de Sousa Soares, C.R.F. Arbuscular mycorrhizal fungi in arsenic-contaminated areas in Brazil. *J. Hazard. Mater.* **2013**, *262*, 1105–1115. [CrossRef] [PubMed]
131. Gonzalez-Chavez, C.; D'haen, J.; Vangronsveld, J.; Dodd, J. Copper sorption and accumulation by the extraradical mycelium of different *Glomus* spp. (arbuscular mycorrhizal fungi) isolated from the same polluted soil. *Plant Soil* **2002**, *240*, 287–297. [CrossRef]
132. Huang, Z.; Zhao, F.; Hua, J.; Ma, Z. Prediction of the distribution of arbuscular mycorrhizal fungi in the metal (loid)-contaminated soils by the arsenic concentration in the fronds of *Pteris vittata* L. *J. Soils Sediments* **2018**, *18*, 2544–2551. [CrossRef]
133. Su, S.; Zeng, X.; Bai, L.; Williams, P.N.; Wang, Y.; Zhang, L.; Wu, C. Inoculating chlamydospores of Trichoderma asperellum SM-12F1 changes arsenic availability and enzyme activity in soils and improves water spinach growth. *Chemosphere* **2017**, *175*, 497–504. [CrossRef]
134. Li, J.; Chen, B.; Zhang, X.; Hao, Z.; Zhang, X.; Zhu, Y. Arsenic transformation and volatilization by arbuscular mycorrhizal symbiosis under axenic conditions. *J. Hazard. Mater.* **2021**, *413*, 125390. [CrossRef]
135. Spagnoletti, F.N.; Balestrasse, K.; Lavado, R.S.; Giacometti, R. Arbuscular mycorrhiza detoxifying response against arsenic and pathogenic fungus in soybean. *Ecotoxicol. Environ. Saf.* **2016**, *133*, 47–56. [CrossRef]
136. Maldonado-Mendoza, I.E.; Harrison, M.J. RiArsB and RiMT-11: Two novel genes induced by arsenate in arbuscular mycorrhiza. *Fungal Biol.* **2018**, *122*, 121–130. [CrossRef]
137. Lomax, C.; Liu, W.J.; Wu, L.; Xue, K.; Xiong, J.; Zhou, J.; McGrath, S.P.; Meharg, A.A.; Miller, A.J.; Zhao, F.J. *Methylated arsenic species in plants originate from soil microorganisms*. *New Phytol.* **2012**, *193*, 665–672. [CrossRef]
138. Zhang, Z.; Guo, G.; Teng, Y.; Wang, J.; Rhee, J.S.; Wang, S.; Li, F. Screening and assessment of solidification/stabilization amendments suitable for soils of lead-acid battery contaminated site. *J. Hazard. Mater.* **2015**, *288*, 140–146. [CrossRef]
139. Li, J.; Sun, Y.; Jiang, X.; Chen, B.; Zhang, X. Arbuscular mycorrhizal fungi alleviate arsenic toxicity to *Medicago sativa* by influencing arsenic speciation and partitioning. *Ecotoxicol. Environ. Saf.* **2018**, *157*, 235–243. [CrossRef] [PubMed]
140. Vodnik, D.; Grčman, H.; Maček, I.; Van Elteren, J.; Kovačevič, M. The contribution of glomalin-related soil protein to Pb and Zn sequestration in polluted soil. *Sci. Total Environ.* **2008**, *392*, 130–136. [CrossRef] [PubMed]
141. Spagnoletti, F.; Carmona, M.; Gómez, N.E.T.; Chiocchio, V.; Lavado, R.S. Arbuscular mycorrhiza reduces the negative effects of *M. phaseolina* on soybean plants in arsenic-contaminated soils. *Appl. Soil Ecol.* **2017**, *121*, 41–47. [CrossRef]
142. Christophersen, H.M.; Smith, F.A.; Smith, S.E. Unraveling the influence of arbuscular mycorrhizal colonization on arsenic tolerance in Medicago: Glomus mosseae is more effective than *G. intraradices*, associated with lower expression of root epidermal Pi transporter genes. *Front. Physiol.* **2012**, *3*, 91. [CrossRef]
143. Zafarzadeh, A.; Rahimzadeh, H.; Mahvi, A.H. Health risk assessment of heavy metals in vegetables in an endemic esophageal cancer region in Iran. *Health Scope* **2018**, *7*, 7. [CrossRef]
144. McBride, M.B. Chemisorption and precipitation of inorganic ions. In *Environmental Chemistry of Soils*; Oxford University Press: New York, NY, USA, 1994; pp. 121–168.
145. Hou, S.; Zheng, N.; Tang, L.; Ji, X.; Li, Y.; Hua, X. Pollution characteristics, sources, and health risk assessment of human exposure to Cu, Zn, Cd and Pb pollution in urban street dust across China between 2009 and 2018. *Environ. Int.* **2019**, *128*, 430–437. [CrossRef]
146. Zhao, F.-J.; Ma, Y.; Zhu, Y.-G.; Tang, Z.; McGrath, S.P. Soil contamination in China: Current status and mitigation strategies. *Environ. Sci. Technol.* **2015**, *49*, 750–759. [CrossRef]

147. Kasemodel, M.; Sakamoto, I.; Varesche, M.; Rodrigues, V. Potentially toxic metal contamination and microbial community analysis in an abandoned Pb and Zn mining waste deposit. *Sci. Total Environ.* **2019**, *675*, 367–379. [CrossRef]
148. Liu, Y.; Wang, X.; Zeng, G.; Qu, D.; Gu, J.; Zhou, M.; Chai, L. Cadmium-induced oxidative stress and response of the ascorbate–glutathione cycle in *Bechmeria nivea* (L.) Gaud. *Chemosphere* **2007**, *69*, 99–107. [CrossRef]
149. Beryllium, I. Cadmium, mercury, and exposures in the glass manufacturing industry. Working Group views and expert opinions, Lyon, 9–16 February 1993. *IARC Monogr. Eval. Carcinog. Risks Hum.* **1993**, *58*, 1–415.
150. Fang, W.; Wei, Y.; Liu, J. Comparative characterization of sewage sludge compost and soil: Heavy metal leaching characteristics. *J. Hazard. Mater.* **2016**, *310*, 1–10. [CrossRef] [PubMed]
151. Dusek, J.; Vogel, T.; Lichner, L.; Cipakova, A. Short-term transport of cadmium during a heavy-rain event simulated by a dual-continuum approach. *J. Plant Nutr. Soil Sci.* **2010**, *173*, 536–547. [CrossRef]
152. Mi, Y.; Zhan, F.; Li, B.; Qin, L.; Wang, J.; Zu, Y.; Li, Y. Distribution characteristics of cadmium and lead in particle size fractions of farmland soils in a lead–zinc mine area in Southwest China. *Environ. Syst. Res.* **2018**, *7*, 14. [CrossRef]
153. Uraguchi, S.; Mori, S.; Kuramata, M.; Kawasaki, A.; Arao, T.; Ishikawa, S. Root-to-shoot Cd translocation via the xylem is the major process determining shoot and grain cadmium accumulation in rice. *J. Exp. Bot.* **2009**, *60*, 2677–2688. [CrossRef] [PubMed]
154. Garg, N.; Aggarwal, N. Effect of mycorrhizal inoculations on heavy metal uptake and stress alleviation of *Cajanus cajan* (L.) Millsp. genotypes grown in cadmium and lead contaminated soils. *Plant Growth Regul.* **2012**, *66*, 9–26. [CrossRef]
155. Wahid, A.; Arshad, M.; Farooq, M. Cadmium phytotoxicity: Responses, mechanisms and mitigation strategies: A review. In *Organic Farming, Pest Control and Remediation of Soil Pollutants*; Springer: Dordrecht, The Netherlands, 2009; pp. 371–403.
156. Vivas, A.; Vörös, A.; Biró, B.; Barea, J.; Ruiz-Lozano, J.; Azcón, R. Beneficial effects of indigenous Cd-tolerant and Cd-sensitive *Glomus mosseae* associated with a Cd-adapted strain of *Brevibacillus* sp. in improving plant tolerance to Cd contamination. *Appl. Soil Ecol.* **2003**, *24*, 177–186. [CrossRef]
157. Li, R.-Y.; Ago, Y.; Liu, W.-J.; Mitani, N.; Feldmann, J.; McGrath, S.P.; Ma, J.F.; Zhao, F.-J. The rice aquaporin Lsi1 mediates uptake of methylated arsenic species. *Plant Physiol.* **2009**, *150*, 2071–2080. [CrossRef]
158. Sun, H.; Xie, Y.; Zheng, Y.; Lin, Y.; Yang, F. The enhancement by arbuscular mycorrhizal fungi of the Cd remediation ability and bioenergy quality-related factors of five switchgrass cultivars in Cd-contaminated soil. *PeerJ* **2018**, *6*, e4425. [CrossRef]
159. Jiang, Q.-Y.; Zhuo, F.; Long, S.-H.; Zhao, H.-D.; Yang, D.-J.; Ye, Z.-H.; Li, S.-S.; Jing, Y.-X. Can arbuscular mycorrhizal fungi reduce Cd uptake and alleviate Cd toxicity of Lonicera japonica grown in Cd-added soils? *Sci. Rep.* **2016**, *6*, 21805. [CrossRef]
160. Zhan, F.; Li, B.; Jiang, M.; Yue, X.; He, Y.; Xia, Y.; Wang, Y. Arbuscular mycorrhizal fungi enhance antioxidant defense in the leaves and the retention of heavy metals in the roots of maize. *Environ. Sci. Pollut. Res.* **2018**, *25*, 24338–24347. [CrossRef]
161. Abdelhameed, R.E.; Metwally, R.A. Alleviation of cadmium stress by arbuscular mycorrhizal symbiosis. *Int. J. Phytoremediat.* **2019**, *21*, 663–671. [CrossRef] [PubMed]
162. Zhan, F.; Li, B.; Jiang, M.; Li, T.; He, Y.; Li, Y.; Wang, Y. Effects of arbuscular mycorrhizal fungi on the growth and heavy metal accumulation of bermudagrass [*Cynodon dactylon* (L.) Pers.] grown in a lead–zinc mine wasteland. *Int. J. Phytoremediat.* **2019**, *21*, 849–856. [CrossRef] [PubMed]
163. Gonzalez-Chavez, M.; Carrillo-Gonzalez, R.; Wright, S.; Nichols, K. The role of glomalin, a protein produced by arbuscular mycorrhizal fungi, in sequestering potentially toxic elements. *Environ. Pollut.* **2004**, *130*, 317–323. [CrossRef] [PubMed]
164. Gil-Cardeza, M.L.; Ferri, A.; Cornejo, P.; Gomez, E. Distribution of chromium species in a Cr-polluted soil: Presence of Cr (III) in glomalin related protein fraction. *Sci. Total Environ.* **2014**, *493*, 828–833. [CrossRef] [PubMed]
165. Gunathilakae, N.; Yapa, N.; Hettiarachchi, R. Effect of arbuscular mycorrhizal fungi on the cadmium phytoremediation potential of *Eichhornia crassipes* (Mart.) Solms. *Groundw. Sustain. Dev.* **2018**, *7*, 477–482. [CrossRef]
166. Audet, P.; Charest, C. Allocation plasticity and plant–metal partitioning: Meta-analytical perspectives in phytoremediation. *Environ. Pollut.* **2008**, *156*, 290–296. [CrossRef]
167. Gabarrón, M.; Faz, A.; Acosta, J. Effect of different industrial activities on heavy metal concentrations and chemical distribution in topsoil and road dust. *Environ. Earth Sci.* **2017**, *76*, 129. [CrossRef]
168. Moosavi, S.; Seghatoleslami, M. Phytoremediation: A review. *Adv. Agric. Biol.* **2013**, *1*, 5–11. Available online: www.pscipub.com/AAB (accessed on 8 November 2021).
169. Cheng, H.; Hu, Y. Lead (Pb) isotopic fingerprinting and its applications in lead pollution studies in China: A review. *Environ. Pollut.* **2010**, *158*, 1134–1146. [CrossRef]
170. Gupta, V.; Jatav, P.K.; Verma, R.; Kothari, S.L.; Kachhwaha, S. Nickel accumulation and its effect on growth, physiological and biochemical parameters in millets and oats. *Environ. Sci. Pollut. Res.* **2017**, *24*, 23915–23925. [CrossRef]
171. Balakhnina, T.I.; Borkowska, A.; Nosalewicz, M.; Nosalewicz, A.; Wlodarczyk, T.M.; Kosobryukhov, A.A.; Fomina, I.R. Effect of temperature on oxidative stress induced by lead in the leaves of *Plantago major* L. *Int. Agrophys.* **2016**, *30*, 285–292. [CrossRef]
172. Adejumo, S.A.; Ogundiran, M.B.; Togun, A.O. Soil amendment with compost and crop growth stages influenced heavy metal uptake and distribution in maize crop grown on lead-acid battery waste contaminated soil. *J. Environ. Chem. Eng.* **2018**, *6*, 4809–4819. [CrossRef]
173. Cao, S.; Duan, X.; Zhao, X.; Wang, B.; Ma, J.; Fan, D.; Sun, C.; He, B.; Wei, F.; Jiang, G. Health risk assessment of various metal (loid) s via multiple exposure pathways on children living near a typical lead-acid battery plant, China. *Environ. Pollut.* **2015**, *200*, 16–23. [CrossRef] [PubMed]
174. Flora, G.; Gupta, D.; Tiwari, A. Toxicity of lead: A review with recent updates. *Interdiscip. Toxicol.* **2012**, *5*, 47–58. [CrossRef] [PubMed]

175. Saputra, H.M.; Mandia, S.; Retnoaji, B.; Wijayanti, N. Antioxidant properties of liverwort (*Marchantia polymorpha* L.) to lead-induced oxidative stress on HEK293 cells. *J. Biol. Sci.* **2016**, *16*, 77–85. [CrossRef]
176. Salazar, M.J.; Menoyo, E.; Faggioli, V.; Geml, J.; Cabello, M.; Rodriguez, J.H.; Marro, N.; Pardo, A.; Pignata, M.L.; Becerra, A.G. Pb accumulation in spores of arbuscular mycorrhizal fungi. *Sci. Total Environ.* **2018**, *643*, 238–246. [CrossRef]
177. González-Chávez, M.d.C.A.; Carrillo-González, R.; Cuellar-Sánchez, A.; Delgado-Alvarado, A.; Suárez-Espinosa, J.; Ríos-Leal, E.; Solís-Domínguez, F.A.; Maldonado-Mendoza, I.E. Phytoremediation assisted by mycorrhizal fungi of a Mexican defunct lead-acid battery recycling site. *Sci. Total Environ.* **2019**, *650*, 3134–3144. [CrossRef]
178. Chen, L.; Hu, X.; Yang, W.; Xu, Z.; Zhang, D.; Gao, S. The effects of arbuscular mycorrhizal fungi on sex-specific responses to Pb pollution in *Populus cathayana*. *Ecotoxicol. Environ. Saf.* **2015**, *113*, 460–468. [CrossRef]
179. Baghaie, A.H.; Aghilizefreei, A. Neighbor presence of plant growth-promoting rhizobacteria (PGPR) and arbuscular mycorrhizal fungi (AMF) can increase sorghum phytoremediation efficiency in a soil treated with Pb polluted cow manure. *J. Hum. Environ. Health Promot.* **2019**, *5*, 153–159. [CrossRef]
180. Khan, M.A.; Ramzani, P.M.A.; Zubair, M.; Rasool, B.; Khan, M.K.; Ahmed, A.; Khan, S.A.; Turan, V.; Iqbal, M. Associative effects of lignin-derived biochar and arbuscular mycorrhizal fungi applied to soil polluted from Pb-acid batteries effluents on barley grain safety. *Sci. Total Environ.* **2020**, *710*, 136294. [CrossRef]
181. Viti, C.; Marchi, E.; Decorosi, F.; Giovannetti, L. Molecular mechanisms of Cr (VI) resistance in bacteria and fungi. *FEMS Microbiol. Rev.* **2014**, *38*, 633–659. [CrossRef] [PubMed]
182. de los Angeles Beltrán-Nambo, M.; Rojas-Jacuinde, N.; Martinez-Trujillo, M.; Jaramillo-López, P.F.; Romero, M.G.; Carreón-Abud, Y. Differential strategies of two species of arbuscular mycorrhizal fungi in the protection of maize plants grown in chromium-contaminated soils. *BioMetals* **2021**, *34*, 1247–1261. [CrossRef] [PubMed]
183. Dhal, B.; Thatoi, H.; Das, N.; Pandey, B. Chemical and microbial remediation of hexavalent chromium from contaminated soil and mining/metallurgical solid waste: A review. *J. Hazard. Mater.* **2013**, *250*, 272–291. [CrossRef] [PubMed]
184. Nigam, H.; Das, M.; Chauhan, S.; Pandey, P.; Swati, P.; Yadav, M.; Tiwari, A. Effect of chromium generated by solid waste of tannery and microbial degradation of chromium to reduce its toxicity: A review. *Adv. Appl. Sci. Res.* **2015**, *6*, 129–136.
185. Dayan, A.; Paine, A. Mechanisms of chromium toxicity, carcinogenicity and allergenicity: Review of the literature from 1985 to 2000. *Hum. Exp. Toxicol.* **2001**, *20*, 439–451. [CrossRef]
186. Kumar, P. Soil applied glycine betaine with Arbuscular mycorrhizal fungi reduces chromium uptake and ameliorates chromium toxicity by suppressing the oxidative stress in three genetically different Sorghum (*Sorghum bicolor* L.) cultivars. *BMC Plant Biol.* **2021**, *21*, 1–16. [CrossRef]
187. Kováčik, J.; Babula, P.; Hedbavny, J.; Kryštofová, O.; Provaznik, I. Physiology and methodology of chromium toxicity using alga *Scenedesmus quadricauda* as model object. *Chemosphere* **2015**, *120*, 23–30. [CrossRef]
188. Gill, R.A.; Zang, L.; Ali, B.; Farooq, M.A.; Cui, P.; Yang, S.; Ali, S.; Zhou, W. Chromium-induced physio-chemical and ultrastructural changes in four cultivars of *Brassica napus* L. *Chemosphere* **2015**, *120*, 154–164. [CrossRef]
189. Hussain, A.; Rizwan, M.; Ali, Q.; Ali, S. Seed priming with silicon nanoparticles improved the biomass and yield while reduced the oxidative stress and cadmium concentration in wheat grains. *Environ. Sci. Pollut. Res.* **2019**, *26*, 7579–7588. [CrossRef]
190. Ullah, S.; Ali, R.; Mahmood, S.; Atif Riaz, M.; Akhtar, K. Differential growth and metal accumulation response of *Brachiaria mutica* and *Leptochloa fusca* on cadmium and lead contaminated soil. *Soil Sediment Contam. Int. J.* **2020**, *29*, 844–859. [CrossRef]
191. Komal, T.; Mustafa, M.; Ali, Z.; Kazi, A.G. Heavy metal uptake and transport in plants. In *Heavy Metal Contamination of Soils*; Springer: Cham, Switzerland, 2015; pp. 181–194.
192. Singh, M.; Srivastava, P.; Verma, P.; Kharwar, R.; Singh, N.; Tripathi, R. Soil fungi for mycoremediation of arsenic pollution in agriculture soils. *J. Appl. Microbiol.* **2015**, *119*, 1278–1290. [CrossRef] [PubMed]
193. Gil-Cardeza, M.L.; Calonne-Salmon, M.; Gómez, E.; Declerck, S. Short-term chromium (VI) exposure increases phosphorus uptake by the extraradical mycelium of the arbuscular mycorrhizal fungus *Rhizophagus irregularis* MUCL 41833. *Chemosphere* **2017**, *187*, 27–34. [CrossRef] [PubMed]
194. Wu, S.; Zhang, X.; Sun, Y.; Wu, Z.; Li, T.; Hu, Y.; Su, D.; Lv, J.; Li, G.; Zhang, Z. Transformation and immobilization of chromium by arbuscular mycorrhizal fungi as revealed by SEM–EDS, TEM–EDS, and XAFS. *Environ. Sci. Technol.* **2015**, *49*, 14036–14047. [CrossRef] [PubMed]
195. Holland, S.L.; Avery, S.V. Chromate toxicity and the role of sulfur. *Metallomics* **2011**, *3*, 1119–1123. [CrossRef] [PubMed]
196. Sobrino-Plata, J.; Meyssen, D.; Cuypers, A.; Escobar, C.; Hernández, L.E. Glutathione is a key antioxidant metabolite to cope with mercury and cadmium stress. *Plant Soil* **2014**, *377*, 369–381. [CrossRef]

Journal of *Fungi*

MDPI

Article

The Edible Gray Oyster Fungi *Pleurotus ostreatus* (Jacq. ex Fr.) P. Kumm a Potent Waste Consumer, a Biofriendly Species with Antioxidant Activity Depending on the Growth Substrate

Raluca A. Mihai [1,*], Erly J. Melo Heras [1], Larisa I. Florescu [2] and Rodica D. Catana [3]

1   CICTE, Department of Life Science and Agriculture, Universidad de Las Fuerzas Armadas—ESPE, Av. General Rumiñahui s/n y, Sangolqui 171103, Ecuador; ejmelo@espe.edu.ec
2   Taxonomy, Ecology and Nature Conservation Department, Institute of Biology Bucharest of Romanian Academy, 296 Splaiul Independentei, 060031 Bucharest, Romania; larisa.florescu@ibiol.ro
3   Developmental Biology Department, Institute of Biology Bucharest of Romanian Academy, 296 Splaiul Independentei, 060031 Bucharest, Romania; rodica.catana@ibiol.ro
*   Correspondence: rmihai@espe.edu.ec

Abstract: Nowadays, climate change is not the only threat facing our planet. There are also other types of pollution such as waste that poisons soils and water and kills plants, harming humans and animals. Sustainability represents a key issue for the actual Global Citizen. For this reason, our article is dedicated to offering biofriendly solutions to decrease wastes, give them a positive meaning, such as a substrate for an edible oyster fungus with nutritive and biological properties usefully for humans. Three types of wastes such as coconut coir, pine sawdust, and paper waste—representative symbols of pollution in Ecuador—have been tested as suitable growing substrate for the edible fungi *Pleurotus ostreatus* (Jacq. ex Fr.) P. Kumm by analyzing parameters such as *Biological Efficiency, Mushroom Yield, and Productive Rate*. The influence of these "waste" substrates on the nutritive (protein content), biological characteristic (antioxidant activity), and the content of human-health-sustaining compounds (phenols, flavonoids) were also evaluated using the Kjeldahal, DPPH, ABTS, FRAP, and Folin–Ciocalteu methods. The results indicate that all the waste products represent desirable substrates for growing the edible fungi, with more focus on coconut coir waste (one of the principal pollution problems in Ecuador), but that also achieved the increase in the fungi's desirable characteristics. Coconut coir waste could be an environmentally friendly solution that also offers for humans additional nutritive and healthy benefits.

Keywords: waste; recyclable substrates; oyster fungi; antioxidants; phenolic compounds

Citation: Mihai, R.A.; Melo Heras, E.J.; Florescu, L.I.; Catana, R.D. The Edible Gray Oyster Fungi *Pleurotus ostreatus* (Jacq. ex Fr.) P. Kumm a Potent Waste Consumer, a Biofriendly Species with Antioxidant Activity Depending on the Growth Substrate. *J. Fungi* **2022**, *8*, 274. https://doi.org/10.3390/jof8030274

Academic Editor: Birgitte Andersen

Received: 31 January 2022
Accepted: 4 March 2022
Published: 9 March 2022

Publisher's Note: MDPI stays neutral with regard to jurisdictional claims in published maps and institutional affiliations.

1. Introduction

All over the world, different types of wastes are generated from different industries and household waste, etc. Since 2014, the waste issue has been recognized as a global environmental problem [1]. An inefficient waste management contributes to air pollution, affecting the health and well-being of ecosystems, species, and humans. Recycling plays important role in finding a sustainable solution, with low impact on the environment, which improves the circular economy and protects natural resources.

An immense amount of waste was produced in the last years due to the intensification of agricultural and industrial activities [2], being estimated at ~998 million tons/year (only the agricultural waste) [3]. Increasing and improper disposal of agro-industrial waste becomes a major source of pollution, affecting the population and environment health and amplifying the global emissions of greenhouse gasses [4]. The use of agro-industrial waste as raw materials for various products (antioxidants through solid-state fermentation) may help to reduce the cost of production, reducing the environmental pollution [5]. Reduction in greenhouse gas emissions, bio-energy, bio-conversion, new jobs, and new green markets are some benefits of recycling agricultural solid wastes, which encourages future studies [6].

In tropical areas, a large amount of unused lignocellulosic by-products is available, being left to rot in the field or burned [7]. Ecuador is one of those areas which has an ideal climate for coconut cultivation, Esmeraldas being the province with the highest coconut production (77.26%), followed by Manabí (18.72%). In Esmeraldas, the cultivation of coconut is concentrated in the border cantons: Eloy Alfaro and San Lorenzo del Pailón [8]. Coconut residue is a by-product waste rich in beneficial nutrients and available in high quantities in Ecuador, without cost [9]. Due to its alimental characteristics, the fiber and shell of the coconut represent wastes with great potential as a component of growing substrates for edible mushrooms [10]. Despite its nutritional benefits (0.52% protein, 0.29% fat, 35.44% fiber, 3.94% ash, and 20.53% humidity, low conductivity, resistance to impact and bacteria [11]), the use of coconut waste in the country is almost null, being considered an environmental problem [12].

Pine wood agroindustry or by-products wastes are suitable for bio-energy production but are difficult to use due to their chemical composition [13], so an alternative for these wastes can be as a substrate for the culture of *Pleurotus* [14], scarcely applied at the moment. At the same time, the pulp and paper industry occupies the place of the third largest industrial polluter of the environment (air, water, and soil). Chlorine-based bleaches used during paper production result in toxic materials being released into the environment. In addition, methane is formed during the rotting of paper waste, and is 25 times more toxic than CO_2. At the moment, the volume of global waste paper has reached 1010 million tons in the last 25 years [15].

In the context of the high amount of different waste sources (coconut, wood, paper), the cultivation of *P. ostreatus* on these waste by-products may be one of the solutions to transform the inedible waste into an edible rich in proteins and with high antioxidant character biomass of high market value [7].

Fungi are of interest in various bioremediation processes due to their ability to grow and develop in different substrates (natural and synthetic solid materials) [16]. Due to their ligninocellulite enzymes [17], fungi are able to transform the lignocellulosic biomass represented by agricultural and forestry wastes into food [18]. This capacity has a high potential application in most industries (paper, chemical, textile, food), agricultural processes, forage production [19], and bioenergy production from agro-industrial waste [20].

Mushroom cultivation represents a useful method to manage environmental pollution. The oyster fungus grows on different materials as by-products or waste from agricultural and agro-industrial activities, which brings an environmental benefit due to the importance of waste management [18]. The genus *Pleurotus* is recognized as species with high nutritional value, having various biotechnological and environmental applications [16]. *Pleurotus* species represent not only a commercially important edible mushroom [21] but also an important component in recycling agricultural wastes turning into protein-rich food. Due to their attractive taste and aroma, nutritional and medicinal value, these species are cultivated globally using numerous agricultural by-products (banana, corn, sugarcane leaves, peanut hull, wheat, rice straw, mango fruits and seeds, leaves) with low-cost production techniques [3]. *Pleurotus* species are considered an important source of dietary fiber, contain numerous important nutrients and polyphenols, which assure their antioxidant character and ability to inhibit free radicals [22]. Although the human body is a balance between free radicals and antioxidants, it is still necessary as a dietary antioxidant to reduce the oxidative stress from the environment [23]. These mushrooms are considered a cheap source of protein since they convert agricultural waste into food [24].

Pleurotus ostreatus is member of the *Pleurotus* genus and represents the second most cultivated edible mushroom after *Agaricus bisporus* due to its economic (edible), ecological (bioremediation agents), and medicinal value (antioxidant activity and biocompounds source) [25]. This fungus is able to colonize a large variety of lignocellulosic substrates and other agricultural, forest, and food-processing wastes [26], and its cultivation is an effective alternative to produce valuable food and nutraceuticals. As a member of basidiomycetes group, it brings to light the presence of ribotoxin—like protein (Ostreatin), a novel specific ribonucleases family. Ostrein could contribute to assigning potential biotechnological applications in agriculture (crops protection towards pathogens or pests) and in medicine (cytotoxic effects) [27].

The advantage of *P. ostreatus* is that it has a shorter growth time (comparing with other edible mushrooms), the substrate used requires pasteurization, it has high profitability (converting a high percentage of the substrate to fruiting bodies), and it is less attacked by diseases and pests [26].

Although there are some studies concerning the cultivation of *P. ostreatus* on different substrates (chestnut, corncobs, beech, oak, linden, potatoes farm wastes, walnut, poplar, peanut wastes, walnut and orange tree sawdust, etc.), to our knowledge, there are still limited studies about the antioxidant character of *P. ostreatus* cultivated on recyclable substrates such as pine wood, coconut coir (fiber, shells), and waste paper.

The study was conducted to evaluate different agricultural wastes (pine sawdust, coconut coir, and paper waste), representative of contamination for Ecuador, as suitable opportunity to diminish the pollution by their recycling into growing substrate for a bio-friendly fungi with desirable organoleptic, phytochemical properties, and antioxidant activity, with final human health and protective planet benefits.

2. Materials and Methods

2.1. Study Area, Experimental Substrates, and Spawn Preparation

Three different contaminant wastes (pine sawdust—T1; coconut coir—T2; and waste paper—T3) were evaluated for biological efficiency, yield, production rate, and antioxidant activity of oyster mushroom. In our study, coconut coir was represented by the mixture of fiber and sawdust from dried coconut shells, collected from local coconut plantations in Esmeraldas, localities Borbón, Limones. Paper waste was collected from the administrative offices of the University, and pine sawdust was purchased from wood factories. It was taken into account that each substrate must have a humidity lower than 12% for its use, securing a predominantly dry and mold-free material [28]. The substrates were ground into 2–5 cm length pellets and soaked in water at 60 °C for 24 h. Excess water was drained off from the materials, mixed with 10% wheat bran, and 2% gypsum (calcium sulfate) as nutrient supplement, at pH 9 [29]. Water content was attained until optimum moisture (75%). The mixture was then filled into $9' \times 14'$ polystyrene bags that were pasteurized for 6 h to eliminate bacteria or other contaminant fungi that would compete with the oyster mushroom for the substrate.

Pure culture of oyster mushroom (*P. ostreatus*) was obtained from the mushroom farm "Fungi Andino" located in the neighborhood La Morita, Tumbaco Valley in Quito, Ecuador. Spawns were prepared in 1 kg polystyrene plastic bags filled with 600 g of oatmeal kernel, which was hydrated with 75% of water. Each bag was supplemented with 2% gypsum, in terms of dry weight basis and pasteurized for 6 h. After cooling at room temperature, each sterilized bag was inoculated with 250 g of primary mycelium. The spawn was incubated for 28 days at 25 °C.

2.1.1. Inoculation, Incubation, and Harvest

The sterilized bags were then allowed to cool at room temperature. The inoculation took place in a tissue culture hood to avoid contamination, so the bags were aseptically inoculated with a piece, 10% in weight, of mycelium culture of 14 days old. The bags were subsequently incubated at 21 ± 3 °C for 28 days until the mycelium fully invaded the substrate and then relocated to a fruiting room with light entry with a mean of 12 h of light per day. After mycelium growth in the bags became abundant, to facilitate the development of fruiting bodies, perforations were created in the bags. Then, the fully colonized substrates were transferred to the growth room and placed on racks at a spacing of 15–20 cm. Proper ventilation of the growth room was assured by opening the door every 2–3 days. To keep the mycelium moist, the inoculated bags were watered 2–3 times a day. Relative humidity (RH) and room temperature were monitored and maintained with thermo-hygrometer hydrometer, and RH was maintained between 80 and 85% by spraying fine mist of water occasionally [30].

2.1.2. Biological Efficiency, Mushroom Yield, and Production Rate

The growth and development of mushrooms were monitored daily. The time (number of days) required from inoculation to completion of mycelium running, time elapsed between opening the plastic bags to pinhead formation, and time required from opening the plastic bags to first round harvesting were recorded. One round of mushroom harvest was made across all substrate types in the course of the experiment.

To evaluate the growth performance of mushrooms on different substrates, yield, biological efficiency, and production rate were calculated for each experimental treatment. The mushroom yield (%) was determined by weighing the total amount of harvested fruiting bodies without taking the base out, which was then divided by the total weight of the dry substrate used. Biological yield (g) was determined by weighing the whole cluster of fruiting bodies without removing the base of stalks, and economic yield (g) was determined by weighing all the fruiting bodies on a substrate after removing the base of stalks. Finally, biological efficiency (%) was calculated by using the following equation:

$$\% \, BE = \frac{FW_m}{DW_s} * 100\% \tag{1}$$

where BE—the Biological Efficiency (%); FW_m—the fresh weight (g) of the harvested mushrooms; and DW_s—the dry weight of the substrate (g) [31].

The production rate (%) was determined by dividing the biological efficiency (%) between the total number of days of the process [TP = EB (%)/number of days of the process] [32].

2.1.3. Preparation of the Fungi Solution

The fresh oyster *Pleurotus* were cleaned with distilled water before oven drying at 40 °C. When heating to constant weight was achieved, the dried material was grinded in a laboratory mill, and 1.0 g was critically weighed and extracted in an ultrasonic cleaner at 50 °C with 40 times of 80% methanol. The solution was then filtered through a Whatmann filter, and the filter extract was concentrated into a dry powder by the rotary evaporator at 50 °C, dissolved in 70% ethanol, and placed in 25 mL volumetric flasks for further antioxidant and metabolic content analysis.

2.1.4. Determination of the Antioxidant Aptness

The DPPH free radical scavenging activity of fungi samples were determined according to the method described by [33]. The DPPH solution was prepared as reported by [34]. After weighting 7.89 mg DPPH on a chemical balance, it was dissolved in 99.5% ethanol to obtain a constant volume by filling 100 mL of a measuring flask (0.2 mM DPPH). The formed solution was kept in the dark for 2 h until the absorbance was stabilized. After this, 1 mL of DPPH solution was added into a test tube, followed by 200 μL of ethanol and 800 μL of 0.1 M Tris·HCl buffer (pH 7.4). After mixing, the absorbance at 517 nm was measured. A mixed solution containing 1.2 mg of ethanol and 800 μL of Tis·HCl buffer was used as a blank. When the absorbance was in a range of 1.00, the prepared solution was used directly for the measurements. If the absorbance exceeded 1.05, ethanol was added to dilute the solution until the absorbance was in the range of 1.00. The solution used for measurements was stored in the dark during the assay. Briefly, the DPPH Radical Scavenging Assay consists of adding an aliquot (40 μL) of fungi extract to 3 mL of methanolic DPPH solution. The change in absorbance at 515 nm was measured after 30 min, and the antiradical activity (AA) was determined using the following formula:

$$AA\% = 100 - [(\text{Abs: sample} - \text{Abs: empty sample})] \times 100)/\text{Abs: control} \tag{2}$$

The optic density of the samples, the control and the empty samples, were measured in comparison with methanol. One synthetic antioxidant represented by Trolox was used as positive control. The antioxidant capacity based on the DPPH free radical scavenging ability of the extract was expressed as μmol Trolox equivalent per gram of dry weight of fungi material.

2.1.5. ABTS Free Radical Scavenging Assay

The ABTS radical cation scavenging activity was performed according to [35], with slight modifications. The ABTS solution (7 mM) was reacted with potassium persulfate (2.45 mM) solution and kept overnight in the dark to yield a dark-green-color solution containing ABTS radical cation. Prior to use in the assay, the ABTS radical cation was diluted with 50% methanol for an initial absorbance of about 0.700 ± 0.02 at 734 nm using UV-VIS spectrophotometer. Free radical scavenging activity was assayed by mixing 100 μL of test sample with 2.9 mL of an ABTS working standard in a microcuvette. The decrease in absorbance was measured at exactly 1 min after mixing the solution and then at 1-min intervals up to 6 min, when final absorbance was recorded. The inhibition % was calculated using the formula: Inhibition% = A (control) − A (test sample)/A (control) × 100.

The antioxidant radical scavenging activity in the mushroom extract is evaluated using DPPH and ABTS radicals. The absorbance was measured against the reagent blank at 515 nm for DPPH and 435 nm for ABTS. Three repetitions were performed. The radical scavenging activity is calculated according to the regression equations for DPPH ($y = -0.8033x + 0.8131$, $r^2 = 0.9542$) and ABTS ($y = -0.4202x + 0.7797$, $r^2 = 0.9903$) obtained from the TROLOX calibration curves.

2.1.6. Reducing Ability (FRAP Assay)

The ability to reduce ferric ions was measured using a modified method of [36]. An aliquot (200 μL) of the fungi extract with appropriate dilution was added to 3 mL of FRAP reagent (10 parts of 300 mM sodium acetate buffer at 3.6 pH, 1 part of 10 mM TPTZ solution, and 1 part of 20 mM $FeCl_3$ $6H_2O$ solution), and the reaction mixture was incubated in a water bath at 37 °C. The increase in absorbance at 593 nm was measured after 30 min. The antioxidant capacity based on the ability to reduce ferric ions of the extract was expressed in μM Fe (II)/g dry mass and compared with ascorbic acid as standard.

The Fe^{2+} calibration curve for reducing activity was used to calculate the Fe^{2+}-TPTZ concentration, which reveals the existence of metabolites with antioxidant capacity in the mushroom extract. The calibration curve was calculated, obtaining the following equation: $y = 1.5431x + 0.0004$ ($n = 3$, $r^2 = 0.9961$).

2.1.7. Determination of Total Phenolic Content (TPC)

Obtaining the TPC content of our fungi extracts was achieved using the methodology described by [37], based on the Folin–Ciocalteu method. Briefly, 10 mg of gallic acid was dissolved in 100 mL of 50% methanol (100 μg/mL) and then further diluted to 6.25, 12.5, 25, or 50 μg/mL [38]. Next, 1 mL aliquots of each dilution were taken in a test tube and diluted with 10 mL of distilled water. Then, 1.5 mL Folin–Ciocalteu's reagent was added and incubated at room temperature for 5 min; 4 mL of 20% (w/w) Na_2CO_3 was added to each test tube, adjusted with distilled water up to the mark of 25 mL, agitated, and left to stand for 30 min at room temperature. Absorbance of the standard was measured at 765 nm using UV/VIS spectrophotometer, with distilled water as blank. Total phenolic content was expressed as gallic acid equivalent (GAE) in the dry sample. Results were expressed percentage w/w and calculated using the following formula: Total phenolic content (% w/w) = GAE × V × D × 10^{-6} × 100/W, where GAE—Gallic acid equivalent (μg/mL); V—Total volume of sample (mL); D—Dilution factor; and W—Sample weight (g).

For the calculation of the TPC values, an experimental calibration curve was taken with the equation $y = 0.0157x + 0.0357$ ($n = 3$, $r^2 = 0.9968$), which was obtained for gallic acid, where y represents the known pyrogallol concentration and x is the registered absorbance.

2.1.8. Total Flavonoid Content Determination

The quantification method for total flavonoids contents followed the method in [36]. The sample (1 mL) consisting of the fungi extract was mixed with $NaNO_3$ (0.3 mL) in a test tube covered with aluminum foil and left for 5 min. Then, 10% $AlCl_3$ (0.3 mL) was added, followed by 1 M NaOH (2 mL). Later, the absorbance was measured at 510 nm using a spectrophotometer with quercetin as a standard (results expressed as mg/g^{-1} quercetin dry sample).

2.1.9. Protein Content Determination

Fresh oyster mushroom samples were cleaned of substrate residues, then dried in a food dehydrator at a low temperature of 40 °C and ground to a fine powder before sifting to remove lumps. This procedure is necessary to prevent lignin artifacts in the powder [39]. The crude protein content in dried mushrooms was determined by the macro-Kjeldahl method using the conversion factor of $N \times 4.38$ [40] and using 3 repetitions for each treatment.

2.1.10. Sensory Evaluation and Organoleptic Properties

The organoleptic properties and sensory evaluation of the oyster mushroom fruiting bodies were conducted through a preference test. The three treatments were assessed by means of 7 sensory attributes and 5 acceptability parameters using the 15 cm labelled magnitude scales (LMS), described in the Table 1. Thirty untrained participants tasted the fruiting bodies following a randomized order and were asked to assess the intensity of each attribute. Acceptability parameters were also evaluated by the same LMS preference test described in Table 1. The evaluated sensory attributes and consumer test were selected by the studies of [41,42], respectively, while acceptability parameters were selected by following the preference test conducted by [43].

Table 1. Sensory attributes and acceptability parameters evaluated in the consumer test. The fruiting bodies were assessed by seven sensory attributes and five acceptability parameters. The 15-cm labelled scale (LMS) was used to conduct the assessment.

Sensory Attributes	Scale
Sweetness Aroma Astringency Bitterness Sourness Umami Chewiness	0 = not perceived at all 15 = strongly perceived
Acceptability Parameters	**Scale**
Appearance Flavor Color Texture Overall Acceptability	0 = bad 15 = excellent

2.2. Statistical Analysis

The experiment was conducted on a completely randomized experimental design of one factor with 3 treatments and 20 repetitions each. The statistical hypotheses were evaluated according to the proposed experimental design, where the means of the treatments are compared in a one-way analysis of variances (ANOVA). The statistics analysis was determined using the statistics software [44].

The one-way analysis of variance (ANOVA) and Tukey post hoc test were performed for the comparison of treatments. The principle of the method allows for a comparison between the averages of two or more data sets, based on the same principles as the Student's *t* test. Tukey post hoc test provides additional information to ANOVA analysis, highlighting the significance of differences between pair groups.

Pearson's correlation matrix was applied to establish the relationships between the variables. Our results contain Pearson's correlation coefficient r and *p*-values, presented in the same table. It is a fast method to highlight simultaneously, through a matrix, the interdependence relations between N variables.

The statistical significances of *p*-value for both the ANOVA and Pearson correlation matrix are classified as follows: $p > 0.05$ is not statistically significant and statistically significant for $p \leq 0.05$ (*), $p \leq 0.01$ (**); $p \leq 0.001$ (***) and $p < 0.0001$ (****).

An easy visualization of the correlations between the variables was obtained by principal component analysis (PCA). The PCA can establish the relationships of a large number of variables depending on the principal components (axes or factors) resulting in the analysis. The first identified component (F1) is assigned to the largest of the data variants. The second component corresponds to the second variant. The analysis was performed with [45].

3. Results

3.1. Biological Efficiency, Mushroom Yield, and Production Rate

The biological efficiency, mushroom yield, and production rate were measured for the first fruiting. An analysis of variance was performed to determine the differences between treatments. The statistical analysis revealed that there were significant differences ($p \leq 0.05$) in the biological efficiency and yield of *P. ostreatus*. The best values for the biological efficiency and yield of *P. ostreatus* were obtained on the waste paper substrate variant, almost double than the value obtained in the pine sawdust (Figure 1).

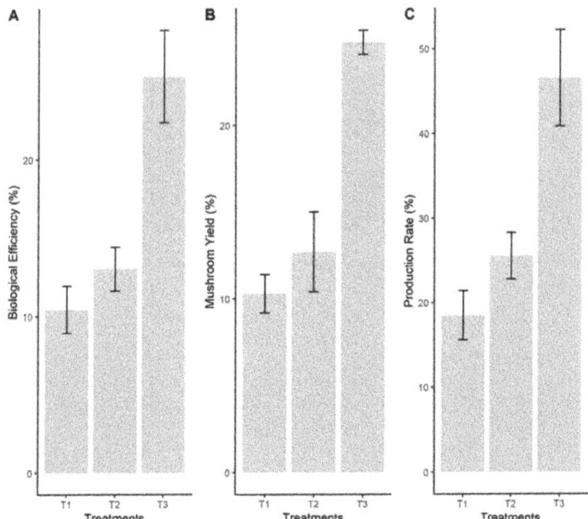

Figure 1. Mushroom Growth analysis in function of the growth media: (**A**) Biological Efficiency (%), (**B**) Mushroom Yield (%), (**C**) Production Rate (%). Legend: T1—pine sawdust; T2—coconut coir; T3—waste paper.

3.2. Antioxidant Aptness

Antioxidant Capacity through DPPH and ABTS Tests

The statistical analysis revealed significant differences between means ($p < 0.05$) (Figure 2). Similar results are observed between treatments in respect of the other tests. The coconut coir substrate (T2) presented the highest antioxidant activity, followed by the pine sawdust (T1) and, finally, the waste paper (T3), with a value similar to that of pine sawdust.

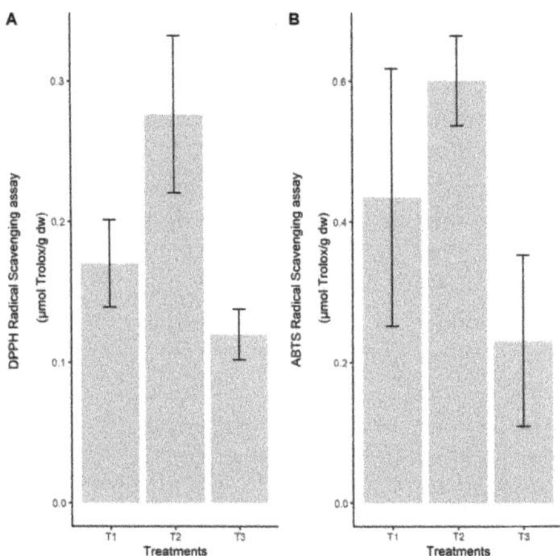

Figure 2. Inhibition percentage (%) and μmol Trolox equivalent per gram of dry weight fungi material, according to the substrate used in the cultivation of *P. ostreatus*: (**A**) DPPH radical scavenging test, (**B**) ABTS radical scavenging test. Legend: T1—pine sawdust; T2—coconut coir; T3—waste paper.

3.3. Reducing Activity

An ANOVA test was conducted to compare the treatments' means, obtaining significant differences between the three treatments ($p < 0.05$). In Figure 3, the results of the test are observed. A higher value for treatment 2, represented by the coconut coir substrate, was obtained in comparison with T1 and T3 that showed almost similar results.

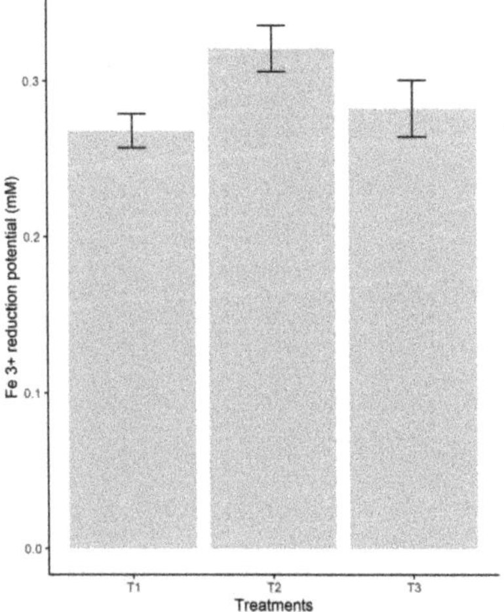

Figure 3. Reducing activity according to the type of substrate used in the cultivation of *P. ostreatus*. Legend: Legend: T1—pine sawdust; T2—coconut coir; T3—waste paper.

3.4. Determination of the Total Phenolic Content (TPC)

Comparing the means reveals a significant difference in the values between the different treatments ($p < 0.05$). The samples' growth on the coconut fiber (T2 substrate) showed a high total polyphenol content (586.60 ± 31.97 mgGAE/g dw), followed by T1 and T3 with close results (397.22 ± 14.87 mgGAE/g dw and 302.95 ± 19.40 mgGAE/g dw) (Figure 4). A One-way ANOVA (F = 38.64, $p = 0.000$) showed significant differences between treatments. The Tukey post hoc test established that the test response was mainly determined by the differences between T1 vs. T2 ($p = 0.003$) and T2 vs. T3 ($p = 0.0005$). The post hoc test confirmed that between T2 and T3 there was no significant differences.

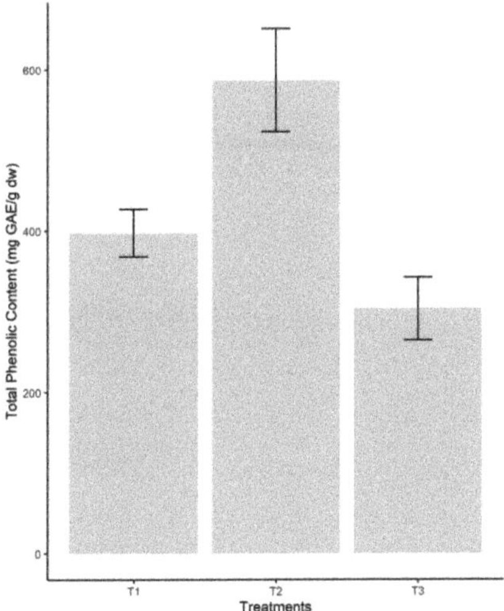

Figure 4. Comparison of the Total Phenolic Content according to the substrate used in the cultivation of *P. ostreatus*. Legend: T1—pine sawdust; T2—coconut coir; T3—waste paper.

3.5. Determination of Total Flavonoid Content

The highest content of total flavonoids in our samples was found in samples cultivated on coconut fiber (T2 variant) (77.54 ± 16.22 mg QE/g dw) being almost double than the total flavonoids content of the samples from recycling paper (T3 variant) (50.79 ± 7.30 mg QE/g dw) (Figure 5). Although a higher value was reached in the T2 treatment, this did not significantly influence the experimental results. Using One-way Anova (F = 0.864, $p = 0.467$), no significant differences were identified between flavonoid contents depending on the different recyclable substrates.

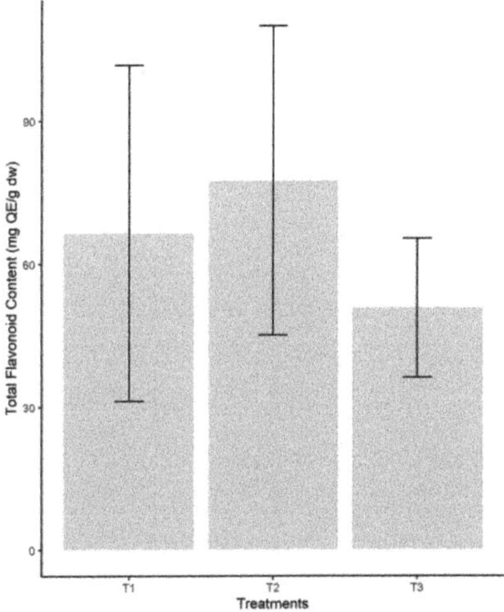

Figure 5. Comparison of the total flavonoid content, according to the type of substrate used in the cultivation of *P. ostreatus*. Legend: T1—pine sawdust; T2—coconut coir; T—waste paper.

The relationships established between TPC, ABTS, FRAP, DPHH, and flavonoid content in our observations were assessed by a principal components analysis (PCA) (Figure 6). The PCA revealed a particular response of flavonoid content (associated with the F2 axis) compared to the other variables, FRAP, TPC, DPPH, and ABTS (which were associated with the F1 axis).

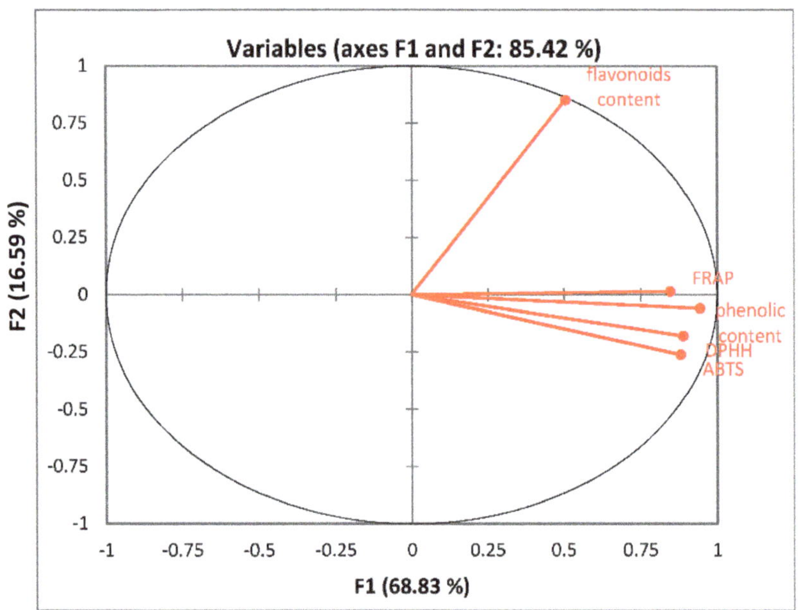

Figure 6. Principal Components Analysis (PCA) between total phenolic content, flavonoid content, and the antioxidant properties DPPH, ABTS, FRAP. Legend: DPPH—radical scavenging test; ABTS—radical scavenging test; FRAP—ferric reducing antioxidant power.

In order to determine the strength of the relationships between them, a Pearson's correlation matrix was applied, and it is presented in Table 2. All the significant results showed positive correlations, which highlights the stimulating influences of the treatments. The total phenolic content, in order of p significance (Table 2 in gray), was correlated with DPPH, FRAP, and ABTS. DPPH presented the strongest influence on ABTS ($r = 0.86$; $p = 0.0076$). Total phenolic content contributes to DPPH and ABTS activity. In addition, the flavonoids content contributes to FRAP, and the total phenolic content contributes to DPPH and ABTS activities.

Table 2. Correlation matrix between the phenolic content (TPC), flavonoid content (FC), radical scavenging test (ABTS), ferric reducing antioxidant power (FRAP), and free radical scavenging (DPHH).

Variables	TPC	FC	ABTS	FRAP	DPHH
TPC		0.2798	**0.0218**	**0.0096**	**0.0029**
FC	0.40		0.4976	0.3074	0.3807
ABTS	**0.74**	0.26		**0.0345**	**0.0076**
FRAP	**0.80**	0.38	0.70		0.1161
DPHH	**0.86**	0.33	**0.81**	0.56	

Legend: bold values are the significant results, Pearson correlation coefficients r—lower white half, *p*-values—upper grey half.

3.6. Protein Content

The dried mushroom powders were analyzed using the macro-Kjeldhal method. The total protein content was measured and recorded in Figure 7, achieving a higher content in T1, the pine sawdust substrate, with significant difference between the three treatments' ANOVAs ($p \leq 0.05$). Coconut coir (T2) showed a lower value than T1 but closer to paper waste substrate (T3), which was the lowest one (Figure 7).

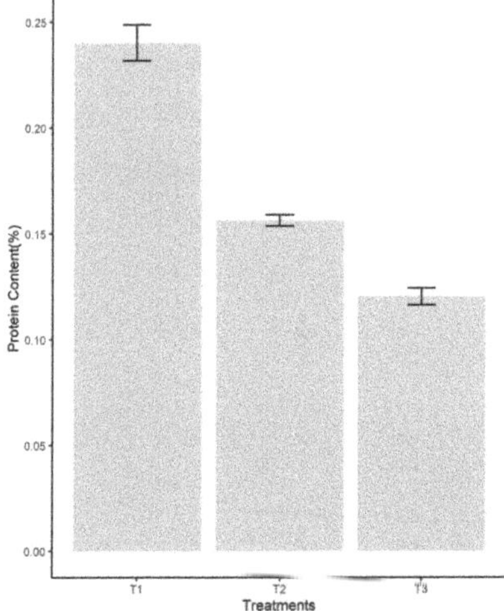

Figure 7. Protein content (%) according to the substrate used in the cultivation of *P. ostreatus*. Legend: T1—pine sawdust; T2—coconut coir; T3—waste paper.

3.7. Sensory Evaluation and Organoleptic Properties

After data collection, statistical analyses were performed to identify the preferences and acceptability of oyster mushrooms grown on different substrates. A one-way ANOVA and a Tukey test were performed to determine significant differences between treatments, checking the hypothesis test with a $p < 0.05$ for each parameter. Table 3 reports the mean value of each sensory attribute and acceptance for each treatment.

Table 3. Sensory attributes and acceptability parameters evaluated in the consumer test.

Treatment	Sensory Attributes		
	T1	**T2**	**T3**
Sweetness	1.933	2.267	1.867
Aroma	11.267	10.867	10.167
Astringency	2.667	2.167	2.700
Bitterness	2.767	2.766	2.967
Chewiness	11.367	11.066	11.000
Sourness	2.400	2.367	2.400
Umami	11.500	11.600	10.600
Treatment	Acceptance		
	T1	**T2**	**T3**
Taste	11.400	11.867	10.567
Color	11.533	11.366	11.233
Texture	11.700	11.967	11.866
Appearance	11.133	11.567	11.566
Overall	10.867	11.400	11.200

Legend: T1—pine sawdust; T2—coconut coir; T3—waste paper.

Regarding the sensory parameters, whose comparison is observed in Figure 8, it was found that the three treatments evaluated showed similar attributes, the coconut coir substrate being the one to produce fruiting bodies with better sensory attributes but with no significant differences between treatments. Some parameters are not relevant for the analyses, such as sourness, astringency, or bitterness, since they were assessed generally by "not perceived at all".

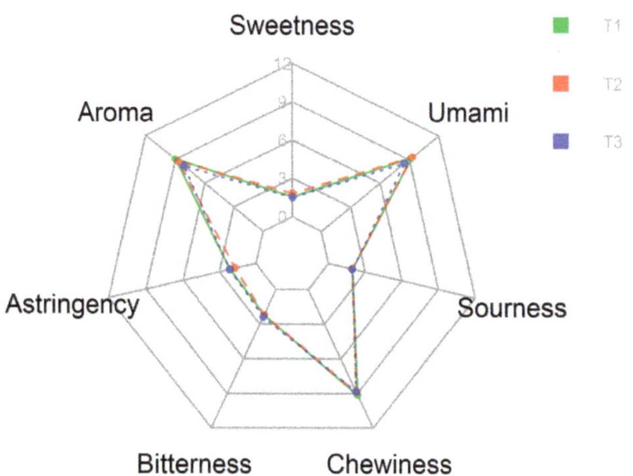

Figure 8. Radar chart for sensory attributes of oyster mushroom grown on different substrates. Legend: T1—pine sawdust; T2—coconut coir; T3—waste paper.

The acceptability test revealed similar results for the three treatments, with little significant differences (Figure 9). For "Taste", T2 (11.867) obtained a significantly higher score than the other two treatments, T3 was the lowest one. Regarding the "Color", there are no significant differences. For the "Texture" attribute, T2 (11.967) shows a higher value than the other treatments. The lowest value of "Appearance" (11.133) was expressed for T1, grown on pine sawdust. The overall acceptability positioned the mushrooms grown in coconut coir (T2) as the product of preference. On the contrary, the mushrooms grown on pine sawdust showed the lowest score, although there are no significant differences between the three treatments.

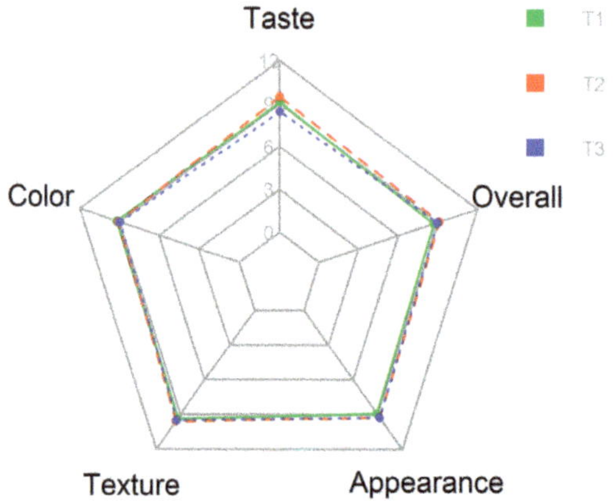

Figure 9. Radar chart for acceptability items for oyster mushroom grown on different substrates. Legend: T1—pine sawdust; T2—coconut coir; T3—waste paper.

4. Discussion

4.1. Biological Efficiency, Mushroom Yield and Production Rate

P. ostreatus, also known as "oyster mushroom", "hiratake", "shimeji", or "houbitake", is able to grow in available waste materials. The different waste by-products of lignocellulose composition tested as substrate for the *P. ostreatus* cultivation represented by pine sawdust, coconut coir (fiber mixed with shell), and waste paper were found to be a good support for the growth of the fungus, with the mycelium fully colonizing the substrates at 28 days. A similar mycelium growth rate, however, did not correspond with yield, indicating that the mycelium growth and yield of mushrooms have different requirements [46]. The results showed that mushroom yield is reliant on biological efficiency, as overall biological efficiency determines the mushroom yield, which is in accordance with the findings of [47], displaying that mushroom yield is dependent on biological efficiency. In our study, the highest yield was harvested from paper waste substrate (T3) with the most elevated percentage of biological efficiency, followed by coconut coir (T2), while the lowest was observed in pine sawdust (T1). Similarly, the biological efficiency (BE) also varied significantly among the different substrates used. The performance of oyster growth and yield in the sawdust substrate was minimal, a result similar with the data obtained by [31]. This could be attributed to the fact that the lignocellulosic materials in pine sawdust are generally low in protein content and insufficient for the mushroom's cultivation [46]. Therefore, the sawdust substrate for mushroom production should undergo a period of composting to breakdown the cellulose and lignin components of the wood in order to release the essential materials for the establishment of mushroom mycelium.

In addition, the mean comparisons (separated using Tukey test) revealed that the biological yield from paper waste was significantly different from the rest of the substrates at a 5% confidence level (Figure 1B). The results of this study are in line with other studies elsewhere (e.g., [6,30]), where paper was identified as an important substrate for significant improvement in the yield of oyster mushroom. The possible justification may be that the paper waste is a high C-content waste material, i.e., PW (C/N = 379) [48], and is accepted as a superior substrate over pine sawdust.

Generally, the present study confirmed that oyster mushrooms can grow on pine sawdust, paper waste, and coconut coir, with varying growth performances. Paper waste was identified as a suitable substrate for oyster mushroom cultivation, since it produced a significantly higher yield, biological efficiency, and production rate compared to the other substrates. The BE, yield, and production rate of both coconut coir and pine sawdust proved similar values, where coconut coir have slightly higher parameters than pine sawdust. Paper waste proved to be better in terms of mycelium density, pin-head formation, and the development of fruiting bodies, and it is a good recommendation as a preferred substrate for oyster mushroom cultivation, serving as a viable solution for the environmental contamination by using the huge paper wastes available. In addition, coconut coir, a contaminant waste product in coconut producing countries, can be used as an alternative substrate given that the growth performance and yield of oyster mushrooms was better than the pine sawdust.

4.2. Antioxidant Activity

Oxidation is essential for the living organisms to produce energy for the biological processes. Free radicals are produced in normal and pathological cell metabolism. The uncontrolled production of oxygen-derived free radicals is involved in the onset of many diseases. The antioxidant components are responsible for defending our body against free radicals, and it is known that low levels of antioxidants cause oxidative stress and may damage or kill cells [49]. Numerous fungi were reported to have antioxidant components, higher than in most vegetables and fruits, and concentrated in fruit bodies and both mycelium and culture. For this reason, our study was also conducted to evaluate the antioxidant activity of the oyster mushroom, depending on the waste by-product used as substrate, using spectrophotometrically methodologies with different mechanisms, one that report the scavenging ability on DPPH and ABTS radicals (DPPH and ABTS assays) and the other monitoring the reducing power of compounds as a significant indication of its potential activity [50]. The presence of reducers (i.e., antioxidants) causes the reduction of the Fe^{3+}/Ferricyanide complex to ferrous form in the case of the FRAP assay.

The antioxidant activities measured in fungi ethanolic extracts were obtained using the three DPPH, ABTS, and FRAP assays that gave a comparable ranking of antioxidant activity among the substrates used with the highest antioxidant capacity revealed by the coconut coir substrate (T2). So, this waste by-product used as substrate for oyster cultivation showed the highest values of scavenging ability on DPPH and ABTS free radicals with the range of 44% to 65%. The results of the FRAP assay indicate that the significantly highest reducing power inhibition could be identified in the extract of oyster fungi grown in the same substrate represented by coconut coir. This substrate has been found to significantly reduce the power in ferric ions of oyster fungi due to its effect on the total phenolic and flavonoid contents of the fruiting body extracts, which play an important role in antioxidant activities. According to [51], the reducing power might be due to their hydrogen-donating ability, and certain mushrooms contain higher amounts of reduction, which could react with free radicals to stabilize and terminate radical chain reactions.

4.3. Total Phenolic Content (TPC)

Phenolic compounds possess a common chemical structure comprising an aromatic ring with one or more hydroxyl substituents that can be divided into several classes, and the main groups of phenolic compounds include flavonoids, phenolic acids, tannins, stilbenes, and lignans [52]. It has been reported that phenolic compounds exhibit antioxidant activity in biological systems, acting as free radical inhibitors, peroxide decomposers, metal inactivators, or oxygen scavengers [53,54]. Phenolic compounds are present in all the mushrooms. These compounds can be pyrogallol, myricetin, caffeic acid, quercetin, and catechin, among others. The fruiting bodies of *Pleurotus* respond dramatically to the chemical composition of the substrate where they grow and develop. The bioactive compounds, including phenolics, can be effectively absorbed by the fruiting bodies of *Pleurotus* [55,56]. Our study could register a significant difference in TPC of *P. ostreatus* grown on various waste by-products substrates. The highest values of TPC were shown in fruiting bodies of the edible fungi obtained from coconut coir containing substrate that exhibited a TPC in the range of 586.60 ± 31.97 mg GAE/g dry weight compared to the lower values of 397.21 ± 14.87 mg GAE/g dry weight on waste paper substrate and followed by 302.95 ± 19.40 mg GAE/g dry weight on the pine sawdust substrate. The changes in the TPC of the oyster fungi fruiting bodies grown in different substrate formulas are explained by the difference in the lignin composition of the substrates. The presence of lignin in substrates reduces the number of some biological active molecules, which is directly linked to a decreased biological activity [57]. This activity also depends on how easily the substrate is decomposed by the mycelium and the quality of nutrients assimilated by the mushrooms [58]. It was observed that the coconut coir substrate significantly increased the TPC and antioxidant activity against DPPH and ABTS+ radicals. It was noticed that the phenolic contents correlate well with DPPH and ABTS assays, with correlation coefficients of 0.8601 and 0.7429, respectively, confirming that phenolic compounds contribute to radical scavenging activity of the fungi extracts, having a redox potential. These results are in agreement with those observed in *Camellia sinensis*, *Annona muricata*, *Zingiber officinale* [59], and *Syzigium aromaticum* and *Allium sativum* [60].

4.4. Total Flavonoid Content (TFC)

Flavonoids represent a group of natural substances with variable phenolic structures, considered an indispensable component in a variety of nutraceutical, pharmaceutical, medicinal, and cosmetic applications. The TFC assay was estimated to extract flavonoids, isoflavonoids, and neoflavonoids, or collectively called bioflavonoids. Several studies have demonstrated that these compounds may act as antioxidant by breaking the radical chains into more stable products in liver microsomal membranes, with the ability to protect low-density lipoprotein or LDL from being demolished by heavy metals and macrophages. They also play an important role in providing instinctive protection against oxidative stress and side effects by its contribution with vitamins.

In the present study, total flavonoid content was calculated using quercetin as a standard and the results are expressed as mg of quercetin equivalents per gram of extract. It showed that the fungi extract with the highest total phenolic content also exhibited the highest total flavonoid content (77.54 ± 16.22 mg QE/g) on the growing substrate represented by coconut coir. The changes in TFC under the influence of substrates can also be explained by the different quantity of lignin, in the same way as it was influenced by the TPC. It could be observed that the flavonoid content in the fungi extracts showed a higher correlation with reducing power FRAP assay than with the radical scavenging activity. This could estimate that the flavonoid compounds present in the extracts act as an antioxidant directly through the mechanism of the reduction of oxidized intermediate in the chain reaction.

4.5. Protein Content

Protein is an essential nutrient in human life activities, helping the formation, growth, constitution, or repair of human tissues. The fungal ones provide a healthy new protein with a low environmental impact [61]. *P. ostreatus* is generally considered to be a good source of digestible proteins, providing all the essential amino acids required by an adult [62]. Protein content of *Pleurotus* depends on the composition of the substrates and mushrooms species [63]. The protein analysis of the fruiting bodies from each growing substrates indicate that pine sawdust substrate offered the highest protein content (24.02%) for the oyster fungi followed by 15.64% on the coconut coir substrate and 12.04% on the waste paper. Nutritional value, in particular the protein content of *P. ostreatus* cultivated on pine sawdust substrate, was higher than those exhibited by another popular edible mushrooms Pioppino (*A. aegerita*) with 20.5% and Champignon (*A. bisporus*) mushrooms with 13.4% [41], the last one being slightly similar with *P. ostreatus* cultivated in coconut coir and waste paper substrates. The amount of nitrogen supplied in the growth media may be used to control the protein content of the mushroom mycelium. The carbon/nitrogen influences the protein content in the mushroom mycelium [64]. The obtained data corroborate the results of [65], which reported that Oyster (*Pleurotus*) mushrooms are considered to be one of the most efficient producers of food protein, producing 30% of its dry weight. Carbon is readily available from cellulose, hemicelluloses, and lignin, so pine sawdust represents a good source of carbon [66]. Sharma et al. [67] noted that the C/N ratio significantly influences the values obtained from the protein composition of the fungus *P. ostreatus*. The C/N ratio of the substrate is critical to the initial development of the fungus, given the value of carbon for the formation of new cells; a low C/N ratio in the substrate will influence the fungus negatively during the mycelium growth stage.

4.6. Sensory Evaluation and Organoleptic Properties

The sensory attributes of *P. ostreatus* cultivated in different substrates were analyzed through a consumer test, assessed by seven sensory attributes and five acceptability parameters. Mushroom samples were hygienically and neatly prepared to be presented for consumption. Panelists gave similar preference scores for all samples from different substrates, which indicated that all were highly satisfactory as judged by appearance, color, taste, and texture, with an overall liking and acceptance for all three. Nonetheless, panelists showed some significant ($p < 0.05$) preferences at taste for the coconut cultivated oyster mushroom, which could be due to the impact of growth substrate on the oyster mushroom fruiting bodies composition, yield, biological efficiency, and nutritional profile [68]. The flavor experienced from eating mushrooms, or any other food, comes from a combination of taste, texture, temperature, spiciness, and aromatic qualities [41].

Our findings showed that consumer acceptability of the fruiting bodies was largely influenced by their sensory characteristics, as well as by their visual appearance. It is well-documented that sensory characteristics such as taste, appearance, freshness, texture, color, and smell are essential motivating factors to lead consumers towards the consumption of food products [69]. Overall, one of the unconventional tools for sensory evaluation of mushrooms is the smell, which is important for both the identification of species and for the oro-sensory sensations one experiences while eating them, among several issues regarding gustation and olfaction [70].

In our study, recorded sensory attributes were similar for the harvested fruiting bodies from three different substrates. As stated in literature, flavor-related characteristics predict the best the consumer preferences for overall eating quality, and therefore, consumer preferences are affected primarily by sensory characteristics [42]. For the present study, aroma, astringency, bitterness, chewiness, sourness, sweetness and umami, flavors found in mushrooms, were evaluated by the panelists. As Du et al. [71] explain, in mushrooms, carboxylic acids contribute to sour taste, while sweetness can be perceived by the presence of sugars, polyols, and several amino acids, and umami taste is elicited by $5'$-nucleotides, monosodium glutamate (MSG), and several other free amino acids and nucleotides. As for texture parameters, chewiness was assessed with a high value and so were liked by the panelists. This feature is related to the mushroom state at which fruiting bodies are harvested, and it is related to good texture and flavor which receive high demand from costumers [72]. Astringency was not perceived at all along with bitterness, sweetness, and sourness; its mouthfeel, as the literature states, is an event induced by tannin interaction and the precipitation of salivary proline-rich proteins (PRPs) in the oral cavity [73].

5. Conclusions

Our study is conducted to demonstrate the transformation of wastes (a contaminant problem) into a biological substrate for the edible fungi *P. ostreatus*. In our case, waste paper (the third largest industrial polluter of the environment) was the growth substrate with a higher yield value and biological efficiency for the harvested oyster mushrooms. The substrate suitable for the production of fungi with a higher concentration of bioactive compounds (phenolics and flavonoids) that offer a high antioxidant capacity for a better medicinal quality was represented by coconut coir. This is a waste product that generates great pollution to the ecosystem, favors the proliferation of insects and rodents, and affects the lives of the inhabitants of the sectors where this fruit is grown, since they tend to throw it into the rivers and estuaries. Our study could offer an economical biotechnology for the organic waste recycling of lignocelluloses that combines the production of protein-rich food with desirable organoleptic properties, therapeutic benefits, and the reduction in environmental pollution.

Author Contributions: Conceptualization, R.A.M. and E.J.M.H.; methodology, R.A.M.; validation, R.A.M., E.J.M.H. and R.D.C.; formal analysis, E.J.M.H. and L.I.F.; investigation, R.A.M.; resources, R.A.M., E.J.M.H. and R.D.C.; data curation, R.A.M. and L.I.F.; writing—original draft preparation, E.J.M.H.; writing—review and editing, R.A.M. and R.D.C.; supervision, R.A.M.; project administration, R.A.M.; funding acquisition, R.A.M. All authors have read and agreed to the published version of the manuscript.

Funding: This research was funded by Universidad de Las Fuerzas Armadas ESPE, grant number CV-GNP-0006-2017 and Romanian Academy, grant number RO1567-IBB08/2021.

Institutional Review Board Statement: Not applicable.

Informed Consent Statement: Not applicable.

Data Availability Statement: Not applicable.

Acknowledgments: We are thankful for the sincere supports of Communities from Costal area of Ecuador for providing all the facilities to supply coconut residues. We are greatly indebted to the CICTE Investigation Center, to the Unit of Management of Linkage with Society form Universidad de Las Fuerzas Armadas ESPE, for providing us with all the necessary materials and financial support to undertake the study, as well as to the partners of this project DIGEIM Ecuador (General Directorate of Maritime Interests) that underlined the negative influence of coconuts waste in the navigation traffic that affects the country. The authors also would like to thank anonymous reviewers for their valuable comment.

Conflicts of Interest: The authors declare no conflict of interest.

References

1. Singh, J.; Laurenti, R.; Sinha, R.; Frostell, B. Progress and challenges to the global waste management system. *Waste Manag. Res.* **2014**, *32*, 800–812. [CrossRef] [PubMed]
2. Yusuf, M. Agro-Industrial Waste Materials and their Recycled Value-Added Applications: Review. In *Handbook of Ecomaterials*; Torres Martínez, L.M., Kharissova, O.V., Kharisov, B.I., Eds.; Springer International Publishing: Cham, Switzerland, 2017; pp. 1–12.
3. Raman, J.; Jang, K.Y.; Oh, Y.L.; Oh, M.; Im, J.H.; Lakshmanan, H.; Sabaratnam, V. Cultivation and Nutritional Value of Prominent *Pleurotus* spp.: An Overview. *Mycobiology* **2021**, *49*, 1–14. [CrossRef] [PubMed]
4. Elbasiouny, H.; Elbanna, B.A.; Al-Najoli, E.; Alsherief, A.; Negm, S.; Abou El-Nour, E.; Nofal, A.; Sharabash, S. Agricultural Waste Management for Climate Change Mitigation: Some Implications to Egypt. In *Waste Management in MENA Regions*; Negm, A.M., Shareef, N., Eds.; Springer Nature: Cham, Switzerland, 2020; pp. 149–169.
5. Sadh, P.K.; Duhan, S.; Duhan, J.S. Agro-industrial wastes and their utilization using solid state fermentation: A review. *Bioresour. Bioprocess.* **2018**, *5*, 1–15. [CrossRef]
6. Scarlat, N.; Dallemand, J.F.; Monforti-Ferrario, F.; Nita, V. The role of biomass and bioenergy in a future bioeconomy: Policies and facts. *Environ. Dev.* **2015**, *15*, 3–34. [CrossRef]
7. Tesfaw, A.; Tadesse, A.; Kiros, G. Optimization of oyster (*Pleurotus ostreatus*) mushroom cultivation using locally available substrates and materials in Debre Berhan, Ethiopia. *J. Appl. Biol. Biotechnol.* **2015**, *3*, 15–20.
8. Esmeraldas Concentra la Palma de Coco. Available online: https://www.revistalideres.ec/lideres/esmeraldas-concentra-palma-coco-negocios.html (accessed on 8 November 2021).
9. Sopit, V. The Feasibility of Using Coconut Residue as a Substrate for Oyster Mushroom Cultivation. *Biotechnology* **2007**, *6*, 578–582.
10. Bernabé, G.T.; Domínguez, R.; Salvador, M.; Baltazar, B.; Antonio, S. Cultivo del hongo comestible *Pleurotus ostreatus* var. *florida* sobre fibra de coco y pulpa de café/Cultivation of the edible mushroom *Pleurotus ostreatus* var. *florida* on coconut fiber and coffee pulp. *Rev. Mex. Micol.* **1993**, *9*, 13–18.
11. Paulitz, T. Biological control in greenhouse systems. *Phytopathology* **2001**, *39*, 103–133. [CrossRef]
12. Lemache, M.; Pacheco, K. Estudio del Procesamiento de la Fibra de Coco Para la Exportación a España y Sus Beneficios en la Economía Solidaria de los Productores del Recinto Tolita Pampa de Oro, Cantón Eloy Alfaro, al Norte de Esmeraldas. Ph.D. Thesis, Universidad de Guayaquil, Guayaquil, Ecuador, 2015.
13. Wagner, A.; Donaldson, L.; Ralph, J. Lignification and lignin manipulations in conifers. In *Advances in Botanical Research, Lignins: Biosynthesis, Biodegradation and Bioengineering*; Jouanin, L., Lapierre, C., Eds.; Academic Press: Burlington, NJ, USA, 2012; Volume 61, pp. 37–76.
14. Piškur, B.; Bajc, M.; Robek, R.; Humar, H.; Sinjur, I.; Kadunc, A.; Oven, P.; Rep, G.; Petkovšek, S.A.S.; Kraigher, H.; et al. Influence of *Pleurotus ostreatus* inoculation on wood degradation and fungal colonization. *Int. Biodeterior. Biodegrad.* **2011**, *102*, 10611–10617. [CrossRef]
15. Ma, Z.; Yang, Y.; Chen, W.Q.; Wang, P.; Chao, W.; Chao, Z.; Gan, J. Material Flow Patterns of the Global Waste Paper Trade and Potential Impacts of China's Import Ban. *Environ. Sci. Technol.* **2021**, *55*, 8492–8501. [CrossRef]
16. Salmones, D. *Pleurotus djamor*, un hongo con potencial aplicación biotecnológica para el neotrópico. *Rev. Mex. Mic.* **2017**, *46*, 73–85. [CrossRef]
17. Geetha, D.; Sivaprakasam, K. *Pleurotus-djamor*—A new edible mushroom. *Curr. Sci.* **1993**, *64*, 280–281.
18. Piña-Guzmán, A.B.; Nieto-Monteros, D.A.; Robles-Martinez, F. Utilización de residuos agrícolas y agroindustriales en el cultivo y producción del hongo comestible seta (*Pleurotus* spp.). *Rev. Int. Contam. Ambient.* **2016**, *32*, 141–151. [CrossRef]
19. Cohen, R.; Persky, L.; Hadar, Y. Biotechnological applications and potential of wood-degrading mushrooms of the genus *Pleurotus*. *Appl. Microbiol. Biotechnol.* **2002**, *58*, 582–594. [CrossRef] [PubMed]
20. Sánchez-Cantú, M.; Ortiz-Moreno, L.; Ramos-Cassellis, M.E.; Marín-Castro, M.; la Cerna-Hernández, D. Solid-state treatment of castor cake employing the enzymatic cocktail produced from *Pleurotus djamor* fungi. *Appl. Biochem. Biotechnol.* **2018**, *185*, 434–449. [CrossRef]
21. Knop, D.; Yarden, O.; Hadar, Y. The ligninolytic peroxidases in the genus *Pleurotus*: Divergence in activities, expression, and potential applications. *Appl. Microbiol. Biotechnol.* **2015**, *99*, 1025–1038. [CrossRef]
22. Vamanu, E. In Vitro Antimicrobial and Antioxidant Activities of Ethanolic Extract of Lyophilized Mycelium of *Pleurotus ostreatus* PQMZ91109. *Molecules* **2012**, *17*, 3653–3671. [CrossRef]

23. Adebayo, E.A.; Oloke, J.K.; Ayandele, A.A.; Adegunlola, C.O. Phytochemical, antioxidant and antimicrobial assay of mushroom metabolite from *Pleurotus pulmonarius*—LAU 09 (JF736658). *J. Microbiol. Biotechnol. Res.* **2012**, *2*, 366–374.

24. García-Oduardo, N.; Bermúdez-Savón, R.C.; Serrano-Alberni, M. Formulaciones de sustratos en la producción de setas comestibles *Pleurotus*. *Tecnol. Química* **2011**, *31*, 272–282.

25. Da Luz, J.M.R.; Paes, S.A.; Nunes, M.D.; Torres, D.P.; da Silva, M.C.S. Lignocellulolytic enzyme production of *Pleurotus ostreatus* growth in agroindustrial wastes. *Braz. J. Microbiol.* **2012**, *43*, 1508–1515. [CrossRef]

26. Sánchez, C. Cultivation of *Pleurotus ostreatus* and other edible mushrooms. *Appl. Microbiol. Biotechnol.* **2010**, *85*, 1321–1337. [CrossRef] [PubMed]

27. Ragucci, S.; Landi, N.; Russo, R.; Valletta, M.; Pedone, P.V.; Chambery, A.; Di Maro, A. Ageritin from Pioppino Mushroom: The Prototype of Ribotoxin-Like Proteins, a Novel Family of Specific Ribonucleases in Edible Mushrooms. *Toxins* **2021**, *13*, 263. [CrossRef] [PubMed]

28. Dulal, S. Oyster Mushroom Farming-an Overview (Issue January). *J. Glob. Ecol. Environ.* **2019**, *11*. [CrossRef]

29. Romero, A.O. Technological Development to Control Green Mold Attack (*Trichoderma* spp.) for Commercial Cultivation of Edible Fungi (*Pleurotus ostreatus* and *Lentinula edodes*) in Mexico. Master's Thesis, Universidad Autónoma Chapingo, Texcoco, Mexico, 2007.

30. Oei, P. *Small-Scale Mushroom Cultivation: Oyster, Shiitake and Wood Ear Mushrooms*; First edition of Digigrafi; Agromisa Foundation and CTA: Wageningen, The Netherlands, 2005.

31. Girmay, Z.; Gorems, W.; Birhanu, G.; Zewdie, S. Growth and yield performance of *Pleurotus ostreatus* (Jacq. Fr.) Kumm (oyster mushroom) on different substrates. *AMB Express* **2016**, *6*, 87. [CrossRef] [PubMed]

32. Vega, A.; Franco, H. Productividad y calidad de los cuerpos fructíferos de los hongos comestibles *Pleurotus pulmonarius* RN2 y P. Djamor RN81 y RN82 cultivados sobre sustratos lignocelulósicos. *Inf. Tecnol.* **2013**, *24*, 69–78. [CrossRef]

33. Joyeux, M.; Lobstein, A.; Mortier, F. Comparative antilipoperoxidant, antinecrotic and scavenging properties of terpenes and biflavones from *Gingko* and some flavonoids. *Planta Med.* **1995**, *61*, 126–129. [CrossRef]

34. Shimamura, T.; Sumikura, Y.; Yamazaki, T.; Tada, A.; Kashiwagi, T.; Ishikawa, H.; Matsui, T.; Sugimoto, N.; Akiyama, H.; Ukeda, H. Applicability of the DPPH Assay for evaluating the antioxidant capacity of food additives—Inter-laboratory evaluation study. *Anal. Sci.* **2014**, *30*, 717–721. [CrossRef]

35. Re, R.; Pellegrini, N.; Proteggente, A.; Pannala, A.; Yang, M.; Rice-Evans, C. Antioxidant activity applying an improved ABTS radical cation decolorization assay. *Free Radic. Biol. Med.* **1999**, *26*, 1231–1237. [CrossRef]

36. Ibrahim, M.H.; Jaafar, H.Z.E. The relationship of nitrogen and C/N ratio with secondary metabolites levels and antioxidant activities in three varieties of malaysian kacip fatimah (*Labisia pumila* Blume). *Molecules* **2011**, *16*, 5514–5526. [CrossRef]

37. Madaan, R.; Bansal, G.; Kumar, S.; Sharma, A. Estimation of Total Phenols and Flavonoids in Extracts of *Actaea spicata* Roots and Antioxidant Activity Studies. *Indian J. Pharm. Sci.* **2011**, *73*, 666–669. [CrossRef]

38. Lin, J.Y.; Tang, C.Y. Determination of total phenolic and flavonoid contents in selected fruits and vegetables, as well as their stimulatory effect on mouse splenocyte proliferation. *Food Chem.* **2007**, *101*, 140–147. [CrossRef]

39. Wan-Mohtar, W.A.Q.I.; Mahmud, N.S.; Supramani, R.A.; Zain, N.A.M.; Hassan, N.A.M.; Peryasamy, J.; Halim-Lim, S.A. Fruiting-body-base flour from an oyster mushroom-a waste source of antioxidative flour for developing potential functional cookies and steamed-bun. *Aims Agric. Food* **2018**, *3*, 481–492. [CrossRef]

40. Landi, N.; Pacifico, S.; Ragucci, S.; Di Giuseppe, A.M.; Iannuzzi, F.; Zarrelli, A.; Piccolella, S.; Di Maro, A. Pioppino mushroom in southern Italy: An undervalued source of nutrients and bioactive compounds. *J. Sci. Food Agric.* **2017**, *97*, 5388–5397. [CrossRef] [PubMed]

41. Kortei, N.K.; Odamtten, G.T.; Obodai, M.; Akonor, P.T.; Wiafe-Kwagyan, M.; Buckman, S.; Mills, S.W.N.O. Sensory evaluation, descriptive textural analysis, and consumer acceptance profile of steamed gamma-irradiated *Pleurotus ostreatus* (Ex. Fr.) Kummer kept in two different storage packs. *Sci. Afri.* **2020**, *8*, e00328. [CrossRef]

42. Caracciolo, F.; El-Nakhel, C.; Raimondo, M.; Kyriacou, M.C.; Cembalo, L.; de Pascale, S.; Rouphael, Y. Sensory attributes and consumer acceptability of 12 microgreens species. *Agronomy* **2020**, *10*, 1043. [CrossRef]

43. Mishra, R.; Mishra, Y.D.; Singh, P.P.; Raghubanshi, B.P.S.; Sharma, R. Nutritional and Sensory Evaluation of Oyster Mushroom Supplemented Daily Food Items. *Int. J. Curr. Microbiol. Appl. Sci.* **2018**, *7*, 1465–1471. [CrossRef]

44. Hammer, O.; Harper, D.A.T.; Ryan, P.D. PAST: Paleontological statistics software package for education and data analysis. *Palaeontol. Electron.* **2001**, *4*, 9.

45. XLSTAT Pro. *Data Analysis and Statistical Solution for Microsoft Excel*; Addinsoft: Paris, France, 2013.

46. Obodai, M.; Cleland-Okine, J.; Vowotor, K.A. Comparative study on the growth and yield of *Pleurotus ostreatus* mushroom on different lignocellulosic by-products. *J. Ind. Microbiol. Biotechnol.* **2003**, *30*, 146–149. [CrossRef]

47. Olasupo, O.O.; Asonibare, A.O.; Nurudeen, T.A. Relative Performance of Oyster mushroom (*Pleurotus florida*) Cultivated on different Indigenous Wood Wastes. *J. Agric. Res. Nat. Resour.* **2019**, *3*, 1–11.

48. Shahbaz, M.; Ammar, M.; Korai, R.M.; Ahmad, N.; Ali, A.; Khalid, M.S.; Zou, D.; Li, X.J. Impact of C/N ratios and organic loading rates of paper, cardboard and tissue wastes in batch and CSTR anaerobic digestion with food waste on their biogas production and digester stability. *SN Appl. Sci.* **2020**, *2*, 1436. [CrossRef]

49. Sharifi-Rad, M.; Anil Kumar, N.V.; Zucca, P.; Varoni, E.M.; Dini, L.; Panzarini, E.; Rajkovic, J.; Tsouh Fokou, P.V.; Azzini, E.; Peluso, I.; et al. Lifestyle, Oxidative Stress, and Antioxidants: Back and Forth in the Pathophysiology of Chronic Diseases. *Front. Physiol.* **2020**, *11*, 694. [CrossRef] [PubMed]
50. Oyaizu, M. Studies on products of browning reaction prepared from glucosamine. *Jpn. J. Nutr.* **1986**, *44*, 307–315. [CrossRef]
51. Shimada, K.; Fujikawa, K.; Yahara, K.; Nakamura, T. Antioxidative properties of xanthan on the autoxidation of soybean oil in cyclodextrin emulsion. *J. Agric. Food Chem.* **1992**, *40*, 945–948. [CrossRef]
52. Ayad, R.; Akkal, S. Phytochemistry and biological activities of algerian *Centaurea* and related genera. In *Studies in Natural Products Chemistry*; Atta-ur-Rahman, Ed.; Elsevier: Amsterdam, The Netherlands, 2019; Volume 63, pp. 357–414.
53. Dziezak, J.D. Antioxidants-The ultimate answer to oxidation. *Food Technol.* **1986**, *40*, 94–102.
54. Yagi, K. A rapid method for evaluation of oxidation and antioxidants. *Agric. Biol. Chem.* **1970**, *34*, 142–145. [CrossRef]
55. Carrasco-González, J.A.; Serna-Saldívar, S.O.; Gutiérrez-Uribe, J.A. Nutritional composition and nutraceutical properties of the *Pleurotus* fruiting bodies: Potential use as food ingredient. *J. Food Compos. Anal.* **2017**, *58*, 69–81. [CrossRef]
56. Da Paz, M.F.; Breyer, C.A.; Longhi, R.F.; Oviedo, M.S.V.P. Determining the basic composition and total phenolic compounds of *Pleurotus sajor-caju* cultivated in three different substrates by solid state bioprocess. *J. Biotechnol. Biodivers.* **2012**, *3*, 11–14. [CrossRef]
57. Stefan, R.I.; Vamanu, E.; Angelescu, C. Antioxidant Activity of Crude Methanolic Extracts from *Pleurotus ostreatus*. *Res. J. Phytochem.* **2015**, *9*, 25–32.
58. Hoa, H.T.; Chun-Li, W.; Chong-Ho, W. The Effects of Different Substrates on the Growth, Yield, and Nutritional Composition of Two Oyster Mushrooms (*Pleurotus ostreatus* and *Pleurotus cystidiosus*). *Mycobiology* **2015**, *43*, 423–434. [CrossRef]
59. Ndomou, S.C.H.; Djikeng, T.F.; Teboukeu, G.B.; Doungue, H.T.; Foffe, K.H.A.; Tiwo, C.T.; Womeni, H.M. Nutritional value, phytochemical content, and antioxidant activity of three phytobiotic plants from west Cameroon. *J. Agric. Food Res.* **2021**, *3*, 100105. [CrossRef]
60. Foffe, K.H.A.; Ndomou, S.C.H.; Djikeng, T.F.; Teboukeu, B.G.; Tsopmo, A.; Womeni, H.M. Effect of *Syzygium aromaticum* and *Allium sativum* spice extract powders on the lipid quality of groundnuts (*Arachis hypogaea*) pudding during steam cooking. *Heliyon* **2020**, *6*, e05166. [CrossRef] [PubMed]
61. Derbyshire, E. Protein guidance—Is it time for an update? *Diet. Today* **2020**, 22–24.
62. Pushpa, H.; Purushothoma, K.B. Nutritional analysis of wild and cultivated edible medicinal mushrooms. *World J. Dairy Food Sci.* **2010**, *5*, 140–144.
63. Erjavec, J.; Kos, J.; Ravnika, M.; Dreo, T.; Sabotic, J. Proteins of higher fungi-from forest to application. *Trends Biotechnol.* **2012**, *30*, 259–273. [CrossRef]
64. Shah, Z.A.; Ashraf, M.; Ishtiaq, M.C. Comparative Study on Cultivation and Yield Performance of Oyster Mushroom (*Pleurotus ostreatus*) on Different Substrates (Wheat Straw, Leaves, Saw Dust). *Pak. J. Nutr.* **2004**, *3*, 158–160.
65. Belewu, M.A. Nutritional qualities of corn cobs and waste paper incubated with edible mushroom (*Pleurotus sajor-caju*). *Niger. J. Anim. Prod.* **2003**, *30*, 20–25. [CrossRef]
66. Upadhyay, R.C.; Verma, R.N.; Singh, S.K.; Yadav, M.C. Effect of organic nitrogen supplementation in *Pleurotus* species. *Mushroom Biol. Mushroom Prod.* **2002**, *105*, 225–232.
67. Sharma, S.; Kailash, R.; Pokhrel, C.H. Growth and yield of oyster mushroom (*Pleurotus ostreatus*) on different substrates. *JNBR* **2013**, *2*, 3–8.
68. Onyeka, E.U.; Okehie, M.A. Effect of substrate media on growth, yield and nutritional composition of domestically grown oyster mushroom (*Pleurotus ostreatus*). *Afr. J. Plant Sci.* **2018**, *12*, 141–147. [CrossRef]
69. Imtiyaz, H.; Soni, P.; Yukongdi, V. Role of sensory appeal, nutritional quality, safety, and health determinants on convenience food choice in an academic environment. *Foods* **2021**, *10*, 345. [CrossRef]
70. Hallock, R.M. The Taste of Mushrooms. *McInvainea* **2007**, *17*, 33–41.
71. Du, X.; Sissons, J.; Shanks, M.; Plotto, A. Aroma and flavor profile of raw and roasted *Agaricus bisporus* mushrooms using a panel trained with aroma chemicals. *LWT* **2021**, *138*, 110596. [CrossRef]
72. Nur Sakinah, M.J.; Misran, A.; Mahmud, T.M.M.; Abdullah, S. A review: Production and postharvest management of *Volvariella volvacea*. *Int. Food Res. J.* **2019**, *26*, 367–376.
73. Pires, M.A.; Pastrana, L.M.; Fucinõs, P.; Abreu, C.S.; Oliveira, S.M. Sensorial Perception of Astringency: Oral Mechanisms and Current Analysis Methods. *Foods* **2020**, *9*, 1124. [CrossRef]

Article

Reappraisal of the Genus *Exsudoporus* (*Boletaceae*) Worldwide Based on Multi-Gene Phylogeny, Morphology and Biogeography, and Insights on *Amoenoboletus*

Alona Yu. Biketova [1,2,3,4,*,†], Matteo Gelardi [5,†], Matthew E. Smith [6], Giampaolo Simonini [7], Rosanne A. Healy [6], Yuichi Taneyama [8], Gianrico Vasquez [9], Ádám Kovács [2], László G. Nagy [2], Solomon P. Wasser [1,10], Ursula Peintner [11], Eviatar Nevo [1], Britt A. Bunyard [12] and Alfredo Vizzini [13,*]

1. Institute of Evolution, University of Haifa, Aba Khoushi Ave. 199, Mt. Carmel, Haifa 3498838, Israel; spwasser@research.haifa.ac.il (S.P.W.); nevo@research.haifa.ac.il (E.N.)
2. Institute of Biochemistry, Biological Research Center, Eötvös Lóránd Research Network, Temesvari Blvd. 62, H-6726 Szeged, Hungary; kovadi90@gmail.com (Á.K.); lnagy@fungenomelab.com (L.G.N.)
3. Mycological Society of Israel, P.O. Box 164, Pardesiya 42815, Israel
4. Jodrell Laboratory, Royal Botanic Gardens, Kew, Richmond TW9 3DS, UK
5. Independent Researcher, Via dei Barattoli 3A, Anguillara Sabazia, I-00061 Rome, Italy; timal80@yahoo.it
6. Department of Plant Pathology, University of Florida, Gainesville, FL 32611, USA; trufflesmith@ufl.edu (M.E.S.); rhealy1@ufl.edu (R.A.H.)
7. Independent Researcher, Via Bell'aria 8, I-42121 Reggio Emilia, Italy; giamsim@tin.it
8. Independent Researcher, 392-3 Yashikida, Nagano 381-0055, Japan; boletus@mx1.avis.ne.jp
9. Laboratorio di Isto-Cito-Patologia s.n.c., Via del Bosco 96, I-95125 Catania, Italy; gianricovasquez@hotmail.com
10. M.G. Kholodny Institute of Botany, National Academy of Sciences of Ukraine, Tereshchenkivska St. 2, 01601 Kiev, Ukraine
11. Institute of Microbiology, University of Innsbruck, Technikerstrasse 25, A-6020 Innsbruck, Austria; ursula.peintner@uibk.ac.at
12. Independent Researcher, P.O. Box 98, Batavia, IL 60510, USA; bbunyard@wi.rr.com
13. Department of Life Sciences and Systems Biology, University of Torino, Viale P.A. Mattioli 25, I-10125 Torino, Italy
*   Correspondence: alyona.biketova@gmail.com (A.Yu.B.); alfredo.vizzini@unito.it (A.V.); Tel.: +39-01-1670-5979 (A.V.); Fax: +39-01-1670-5962 (A.V.)
†   These authors contributed equally to this work.

Citation: Biketova, A.Yu.; Gelardi, M.; Smith, M.E.; Simonini, G.; Healy, R.A.; Taneyama, Y.; Vasquez, G.; Kovács, Á.; Nagy, L.G.; Wasser, S.P.; et al. Reappraisal of the Genus *Exsudoporus* (*Boletaceae*) Worldwide Based on Multi-Gene Phylogeny, Morphology and Biogeography, and Insights on *Amoenoboletus*. *J. Fungi* 2022, 8, 101. https://doi.org/10.3390/jof8020101

Academic Editor: José Francisco Cano-Lira

Received: 28 December 2021
Accepted: 14 January 2022
Published: 21 January 2022

Publisher's Note: MDPI stays neutral with regard to jurisdictional claims in published maps and institutional affiliations.

Abstract: The boletoid genera *Butyriboletus* and *Exsudoporus* have recently been suggested by some researchers to constitute a single genus, and *Exsudoporus* was merged into *Butyriboletus* as a later synonym. However, no convincing arguments have yet provided significant evidence for this congeneric placement. In this study, we analyze material from *Exsudoporus* species and closely related taxa to assess taxonomic and phylogenetic boundaries between these genera and to clarify species delimitation within *Exsudoporus*. Outcomes from a multilocus phylogenetic analysis (ITS, nrLSU, *tef1-α* and *rpb2*) clearly resolve *Exsudoporus* as a monophyletic, homogenous and independent genus that is sister to *Butyriboletus*. An accurate morphological description, comprehensive sampling, type studies, line drawings and a historical overview on the nomenclatural issues of the type species *E. permagnificus* are provided. Furthermore, this species is documented for the first time from Israel in association with *Quercus calliprinos*. The previously described North American species *Exsudoporus frostii* and *E. floridanus* are molecularly confirmed as representatives of *Exsudoporus*, and *E. floridanus* is epitypified. The eastern Asian species *Leccinum rubrum* is assigned here to *Exsudoporus* based on molecular evidence, and a new combination is proposed. Sequence data from the original material of the Japanese *Boletus kermesinus* were generated, and its conspecificity with *L. rubrum* is inferred as formerly presumed based on morphology. Four additional cryptic species from North and Central America previously misdetermined as either *B. frostii* or *B. floridanus* are phylogenetically placed but remain undescribed due to the paucity of available material. *Boletus weberi* (syn. *B. pseudofrostii*) and *Xerocomus* cf. *mcrobbii* cluster outside of *Exsudoporus* and are herein assigned to the recently described genus *Amoenoboletus*. Biogeographic distribution patterns are elucidated, and a dichotomous key to all known species of *Exsudoporus* worldwide is presented.

Keywords: *Boletales*; bolete diversity; biogeography; molecular phylogeny; taxonomy

1. Introduction

Red-pored boletes were originally placed in *Boletus* Fr. sect. *Luridi* Fr. emend. Lannoy and Estadès, a species-rich complex typified by *Boletus luridus* Schaeff. of apparently similar taxa mostly sharing an orange to reddish colored hymenophoral surface, variably bluing oxidation reaction of tissues and mild taste [1–6]. Several informal infrageneric groupings (subsections, stirps, series, etc.) were mainly proposed by North American and European authors based on different combinations of morphological, chemotaxonomical and ecological traits [4,7–12]. At least two additional validly published sections, *Boletus* sect. *Rubropori* M. Zang [13] and *Boletus* sect. *Erythropodes* Galli [14], were erected to include boletes with reddish tube dissepiments. However, due to the obvious diversity and phenotypic variability of this large assemblage of boletoid mushrooms, it has long been challenging to address their true phylogenetic affinities based solely on conventional research techniques. Molecular phylogenetic data are essential to determine the evolutionary relationships within the family *Boletaceae*. Based on recent phylogenetic analyses, the historical and long-established genera of *Boletaceae*, including *Boletus, Leccinum, Pulveroboletus, Tylopilus, Xerocomus*, etc., have been shown to be polyphyletic [15–18]. These genera contain species from multiple unrelated lineages and therefore have recently undergone dramatic taxonomic reassessments.

Among several newly described genera recently segregated from *Boletus* s.l. [18–25], *Exsudoporus* Vizzini, Simonini and Gelardi, typified by *Boletus permagnificus* Pöder, was established to accommodate species sharing bright red colors overall, reddish pore surface typically beaded with golden droplets when young and fresh, prominently reticulate to reticulate-alveolate stipe surface, tissues bruising blue on injury, mild to acidic taste, olive-brown spore print, ellipsoid-fusiform, smooth basidiospores, trichodermal to ixotrichodermal or ixocutis pileipellis, hymenophoral trama bilaterally divergent of the "*Boletus*-type", fertile caulohymenium and gymnocarpic ontogenetic development [26–28]. Prior to the establishment of *Exsudoporus* as a genus on its own, the clade encompassing the American species *Boletus frostii* J.L. Russell and *B. floridanus* Singer was already inferred as distantly related to *Boletus s. str.* and separate from other boletoid clades [16,17,29–33]. In addition, further evidence supporting *Exsudoporus* was provided by Zhao et al. [23,34], Gelardi et al. [25], Smith et al. [35], Henkel et al. [36], Crous et al. [37], Bozok et al. [38] and Loizides et al. [39]. Based on the most recent comprehensive phylogenetic classification, *Exsudoporus* is nested within the informal "*Pulveroboletus* group" [17], an unresolved heterogeneous assemblage dominated by boletoid species but also including sequestrate and lamellate taxa. The taxonomic status of *Exsudoporus* has recently been disputed and Wu et al. [18] proposed that *Exsudoporus* be synonymized with *Butyriboletus* D. Arora and J.L. Frank. Moreover, the genus was not evaluated in recent taxonomic overviews of the *Basidiomycota* by He et al. [40] and Wijayawardene et al. [41].

Wu et al. [42] recently described a new genus of *Amoenoboletus* G. Wu, E. Horak and Zhu L. Yang, which has certain morphological similarity with *Exsudoporus* based on size and color of basidiomes and stipe surface ornamentation.

In the present study, several DNA sequences have been generated from representative voucher specimens from across the Northern Hemisphere, including the holotype material of the type of the genus *E. permagnificus*, in order to (1) establish the generic limits of *Exsudoporus* and resolve the taxonomic issue related to the proposed synonymy of *Exsudoporus* with *Butyriboletus* by inferring their phylogenetic relationship, (2) ascertain species-level diversity within *Exsudoporus* on a global scale, (3) elucidate the phylogenetic infrageneric affiliations of species within *Exsudoporus*, (4) clarify the morphological variability of *E. permagnificus* and (5) assess the ecological requirements and biogeographic distribution of *Exsudoporus* species worldwide.

2. Materials and Methods

2.1. Collection Site and Sampling

Specimens were collected at several different localities in Italy, Israel, the USA and Japan and are deposited in BTS, EMAC, F, FH, FLAS, HAI, IB, K-M, LE, MCVE, NY, TO and TNS-F (acronyms from Thiers [43]), while "MG", "AB", "GS" and "FM" refer to the personal herbaria of Matteo Gelardi, Alona Yu. Biketova, Giampaolo Simonini, and Francesco Mondello, respectively. With the only exception of a single collection of *E. permagnificus* from Italy, herbarium numbers are cited for all samples from which morphological features were examined. Author citations follow the Index Fungorum, Authors of Fungal Names [44]. Novel combinations are registered with MycoBank [45] and the epitype of *E. floridanus*—with Index Fungorum [44]. The distribution range of North American species has also been checked on MyCoPortal [46]. Since *E. ruber* is a rare species in Japan, geographic grid references are not reported in the examined material of that species in order to better preserve its occurrence localities.

2.2. Morphological Study

Macroscopic descriptions, macro-chemical reactions (25% NH_3, 30% KOH, $FeSO_4$) and ecological information, such as habitat notations, time of fruiting and associated plant communities accompanied the detailed field notes of the fresh basidiomes. For some collections, macro-morphological characteristics of the specimens were also examined using Carl Zeiss Stemi DV4 stereo microscope. In the field, latitude, longitude and elevation were determined with a global positioning system (GPS) receiver. Color terms in capital letters (e.g., White, Plate LIII) are from Ridgway [47]. Microscopic anatomical features were observed and recorded from revived dried material; sections were rehydrated either in water, 5% potassium hydroxide (5% KOH), 3% NH_3 or in anionic solution saturated with Congo red. All anatomical structures were measured from preparations in anionic Congo red. Colors and pigments were described after examination in water and 5% KOH. Measurements were made at $1000\times$ using a calibrated ocular micrometer (Nikon Eclipse E200 (Tokyo, Japan) and Carl Zeiss Axiostar 1122-100 (Germany) light microscopes). Basidiospores were measured directly from the hymenophore of mature basidiomes, average sizes were calculated for each collection and used in the description, dimensions are given as (minimum) average $\pm$ standard deviation (maximum), Q_m = average quotient (length/width ratio) $\pm$ standard deviation with extreme values (minimum and maximum) in parentheses, and average spore volume was approximated as a rotation ellipsoid ($V = (\pi \times L \times W^2)/6 \pm$ standard deviation). The notation (n/m/p) indicates that measurements were made on "n" randomly selected basidiospores from "m" basidiomes of "p" collections. The morphometric variables "spore length" and "spore width" were measured and statistically analyzed. Spore size distribution (length and width) with Gauss' bivariate confidence ellipse of *Exsudoporus permagnificus*. In the field of the variables "spore length" and "spore width" the ellipse represents the pairs of the variable values that have an identical probability (equal to 68%) to occur. All the points inside the ellipse represent pairs of the variables having a probability of occurrence greater than 68%. As for the other anatomical elements aside from spores, absolute sizes are given. The width of each basidium was measured at the widest part, and the length was measured from the apex (sterigmata excluded) to the basal septum. Radial and/or vertical sections of the pileipellis were taken midway between the center and margin of the pileus. Sections of the stipitipellis were taken from the middle part along the longitudinal axis of the stipe. Metachromatic, cyanophilic and iodine reactions were tested by staining the basidiospores in Brilliant Cresyl blue, Cotton blue and Melzer's reagent, respectively. Line drawings of microstructures were traced in free hand based on digital photomicrographs of rehydrated material.

2.3. DNA Extraction, PCR Amplification and DNA Sequencing

Total genomic DNA was extracted from dried basidiomes using NucleoSpin Plant II kit with minor modifications. The following primers were used: ITS1F, ITS1, ITS4B, ITS4 and ITS2 for internal transcribed spacer (ITS) [48,49], LROR, LR5 and LR7 for nuclear large subunit ribosomal DNA (nrLSU) [50,51], EF1-983F, EF1-1567R and EF1-2218R for translation elongation factor 1-α gene (*tef1-α*) [52] and RPB2-BF2, bRPB2-7R2 and RPB2-BR for DNA-directed RNA polymerase II subunit 2 gene (*rpb2*) [53].

For ITS and nrLSU, PCR was carried out under the following cycling parameters: initial denaturation: 95 °C for 5 min, followed by 35 cycles: 95 °C for 30 s, 55 (or 53) °C for 30 s, 72 °C for 1 min, and final extension at 72 °C for 7 min. For tef1-α and rpb2, PCR conditions were as follows: initial denaturation at 95 °C for 5 min, followed by 40 cycles: 95 °C for 30 s, 55 °C for 45 s, 72 °C for 45 s, and final extension at 72 °C for 7 min.

Sequences were manually edited and assembled using Sequencer 4.1.4 program (Gene Codes Corporation, Ann Arbor, MI, USA). Sequences generated for this study were submitted to GenBank and their accession numbers are cited in Table 1.

Table 1. Information on specimens used in multilocus phylogenetic analysis and their GenBank accession numbers. Newly generated sequences are in boldface.

Species	Voucher	Locality	GenBank Accession Number				Notes
			ITS	nrLSU	*tef1-α*	*rpb2*	
Amoenoboletus cf. *granulopunctatus* 1	KUN-HKAS 56280	China	MZ708840	KF112418	KF112265	KF112708	-
A. cf. *granulopunctatus* 2	MHHNU 9490	China	MW520189	MW520186	MW566747	MW560081	-
A. cf. *granulopunctatus* 2	KUN-HKAS 80250	China	MW520191	MW520185	MW566746	MW560080	-
A. cf. *granulopunctatus* 2	KUN-HKAS 86007	China	MW520190	MW520187	MZ741478	MW560079	-
A. mcrobbii	PDD 97418	New Zealand	MZ708841	JQ924329	MZ708841	-	-
A. cf. *mcrobbii*	**NY 2686023 (Halling 9916)**	**Australia**	**OL960511**		-	-	-
A. miraculosus	Z-ZT 14046	Malaysia	MZ708842	MZ708842	MZ708842	-	holotype
A. *weberi*	**FLAS-F-61525**	**USA**	**MH211950**	-	-	-	-
A. *weberi*	**FLAS-F-68076**	**USA**	**OL960512**	-	-	-	**topotype**
A. weberi	MO 179586	USA	MH251719	MH249987 MH249988	-	-	-
Boletus billieae	MO 333089	USA	MK542835	MK542836	-	-	-
B. *kermesinus*	**BTS031J**	**Japan**	**OL960532**		-	-	-
B. *kermesinus*	**TNS-F-37407**	**Japan**	**OL960531**		-	-	holotype
B. pseudofrostii	CFMR BZ-1611 (BOS-266)	Belize	MN250201	MN250176	-	-	holotype
B. subsplendidus	KUN-HKAS 50444	China	KM388725	KT990540	KT990742	KT990379	-
B. subsplendidus	KUN-HKAS 52661	China	KM388726	KF112339	KF112169	KF112676	-
B. subsplendidus	KUN-HKAS 82375	China	KM388727	-	-	-	-
"*Boletus*" sp. 1	KUN-HKAS 52525	China	KU317760	KF112337	KF112163	KF112671	-
"*Boletus*" sp. 2	KUN-HKAS 57774	China	KU317761	KF112330	KF112155	KF112670	-
Butyriboletus abieticola	JLF2654	USA	KC184418		-	-	-
Bu. appendiculatus	MB000286	Germany	KT002599	KT002610	KT002634	-	-
Bu. autumniregius	JLF2275	USA	KC184430		-	-	paratype
Bu. brunneus	NY 00013631	USA	KT002600	KT002611	KT002635	-	-
Bu. fechtneri	AT2003097	USA	KC584784	KF030270	-	-	-
Bu. fuscoroseus	BR50201618465-02	Belgium	KT002602	KT002613	KT002637	-	-
Bu. fuscoroseus	HR:86133	Czech Rep.	KJ419926		-	-	neotype
"*Bu.*" *hainanensis*	N.K.Zeng1197	China	KU961653	KU961651	-	KU961658	paratype
"*Bu.*" *hainanensis*	N.K.Zeng2418	China	KU961654	KU961652	KU961656	KX453856	paratype
"*Bu.*" *hainanensis*	KUN-HKAS 59814	China	KU317762	KF112336	KF112199	KF112699	paratype

Table 1. *Cont.*

Species	Voucher	Locality	GenBank Accession Number				Notes
			ITS	nrLSU	*tef1-α*	*rpb2*	
"Bu." hainanensis	EMF11	China	JF273514	-	-	-	-
Bu. huangnianlaii	FHMU 2207 (N.K.Zeng3246)	China	MH885351	MH879689	MH879718	MH879741	holotype
Bu. peckii	3959	USA	-	JQ326999	JQ327026	-	-
Bu. persolidus	Arora11102	USA	KC184441		-	-	paratype
Bu. primiregius	JLF2030	USA	KC184455		-	-	holotype
Bu. pseudospeciosus	KUN-HKAS 63596	China	-	KT990542	KT990744	KT990381	paratype
Bu. pseudospeciosus	KUN-HKAS 63513	China	KM388728	KT990541	KT990743	-	holotype
Bu. pulchriceps	R. Chapman 0945	USA	KT002604	KT002615	KT002639	-	-
Bu. quercireguis	Arora11100	USA	KC184461		-	-	holotype
Bu. regius	KUN-HKAS 84878	Germany	-	MT264910	MT269659	MT269661	-
Bu. regius	MB000287	Germany	KT002605	KT002616	KT002640	-	-
Bu. roseoflavus	KUN-HKAS 63593	China	KJ909517	KJ184559	KJ184571	-	-
Bu. roseogriseus	PRM 923479	Czech Rep.	KJ419928		-	-	paratype
Bu. roseopurpureus	E.E. Both 3765	USA	KT002606	KT002617	KT002641	-	-
Bu. sanicibus	Arora99211	China	KC184469	KC184470	-	-	holotype
Bu. subappendiculatus	MB000260	Germany	KT002607	KT002618	KT002642	-	-
Bu. taughannockensis	MO 250839	USA	MH234472	MH234473	-	-	holotype
Bu. yicibus	KUN-HKAS 57503	China	KT002608	KT002620	KT002644	-	-
***Butyriboletus* sp. 1**	**K-M000257266**	**Japan**	**OL960513**	-	-	-	-
Butyriboletus sp. 2	MHHNU7456	China	-	KT990539	KT990741	KT990378	-
Butyriboletus sp. 2	Zhangping956	China	KU317759	KU317764	-	-	-
Butyriboletus sp. 3	KUN-HKAS 59467	China	KM388729	-	-	-	-
Caloboletus peckii	MO 246697	USA	-	MH220330	MH318614	-	-
Exsudoporus floridanus	MO 320467	USA	-	MN114633	-	-	-
E. floridanus	MO 320294	USA	-	MK533764	-	-	-
E. floridanus	25A_T2_C5	USA	KX899261	-	-	-	environmental
E. floridanus	25B_Y1_B2	USA	KX899270	-	-	-	environmental
E. floridanus	25D_W1_G3	USA	KX899279	-	-	-	environmental
E. floridanus	**FLAS-F-59069**	**USA**	**OL960514**	**OL960488**	**OL960496**	**OL960503**	**epitype**
E. floridanus	**FLAS-F-59142**	**USA**	**OL960515**	-	-	-	-
E. floridanus	**FLAS-F-61008**	**USA**	**OL960516**	**OL960489**	**OL960497**	**OL960504**	-
E. floridanus	**FLAS-F-61189**	**USA**	**MH211799**	**OL960490**	**OL960498**	-	-
E. floridanus	Farid 499	USA	-	-	MW737484	MW737459	-
E. cf. *floridanus*	BD368	Costa Rica	JN020981	HQ161859	-	-	-
E. cf. *floridanus*	CFMR BZ-3170	Belize	MN250222	MK601725	MK721079	MK766287	-
E. cf. *frostii* 1	JLF2548	USA	KC812303	KC812304	-	-	-
E. cf. *frostii* 1	TENN067311	USA	KT002601	KT002612	KT002636	-	-
E. cf. *frostii* 1	TENN:SAT1221511	USA	-	KP055021	KP055018	KP055027	-
E. cf. *frostii* 1	B1789	USA	KY826056	-	-	-	-
***E.* cf. *frostii* 1**	**TO AVBB10**	**USA**	**OL960517**	**OL960491**	-	-	-
***E.* cf. *frostii* 1**	**TO AVBB11**	**USA**	**OL960518**	**OL960492**	-	**OL960505**	-
E. cf. *frostii* 2	MHM069	Mexico	EU569285	-	-	-	-
E. cf. *frostii* 2	BDCR0418	Costa Rica	-	HQ161855	-	-	-
E. cf. *frostii* 2	NY 815462	Costa Rica	-	JQ924342	KF112164	KF112675	-
E. cf. *frostii* 2	M39B4	Mexico	FJ196902		-	-	environmental
E. cf. *frostii* 3	JLF6850	USA	MN263010	MN258884	-	-	-
E. cf. *frostii* 3	JLF5376	USA	MN263009	MN258883	-	-	-
E. cf. *frostii* 4	man3_soil_G05	USA	GU328546	-	-	-	environmental
E. cf. *frostii* 4	**FLAS-F-60742**	**USA**	**MH016833**	**OL960493**	**OL960499**	**OL960506**	-
E. cf. *frostii* 4	**iNat35326745**	**USA**	**OL960519**	**OL960494**	**OL960500**	**OL960507**	-

Table 1. *Cont.*

Species	Voucher	Locality	GenBank Accession Number				Notes
			ITS	nrLSU	*tef1-α*	*rpb2*	
E. cf. *frostii* 4	iNat30897161	USA	OL960520 OL960521	OL960495	OL960501	OL960508	-
E. permagnificus	IB 19800750	Italy	OL960522	-	-	-	holotype
E. permagnificus	AB B11-03	Israel	OL960523		-	-	-
E. permagnificus	AB B15-254	Israel	OL960524	-	-	-	-
E. permagnificus	AB B15-271	Israel	OL960525		-	OL960509	-
E. permagnificus	GS1001	Italy	OL960526	-	-	-	-
E. permagnificus	GS1275	Italy	OL960527	-	-	-	-
E. permagnificus	MG558	Italy	OL960528	-	-	-	-
E. permagnificus	MG637	Italy	OL960529	-	-	-	-
E. permagnificus	MG829	Italy	OL960530		OL960502	OL960510	-
E. permagnificus	ML61992EP	Cyprus	MH011858	-	-	-	-
E. permagnificus	FR2011120	France	KR782301	-	-	-	-
E. ruber	KUN-HKAS 106891	China	-	MN930518	MT063123	MT063120	-
E. ruber	KUN-HKAS 103513	China	-	MN930519	MT063124	MT063121	-
E. ruber	KUN-HKAS 103122	China	-	MN930520	-	MT063122	-
Rubroboletus sinicus	KUN-HKAS 68620	China	KJ951991	KF112319	KF112146	KF112661	-

2.4. Sequence Alignment, Data Set Assembly and Phylogenetic Analysis

A total of 208 sequences from 95 specimens, including 48 newly generated (19 ITS, 8 nrLSU, 6 ITS-nrLSU, 7 *tef1-α*, 8 *rpb2*) and 160 retrieved from GenBank database (49 ITS, 52 nrLSU, 8 ITS-nrLSU, 36 *tef1-α*, 23 *rpb2*), were used in the phylogenetic analysis (Table 1). *Rubroboletus sinicus* was chosen as an outgroup taxon.

Three alignments were generated for combined ITS-nrLSU (algorithm Q-INS-i), *tef1-α* (algorithm E-INS-i) and *rpb2* (algorithm E-INS-i) datasets with MAFFT 7 [54]. Data for each locus were manually adjusted and combined in MEGA 6.06 [55]. Sites with 99% gaps (59 positions in total) were removed using trimAl v.1.2 program [56]. Phylogenetic reconstructions were performed using the maximum likelihood (ML) and Bayesian inference (BI) methods of analysis. In both ML and BI analyses, the general time reversible model with rates that vary over sites according to gamma (GTR+G) was employed.

The ML phylogenetic analysis was run in raxmlGUI 2.0 [57], which implements the search protocol of Stamatakis et al. [58], under a partitioned model (ITS1, 5.8S, ITS2, nrLSU, *tef1-α*, and *rpb2*), estimating unique parameters for each partition, using 1000 rapid bootstrap replicates and branch lengths saved in the bootstrap trees (BS brL enabled).

BI was performed with MrBayes 3.2.6 software [59], under the described model. The BI analysis was performed with two parallel searches and four chains, with three million generations and a sampling frequency of every 100th generation. Tracer 1.6.0 [60] was used to evaluate the quality of a sample from the posterior and the continuous parameters, using effective sample size (ESS). To remove the pre-stationary posterior probability distribution, a burn-in of 25% was applied.

Only bootstrap support (BS) from 50% and posterior probability (PP) values exceeding 0.7 are reported in the resulting tree (Figure 1). A clade was considered strongly supported if it received bootstrap support (BS) greater than 70% and/or posterior probability (PP) equal to or greater than 0.95. Branch lengths were estimated as mean values over the sampled trees. The final tree was edited in Inkscape v.0.92.

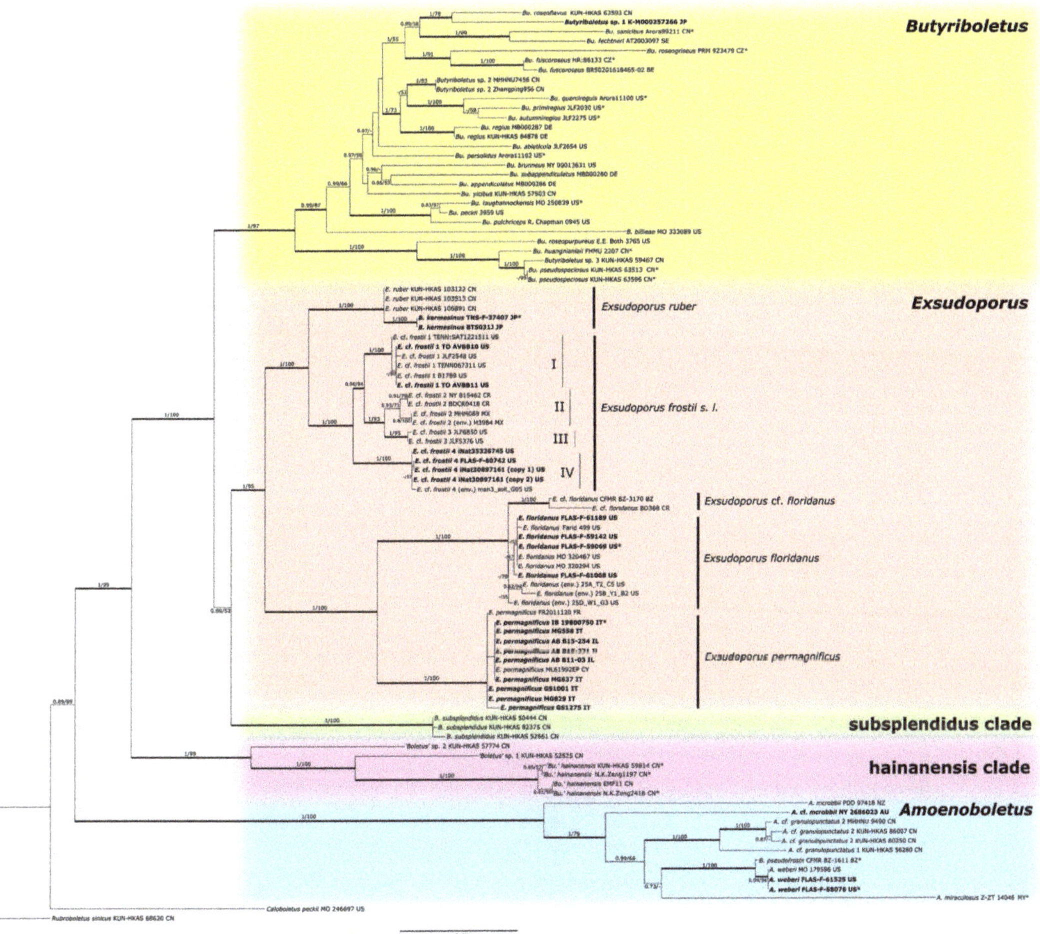

Figure 1. ML phylogenetic tree of *Exsudoporus*, *Amoenoboletus* and allied genera generated from a multilocus (ITS + nrLSU + *tef1-α* + *rpb2*) dataset. PP values ≥ 0.7 and BS support values ≥ 50% are shown at the nodes. Thickened branches indicate PP ≥ 0.95 and BS support ≥ 70%. Newly sequenced collections are indicated in bold; type specimens are indicated with an asterisk (*). Two-letter country codes (ISO 3166-1 alpha-2) reflecting origin of specimens are given.

3. Results

3.1. Molecular Phylogenetic Analysis

The aligned multigene matrix contained a total of 96 samples and 3302 aligned bases with gaps (Supplementary File S1). The ML and BI analyses generated almost identical tree topologies with minimal variation in statistical support values; thus, an ML tree with both BS and PP values was selected for the purposes of display (Figure 1).

Our phylogenetic analysis indicated that genera *Exsudoporus* and *Butyriboletus* form two different generic clades, both with high support and showing a considerable genetic difference. *Exsudoporus* (BS = 95%, PP = 1.00) and *Boletus subsplendidus* (BS = 100%, PP = 1.00) clades form a sister superclade (BS = 52%, PP = 0.89) to the *Butyriboletus* clade (BS = 97%, PP = 1.00). There are two other generic clades with strong statistical support: *"hainanensis"* clade (BS = 99% and PP = 1.00) and *Amoenoboletus* (BS = 100% and PP = 1.00).

3.2. Taxonomy

Exsudoporus Vizzini, Simonini and Gelardi 2014, emend. Biketova and Gelardi.
MYCOBANK MB 550708.

Diagnosis: Basidiome stipitate-pileate with tubular hymenophore, epigeal, evelate; pileus convex to applanate, bright blood red, crimson-red, purplish-red, reddish-pink or reddish-brown, opaque to shiny, dry to subviscid with moist weather, glabrous to subpruinose or subtomentose; hymenophore poroid, adnate or slightly depressed around stipe apex; tubes yellow to olivaceous-brown; pores pinkish-red, reddish-orange, blood red to dark red, rarely yellowish-orange or yellow, often beaded with golden yellow or amber yellow droplets when young and fresh; stipe central, solid, yellowish to concolorous with the pileus, conspicuously reticulate with elongated, red meshes or deeply reticulate-alveolate or can be covered with scaly patches; context pale yellow to bright yellow; tissues quickly turning dark blue or more rarely light blue or even unchanging when injured or exposed, then fading blackish; taste mild to acidic; spore print olive-brown; basidiospores smooth, subfusiform to ellipsoid or ellipsoid-fusoid; pleuro-, cheilo- and caulocystidia present; pileipellis an interwoven (ixo)trichoderm tending to a cutis; hymenophoral trama bilateral-divergent of the "*Boletus*-type"; clamp connections absent; stipe context varies from inamyloid to amyloid or dextrinoid; ontogenetic development gymnocarpic.

Generic type: *Boletus permagnificus* Pöder 1982.

Exsudoporus permagnificus (Pöder) Vizzini, Simonini and Gelardi, Index Fungorum 183:1, 2014.

Figures 2–5.

MYCOBANK MB 550709.

≡ *Boletus permagnificus* Pöder, Sydowia 34:151, 1982 ("1981"). (Basionym).

≡ *Suillellus permagnificus* (Pöder) Blanco-Dios, Index Fungorum 211:1, 2015.

Misapplied names:

– *Boletus flammans* E.A. Dick and Snell, Mycologia 57(3):453, 1965 s. Angarano non s. E.A. Dick and Snell.

– *Boletus frostii* J.L. Russell, Bulletin of the Buffalo Society of Natural Sciences 2:102, 1874 s. Alessio non s. J.L. Roussell.

– *Boletus siculus* Inzenga, Funghi Siciliani. Centuria 2: 57, 1869 s. Alessio non s. Inzenga (? = *Alessioporus ichnusanus* (Alessio, Galli and Littini) Gelardi, Vizzini and Simonini).

Holotype: Italy, Sardinia, Arzachena (OT), Cannigione, under *Quercus suber* with the presence of *Cistus* sp. And *Eryngium* sp., 1 November 1980, R. Pöder, IB 19800750.

Basidiomes small to medium. *Pileus* (2.0) 3.2–9.5 (10.0) cm broad, at first hemispherical then persistently convex and finally broadly pulvinate-flattened to slightly depressed, regularly to sometimes unevenly shaped by shallow depressions, moderately fleshy, firm at the beginning but progressively softer with age, flabby in old basidiomes; margin generally obtuse, steady to faintly wavy-lobed especially in young specimens, initially involute then curved downwards and finally completely plane or even uplifted, extending beyond the tubes up to 2 mm in primordia; surface matt, dry but slightly greasy and polished with moist weather, very finely pubescent in the early stage of development but later smooth and glabrous, not cracked; cuticle somewhat variable in color, ranging from pale pinkish, pinkish-violet to pinkish-red (Shrimp Pink, Geranium Pink, Rose Doree, pl. I; Pale Amaranth Pink, Rose Pink, pl. XII; Rose-Purple, pl. XVI) in young specimens due to the presence of a whitish pubescence which soon tends to dissolve revealing the true color below, blood red to dark carmine red, coral red or garnet red to reddish-purple (Spinel Pink, Spinel Red, pl. XXVI; Spectrum Red, Scarlet-Red, Carmine, pl. I) at maturity, often with scattered orange shades (Orange Chrome, pl. II; Scarlet, pl. I) especially towards the margin, gradually fading with age and becoming copper red to ochraceous red or even ochraceous brown (Madder Brown, Brick Red, pl. XIII; Buckthorn Brown, pl. XV) in senescence or if exposed to direct sunlight; slowly but pronouncedly turning bluish-black (Berlin Blue, pl. VIII) on handling or when injured, particularly in young and fresh basidiomes, then fading dull brownish-red (Indian Lake, Dahlia Carmine, pl. XXVI; Bordeaux, pl. XII); subcuticular layer reddish-purple (Amparo Purple, pl. XI; Lobella Violet, pl. XXXVII). *Tubes* at first thin then increasingly broader and shorter than the thickness of the pileus context (up to 1.8 cm long), adnate-subdecurrent at first but soon adnexed, finally depressed around the stipe apex and shortly decurrent with a tooth, pale yellow (Barita Yellow, pl. IV) at first, yellowish-olive (Yellowish Citrine, pl. XVI) to brownish-olive (Saccardo Olive, pl. XVI; Light Brownish Olive, pl. XXX; Dark Olive-Buff, pl. XL) in old fruiting bodies, bluing (Blanc's Blue, Dusky Greenish Blue, pl. XX) when cut. *Pores* initially forming a flat surface, later convex to ascendant, at first very small then gradually wider (up to 2 mm in diam.), simple, roundish to barely angular and radially stretched at maturity, at first more or less evenly yellow (Lemon Chrome, pl. IV) but soon orange pinkish to blood red (Vinaceous, Deep Vinaceous, pl. XXVII; Eosine Pink, Spectrum Red, Scarlet-Red, pl. I), sometimes yellowish-orange (Flame Scarlet, Orange Chrome, pl. II) towards the margin and finally rusty red (Carmine, pl. I), quickly and intensely turning dark blue (Blanc's Blue, Dusky Greenish Blue, pl. XX) on bruising or when injured and finally fading to blackish (Black, pl. LIII); the hymenophore of young and fresh specimens exude abundant golden yellow (Light Orange-Yellow, pl. III), markedly salty-tasting droplets which tend to disappear with age. *Stipe* (4.0) 4.6–11.0 (12.0) × (0.6) 0.8–3.1 (4.0) cm, slightly longer than or as long as the pileus diameter at maturity, central to slightly off-center, solid, firm, dry, straight or faintly curved, cylindrical to fusiform, rarely progressively attenuated downwards, usually swollen in the middle part and tapering towards both the apex and the base, never clavate, ending with a long pointed taproot at the very base, frequently deeply rooting; surface showing a fine to pronounced reticulum at least in the upper half or over the upper three fourths, sometimes extending down to the base, smooth and glabrous elsewhere, evelate; lemon yellow or bright yellow to occasionally pinkish (Barita Yellow, pl. III; Light Orange-Yellow, pl. II) in the upper third or in the upper half in young specimens, orange yellowish to pinkish (Capucine Yellow, Orange, pl. II; Ochraceous-Salmon, pl. XV) elsewhere, reddish-purple to purplish-brown (Aster Purple, Dahlia Purple, Pansy Purple, pl. XII) in the lower portion, entirely blood red to carmine red or garnet red (Spinel Pink, Spinel Red, pl. XXVI; Spectrum Red, Scarlet-Red, Carmine, pl. I) in mature specimens; reticulum consisting of fine to well-defined, narrow and longitudinally stretched polygonal meshes, concolorous to the ground in early stages of development but soon blood red to carmine red (Spectrum Red, Scarlet-Red, Carmine, pl. I) lengthwise; bruising dark blue (Berlin Blue, pl. VIII) when pressed then fading sordid reddish-purple to blackish (Claret Brown, pl. I; Black, pl. LIII); basal mycelium cream yellowish (Barita Yellow, pl. IV), rhizomorphs brownish (Sayal Brown, pl. XXIX). *Context* firm and tough when young, later soft textured and eventually flabby in the pileus (up to 3.2 cm thick in the central zone and gradually becoming thinner towards the edge), a little more fibrous in the stipe, evenly watery yellow (Citron Yellow, pl. XVI) throughout with darker tones towards the base, with a thin reddish-purple line (Amparo Purple, pl. XI; Amaranth Purple, pl. XII; Lobella Violet, pl. XXXVII) beneath the cuticle and sometimes with reddish-purple scattered spots or shades in the stipe; turning extensively blue (Yale Blue, Vanderpoel's Blue, Blanc's Blue, Dusky Greenish Blue, pl. XX; Grayish Violaceous Blue, pl. XXII) when exposed to air with the only exception of the stipe base in young specimens, which remains practically unchangeable, finally fading dull yellowish (Aniline Yellow, Pyrite Yellow, pl. IV) to dirty yellowish-orange (Yellow Ocher, pl. XV) or occasionally reddish (Etruscan Red, pl. XXVII); reddish-purple (Pomegranate Purple, pl. XII) where eroded by maggots and bright yellow (Strontian Yellow, pl. XVI) to reddish (Etruscan Red, pl. XXVII) in places eaten by slugs; subhymenophoral layer bright yellow (Strontian Yellow, pl. XVI); exsiccate pinkish-red, wine red to dark reddish-purple (Spinel Pink, Spinel Red, pl. XXVI; Spectrum Red, Scarlet-Red, Carmine, pl. I) on pileus and stipe, brownish (Cinnamon-Brown, pl. XV) on hymenophore, beige to dull ochraceous brown on context (Ivory Yellow, pl. XXX; Clay Color, pl. XXIX). *Odor* intensely fruity, agreeable. *Taste* mild then with a weakly acidic aftertaste. *Spore print* olive-brown. *Macrochemical spot-test reactions*: 25% NH$_3$: Pileus cuticle fades ochraceous, no reaction elsewhere (but bluing oxidation on tissues disappears); 30% KOH: staining dark wine red on hymenophore, bright orange to wine red elsewhere; $FeSO_4$: blackish on hymenophore, no response or pale olivaceous on context, brownish-olive on pileus and stipe. *Ontogenetic development* gymnocarpic.

Basidiospores (768/55/23) (12.7) 14.4 ± 0.7 (15.4) × (5.5) 5.9 ± 0.2 (6.1) μm, Q_m = (2.23) 2.47 ± 0.12 (2.61), V = 263 ± 22 $μm^3$, inequilateral, ellipsoid to ellipsoid-fusiform or less frequently broadly ellipsoid in side view, broadly ellipsoid to ellipsoid in face view, smooth, apex rounded, with a short apiculus and usually with a shallow suprahilar depression, moderately thin-walled (0.3–0.5 μm), straw-yellow colored in water and 5% KOH, having one, two or three large oil droplets when mature, less frequently pluri-guttulate, inamyloid, acyanophilic and with an orthochromatic reaction. *Basidia* (28) 31–47 × 9–15 μm (n = 31), subclavate to clavate, moderately thick-walled (0.3–0.8 μm), predominantly four-spored but also one-, two- or three-spored, usually bearing relatively long sterigmata (3–6 μm) (sterigmata up to 7 μm long in one- and two-spored basidia), hyaline to yellowish and containing scattered straw-yellow oil guttules in water and 5% KOH, bright yellow (inamyloid) in Melzer's, without basal clamps; basidioles cylindrical, subclavate to clavate, similar in size to basidia. *Cheilocystidia* (28) 31–60 (62) × 8–13 μm (n = 17), common, moderately slender, projecting straight to sometimes flexuous, fusiform, ventricose-fusiform to sublageniform or lageniform, showing a narrow and long neck with rounded to subacute tip, smooth, moderately thick-walled (0.4–0.9 μm), nearly hyaline to yellowish in water and 5% KOH, bright yellow (inamyloid) in Melzer's, without epiparietal incrustations. *Pleurocystidia* (43) 47–59 (61) × 7–11 μm (n = 12), infrequent, irregularly cylindrical or cylindrical-fusiform to narrowly fusiform or sublanceolate, on average usually slightly longer and narrower than cheilocystidia, color and chemical reactions similar to cheilocystidia. *Pseudocystidia* not recorded. *Pileipellis* an ixocutis consisting of mostly repent, subparallel to interwoven, elongated, frequently branched, filamentous and sinuous to cylindrical hyphae embedded in gelatinous matter; terminal elements 14–90 × 4–18 μm, versiform, short cystidioid or rarely slightly clavate to long and slender cylindrical, apex rounded-obtuse, moderately thick-walled (up to 1 μm), yellowish to pale pinkish-red in water and pale yellowish to bright yellow in 5% KOH, golden yellow to orange yellow (inamyloid) in Melzer's, smooth to sometimes ornamented by a very subtle granular epiparietal incrustation; subterminal elements similar in shape, size and color to terminal elements; the subpellis consists of irregularly and loosely arranged, short, predominantly inflated to nearly globose, 6–18 μm broad, hyaline to pale yellowish cells. *Stipitipellis* a layer of slender, parallel to loosely intermingled and longitudinally running, smooth-walled, adpressed hyphae, 2–13 μm wide, hyaline to very pale yellowish in water and 5% KOH; the stipe apex covered by a well-developed caulohymenial layer consisting of sterile caulobasidioles, very common, predominantly four-spored, fertile caulobasidia, (24) 26–39 (41) × 9–11 (13) μm (n = 15), sterigmata up to 6 μm and very scattered projecting fusiform, ventricose-fusiform to lageniform *caulocystidia* similar in shape and color to hymenial cystidia but distinctly shorter, (25) 31–51 (55) × 6–14 μm (n = 10), having a wall up to 0.5 μm thick. *Lateral stipe stratum* under the caulohymenium present and well differentiated from the stipe trama, of the "boletoid type", at the stipe apex a (10) 20–40 (50) μm thick layer consisting of divergent, inclined and running towards the external surface, loosely intermingled and branched hyphae remaining separate and embedded in a gelatinous substance; the stratum is clearly visible in earlier developmental stages but tends to disappear with age. *Stipe trama* composed of confusedly and densely arranged, subparallel to moderately interwoven, filamentous, smooth, inamyloid hyphae, 3–24 μm broad. *Hymenophoral trama* bilateral divergent of the "*Boletus*-type", with slightly to strongly divergent, recurved-arcuate and loosely arranged, not-branched, distantly septate and generally not restricted at septa, gelatinous hyphae (lateral strata hyphae in transversal section not touching each other, (3) 4–8 (10) μm apart, 3–7 μm broad), hyaline to very pale yellowish in water and 5% KOH, inamyloid in Melzer's; lateral strata (30) 40–60 (70) μm thick, mediostratum (10) 20–30 (40) μm thick, axially arranged, consisting of a tightly adpressed, non-gelatinous bundle of hyphae, 3–7 μm broad, distantly septate; in Congo Red the mediostratum is darker than the lateral strata.

Oleipherous hyphae scattered although more frequently observed in the basal stipe trama, golden yellow to brownish in 5% KOH and Melzer's. *Clamp connections* absent in all tissues. *Hyphal system* monomitic.

Edibility: Edible after prolonged cooking.

Ecology and phenology: Solitary or more frequently gregarious or subcaespitose to truly caespitose, sometimes with basidiomes emerging from the stipe of adjacent fruiting bodies, growing in warm Mediterranean regions preferably on slightly to decidedly acidic, clayey or sandy soil among litter associated exclusively with hardwoods in dry, warm, exposed groves of *Quercus* (*Fagaceae*), occasionally also with *Castanea sativa* (*Fagaceae*) and *Cistus* (*Cistaceae*), sometimes in mixed woods with the presence of *Pinus halepensis* and *P. pinea* (*Pinaceae*), *Erica* spp., *Arbutus unedo* (*Ericaceae*) and *Eryngium* sp. (*Apiaceae*). Summer to late fall (August to November).

Known distribution: Previously reported from southern Europe in warm countries at low altitudes, bordering the Mediterranean basin (Portugal, Spain, Italy, Slovenia, Bulgaria, Greece) east to the Asian Middle East (Cyprus, Israel) and likely extending as far north as France. It probably also occurs in Mediterranean northern Africa. Apparently widespread throughout mainland and insular Mediterranean basin but rare and localized. Northern, eastern and southern distribution limits yet to be established.

Examined material: ISRAEL, Mount Carmel: Mount Carmel National Park, near the crossroad Damon, on the ground in a clearing near *Quercus calliprinos*, 32°44′03″ N, 35°02′22″ E, 505 m, 30 October 2015, legit. Sh. Alter, det. A. Yu. Biketova, AB B15-254 (HAI B15-254); same loc., Nahal Oren valley, Henyon Ha'Agam, under *Quercus calliprinos*, 32°43′26″ N, 35°00′55″ E, 260 m, 13 November 2015, legit. Z. Shafranov, det. A. Yu. Biketova, AB B15-274 (HAI B15-274); Golan Heights: Odem Forest Reserve, under *Quercus* sp., 33°12′56″ N, 35°45′26″ E, 1028 m, 10 October 2011, legit. R. Kuznetsov and Z. Shafranov, det. A. Yu. Biketova, AB B11-06 (HAI B11-06); same loc., under *Quercus calliprinos*, 26 September 2013, legit. Z. Shafranov, det. A. Yu. Biketova, AB B13-165 (HAI B13-165); same loc., under *Quercus calliprinos*, 07 November 2015, legit. R. Kuznetsov and Z. Shafranov, det. A. Yu. Biketova, AB B15-271 (HAI B15-271); Upper Galilee: Hanita Forest, in a mixed wood of *Quercus calliprinos* and *Pinus halepensis*, 33°04′44″ N, 35°09′57″ E, 160 m, 6 October 2011, legit. Y. Cherniavsky, det. A. Yu. Biketova, AB B11-03 (HAI B11-03); same loc., Bar'am Forest, on the ground near *Quercus calliprinos*, 33°02′20″ N, 35°25′33″ E, 679 m, 27 November 2020, legit. Y. Segal, det. A. Yu. Biketova and G. Simonini, AB B20-370; ITALY, Sardinia: Arzachena (OT), Cannigione, under *Quercus suber* with the presence of *Cistus* sp. and *Eryngium* sp., 1 November 1980, legit. And det. R. Pöder, IB 19800750 (holotype); Piedmont: Ceresole d'Alba (CN), under *Quercus* sp., 22 August 1980, legit. M. Strani, det. R. Pöder, IB 19800751a and IB 19800751b (paratypes); Lazio: Nettuno (LT), Tre Cancelli, on slightly acidic, sandy soil, along a trackside in a coastal mixed broadleaved woodland dominated by *Quercus robur* with the presence of *Quercus cerris*, *Quercus frainetto* and *Phyllirea angustifolia*, 41°28′39″ N, 12°43′54″ E, 36 m, 26 September 2009, legit. M. Gelardi, G. Gelardi and C. Aita, det. M. Gelardi, MG238 (MCVE25596); same loc., 18 September 2013, legit. And det. M. Gelardi, MG558; same loc., on a clearing side, 2 October 2014, legit. And det. M. Gelardi and F. Costanzo, MG662; same loc., 22 September 2006, legit. And det. M. Gelardi (no voucher material preserved); Nettuno (LT), Torre Astura, on slightly acidic, sandy soil, in a coastal mixed broadleaved woodland under *Quercus robur*, *Quercus frainetto*, *Quercus cerris* with the presence of *Pinus pinea*, 41°25′15″ N, 12°45′33″ E, 5 m, 9 August 2014, legit. M. Gelardi, M. Tullii and R. Polverini, det. M. Gelardi, MG637; Manziana (RM), Cerreta di Manziana, on acidic soil, along a trackside in a pure stand of *Quercus cerris*, 42°07′12″ N, 12°06′53″ E, 320 m, 4 August 2011, legit. M. Gelardi and V. Migliozzi, det. M. Gelardi, MG418; Mounts Tolfa, Allumiere (RM), Capo Nord, on acidic volcanic soil, along a trackside in a mixed broadleaved woodland under *Quercus cerris*, *Q. Petraea*, *Arbutus unedo*, *Erica arborea* and *Fraxinus ornus*, 42°11′12″ N, 11°55′21″ E, 426 m, 3 October 2020, legit. And det. M. Gelardi, F. Costanzo and O. Gelardi, MG829; Emilia Romagna: Pulpiano, Viano (RE), on acidic soil under *Quercus cerris*, 44°31′08″ N, 10°33′13″ E, 520 m, 06 September 1983, legit. U. Bonazzi, det. G. Simonini, GS143 (MCVE17281); same loc., 3 October 1986, legit. And det. G. Simonini, GS336 (MCVE17298); same loc., 12 September 1987, legit. And det. G. Simonini, GS488 (MCVE29227); same loc., 29 October 1990, legit. And det. G. Simonini, GS784 (MCVE17556); same loc., 24 September 1994, legit. U. Bonazzi, det. G. Simonini, GS1275 (MCVE17763); same loc., 30 August 2015, legit. And det. G. Simonini, GS10151; Quattro Castella (RE), Parco di Roncolo, on basic soil under *Quercus pubescens* and *Ostrya carpinifolia*, 320 m, 19 September 1993, 44°37′20″ N, 10°29′23″ E, legit. And det. G. Simonini, GS1001 (MCVE17523); Calabria: Santa Sofia d'Epiro (CS), Serra di Zoto, on acidic soil under *Quercus pubescens* and *Cistus* sp., 550 m, 01 Sep 1995, 39°33′ N, 16°25′ E, legit. C. Lavorato, det. G. Simonini, GS1599 (MCVE18037); Santa Sofia d'Epiro (CS), Contrada Calamia, on acidic soil under *Castanea sativa* and *Cistus* sp., 800 m, 1 September 1995, 39°33′ N, 16°25′ E, legit. C. Lavorato, det. G. Simonini, GS1570 (MCVE17978); Acri (CS), Cozzo S. Angelo, on acidic soil under *Castanea sativa*, 900 m, 8 September 1996, 39°33′ N, 16°25′ E, legit. C. Lavorato, det. G. Simonini, GS1717 (MCVE18132); Sicily: Mounts Nebrodi, Cesarò (ME), Torti II creek, under *Fagus sylvatica* and *Quercus pubescens*, 1400 m, 1 July 2002, 37°53′32″ N, 14°38′51″ E, legit and det. A. Pappalardo, EMAC300-020701; same loc., Mount Soro, under *Fagus sylvatica* and *Quercus pubescens*, 1500 m, 11 August 2002, 37°55′12″ N, 14°39′25″ E, legit and det. S. Silviani, EMAC311-020811; same loc., Randazzo (ME), along river Flascio, under *Quercus cerris*, 1250 m, 20 September 2020, 37°56′29″ N, 14°52′36″ E, legit and det. G. Vasquez, EMAC1511-200920; Mounts Peloritani, Antillo (ME), on soil under *Castanea sativa*, 500 m, 25 September 2015, 37°58′56″ N, 15°12′57″ E, legit. And det. F. Mondello, FM 20150925_100.

Additional material examined: Unidentified *Boletaceae* sp. (initially identified as *Boletus permagnificus*): RUSSIAN FEDERATION: Belgorod Region, Krasnogvardeysky district, Valuychik village, under *Quercus robur* in oak grove on chalky soil, 50°23′ N, 138°15′ E, 180 m, August 1979, legit. E.P. Bedenko, det. T. Yu. Svetasheva and A. Yu. Biketova, LE 17906.

Comments: The Italian mycologists C.L. Alessio and A. Angarano were the first to struggle with the placement of *Exsudoporus permagnificus* in the early 1980s, assigning it three misapplied species epithets [61,62]. Alessio wrote at length and repeatedly about this species based on several collections recorded in Piedmont (northwestern Italy) in 1973 [61]. Two years later, in 1975, the species was also recorded with several specimens emerging from a common base under a single oak tree near Bologna, Emilia Romagna [63]. Alessio's first contribution was devoted to this species which, following A. Marchand's suggestion, was at first believed to represent the eastern North American species *Boletus frostii* Roussell [61]. At the same time, Angarano misidentified the species as another North American taxon, *B. flammans* E.A. Dick and Snell, based on his collections from Cannigione (northern Sardinia) [62]. Alessio continued to use the name *B. frostii* for collections of *B. permagnificus* [64]. However, Alessio reconsidered the best name for these collections and instead applied the binomial *Boletus siculus* Inzenga [65], a species described from Sicily in the second half of the 19th century [66]. However, as already argued by the French mycologists M. Bon [61] and G. Redeuilh [67], and the Italian F. Bellù [68], the collections of Alessio [61,64,65] and Angarano [62] represented a taxon without a valid name.

Figure 2. Basidiomes of *E. permagnificus*: (**A**) IB 19800750, holotype collection; (**B**) GS1275; (**C**) MG558; (**D**) MG662; (**E**) MG662; (**F**) MG829; (**G**) AB B11-03. Photos by: (**A**) R. Kuhnert; (**B**) G. Simonini; (**C–F**) M. Gelardi; (**G**) R. Kuznetsov.

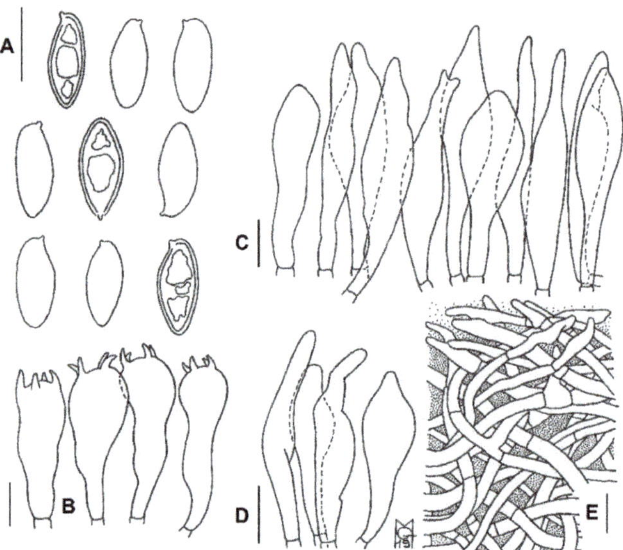

Figure 3. Microscopic features of *E. permagnificus*: (**A**) basidiospores; (**B**) basidia; (**C**) cheilocystidia and pleurocystidia; (**D**) caulocystidia; (**E**) pileipellis. Bars: (**A–D**) = 10 μm; (**E**) = 20 μm. Drawings by M. Gelardi.

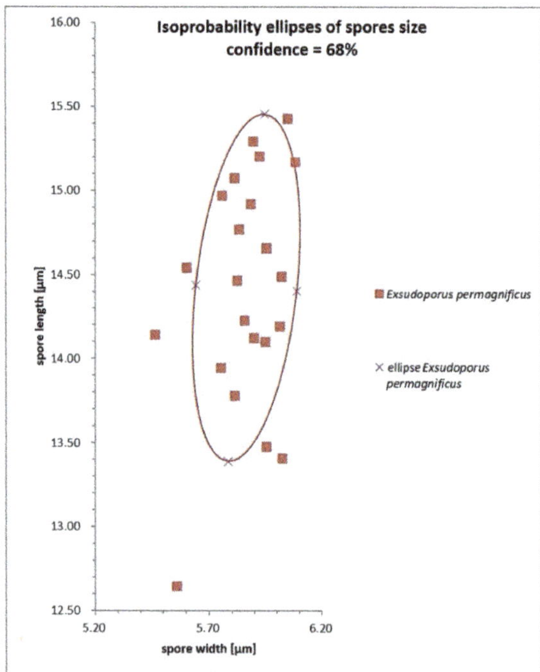

Figure 4. Distribution of spore size of *E. permagnificus* (23 collections) using "isoprobability ellipse". Shown is the distribution of the average values of spore size of any of the collections, at the confidence of 68% (corresponding to one standard deviation).

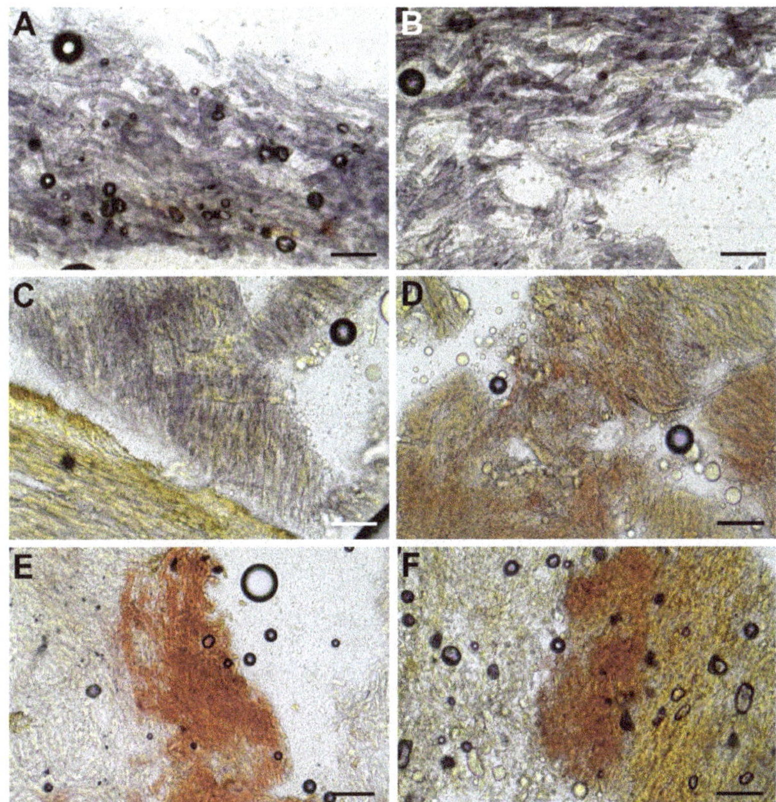

Figure 5. Stipe base hyphae of *E. permagnificus* in Melzer's reagent. (**A**) GS1717, (**B**) GS1717, (**C**) GS336, (**D**) GS784, (**E**) GS1001, (**F**) GS 1275. Bars: 20 μm. Photos by G. Simonini.

Indeed, Pöder [69], working mainly on the same Sardinian specimens previously misidentified by Angarano, described this taxon as new to science under the binomial *Boletus permagnificus* Pöder (see also Pöder [67] for an Italian translation of Pöder's original manuscript by F. Bellù). Despite the convincing evidence of a novel species provided by the Austrian mycologist, in the subsequent years Alessio continued to reiterate his taxonomic view by considering *B. permagnificus* a later heterotypic synonym of *B. siculus* [70–74] and apparently did not change his opinion throughout his life. Conversely, Bellù did not concur with Alessio's interpretations and instead agreed with Pöder that *E. permagnificus* was indeed a new taxon [67,75]. Furthermore, as already pointed out by Lavorato [76] and M. Contu (pers. Comm.), *B. siculus* might represent an older name for *Alessioporus ichnusanus* (Alessio, Galli and Littini) Gelardi, Vizzini and Simonini—a species described by the same Alessio and co-workers in 1984 [77]—even though the exact identity of *B. siculus* remains unclear [78].

In more recent times, *B. permagnificus* was subjected to preliminary phylogenetic analysis which led to the erection of the genus *Exsudoporus* encompassing *E. permagnificus* and its closest relatives [26]. Finally, Blanco-Dios recombined nearly all European red-pored boletes in *Suillellus* Murrill without providing any supporting phylogenetic evidence to justify this placement [79].

Exsudoporus permagnificus is an easily recognized species in the warm regions of southern Europe and the eastern Mediterranean region of western Asia, but it appears to be practically unknown elsewhere, especially outside Europe. It is easily recognizable based on the following diagnostic key features, which place this species in a morphologically unique position among red-pored European boletes: small to medium-sized basidiomes, overall bright red coloration, cylindrical to fusiform stipe that is never clavate but is always covered by a well-developed raised reticulum, hymenophore exuding golden or amber-yellow droplets in young and fresh specimens, yellowish context and basal mycelium, deep blue staining reaction upon bruising or injury, ellipsoid-fusiform, smooth basidiospores, ixocutis pileipellis mainly consisting of filamentous to cylindrical hyphae, subcaespitose to caespitose growth and the occurrence under broadleaved trees or shrubs in warm environments. As far as the pileipellis structure is concerned, inflated to nearly globose cells up to 18–20 μm wide are sometimes observed in the subpellis (as broad as 25 μm) [80–82]. Another peculiar organoleptic trait that has likely gone unnoticed but is worthy of mention is the salty taste of the droplets; this neglected feature has not been previously reported for *E. permagnificus* but should be examined in the remaining extraeuropean species to evaluate its taxonomic significance.

The amyloidity of tissues with Melzer's reagent is a point that deserves further discussion. Bozok et al. reported the amyloid reaction of the hyphae of the stipe context as an additional attribute of *E. permagnificus* [38]. The holotype material and all collections from Israel studied in the present paper also exhibited the same positive reaction of the stipe base context: moderately amyloid in AB B11-03, AB B13-165, AB B15-271 and AB B20-370, weakly amyloid in AB B11-06 and IB 19800750 and variably amyloid and dextrinoid in AB B15-254 and AB B15-274 (Figure 5). As a matter of fact, in the original diagnosis, Pöder noted that the stipe trama was almost black with Melzer's reagent, suggesting amyloid tissues [69]. Icard and Hurtado [80], Lannoy and Estades [5], and Horak [83] also observed a positive amyloid reaction in this species. However, other authors reported a negative reaction with Melzer's reagent [14,82,84–86]. As already suggested by Assyov, contrasting results might be due to different procedures followed in testing the amyloidity of fungal tissues [84]. However, we have tested the iodine reaction of the context at the stipe base according to Imler's procedure [4] on several of our Italian collections using the same techniques and surprisingly we observed dissimilar results depending on the studied samples; some of them showed a strong (GS1717) to weak (MG238, MG662, GS336) amyloid reaction, whereas some others exhibited an inamyloid reaction (MG558, MG637, GS784, GS10151), and a few others even displayed a dextrinoid reaction (MG418, GS1001, GS1275). Consequently, this macro-chemical reaction should be carefully re-evaluated, as it seems to be quite inconstant and variable.

A chromatographic analysis of the pigments in *E. permagnificus* was carried out by Davoli and Weber, which led to the isolation of variegatic acid, xerocomic acid and variegatorubin [87,88].

With regards to its ecological requirements, this species is primarily found in oak-dominated woodlands. The species occurs with deciduous and evergreen *Quercus* species (including *Q. cerris, Q. calliprinos, Q. pubescens, Q. petraea, Q. pyrenaica, Q. virgiliana, Q. ilex, Q. suber, Q. alnifolia, Q. frainetto* and the introduced-in-Europe *Q. rubra*), sometimes in mixed forests with the presence of pines (*Pinus pinea* and *P. halepensis*) and also sporadically under sweet chestnut (*Castanea sativa*), common beech (*Fagus sylvatica*), some *Ericaceae* (*Erica scoparia, E. arborea* and *Arbutus unedo*), and rockroses (*Cistus* spp.), preferably on acidic soil [30,38,76,82,85,89,90]. The occurrence of *E. permagnificus* with several host trees indicates that it is not selective in its plant symbiotic partners. Concerning geographical distribution, the natural range of *E. permagnificus* covers most of the warm and dry regions of southern Europe and Levant in western Asia. Aside from Italy (including Sardinia and Sicily) [14,63,69,72,76,85,89–105], the species has also been reported from Portugal [106,107], Spain [81,82,86,108–115], France (including Corsica) [5,80,116–119], Bulgaria [38,84,120,121], Slovenia [122], Greece [123] and Cyprus [39]; and here it is also reported for the first time from Israel. *Exsudoporus permagnificus* is relatively widespread throughout the Western Mediterranean basin, more rarely found in the East Mediterranean basin, and we assume it might also occur in northern and northwestern Africa. We suspect that it has not been reported from this region because it may have been overlooked. Based on the International Union for Conservation of Nature (IUCN) Red Listing protocol, *E. permagnificus* has most recently been assessed as "Vulnerable" (VU) in central Italy at the ecoregion scale [124]. Finally, *E. permagnificus* is an edible species that is commonly harvested and consumed by local mushroom pickers in the coastal areas south of Rome, although it is rarely prolific enough to be widely eaten.

A bolete sample (LE 17906) identified by Bedenko [125] as *B. permagnificus* from the Belgorod Region in the Russian Federation has been morphologically re-examined by T. Yu. Svetasheva and I. V. Zmitrovich and was confirmed to be another species of *Boletaceae* (likely a member of either the genus *Rubroboletus* or *Neoboletus*). Unfortunately, the specimen is poorly preserved and no DNA sequences could be generated from this material, so it is not possible to identify this specimen to the species level.

The most similar species to *E. permagnificus* is probably *E. floridanus* (Singer) Vizzini, Simonini and Gelardi, which is its closest relative based on phylogenetic data. However, *E. floridanus* is distinguished by the larger size (pileus up to 15 cm diam.), the pileus cuticle staining olivaceous black with NH$_3$, the typically clavate stipe, slightly narrower basidiospores (13.2–16.7 × 4.5–5.0 μm) and the different geographic distribution in eastern and southeastern North America [28,126–129].

The eastern North American *E. frostii* differs from *E. permagnificus* in the larger size (pileus up to 15.5 cm diam.), stouter basidiomes, polished and shining cuticle that is viscid when wet, blood-red to candy-apple-red pileus surface, usually clavate stipe, raised and much coarser, deeply pronounced reticulate-alveolate pattern over the entire length of the stipe, yellowish reticulum on a reddish background, tissues turning light blue slowly and erratically, narrower basidiospores (12–15 × 3.7–5.0 μm, Q_m = 3.1–3.6), smaller basidia (23–32 × 8–10 μm), narrower pileipellis hyphae (up to 6 μm broad) and the occurrence in North America [7,28,69,127–131].

Exsudoporus ruber is consistently separated from *E. permagnificus* by the slightly larger size (pileus up to 13 cm diam.), polished and shining cuticle that is viscid when wet, dark red, brownish-red to violet-brown pileus surface, sometimes unchangeable or very slowly and lighter bluing context on exposure, overall reddish stipe surface (even in young specimens) which tends to become disrupted into irregular scaly patches in ageing specimens especially on the lower half, longer basidiospores (15.5–18.6 × 5.6–6.6 μm, Q_m = 2.8), subcylindrical, slightly narrower cheilocystidia (35–76 × 4–9.5 μm), narrower pileipellis hyphae, 2.0–6.4 μm wide, longer caulocystidia (37–85 × 4–10 μm), narrower stipe trama hyphae (up to 11 μm broad) and the occurrence in subalpine coniferous forests in association with *Pinaceae* (*Abies, Pinus, Tsuga*) in eastern Asia ([132–134], as "*Boletus kermesinus*"; [135,136]).

Another species that is similar to *E. permagnificus* is *Boletus weberi* Singer, which appears to be conspecific with *Boletus pseudofrostii* B. Ortiz, based on our phylogenetic reconstruction. This taxon belongs to another genus (see below) and differs by the smaller size (pileus up to 6.5 cm diam., stipe up to 5.3 × 1.7 cm), unchanging tissues on bruising, white basal mycelium, slightly smaller and shorter basidiospores (8.1–10.7 × 4.1–5.5 µm, Q_m = 2.0), slightly narrower basidia (33–45 × 7–10 µm), smaller cheilocystidia (14–60 × 4–8 µm) and caulocystidia (14–29 × 8–9 µm), cylindrical, narrower pileipellis terminal cells (up to 10 µm wide) and the occurrence in *Pinus*-dominated or in mixed forests with *Fagaceae* in North (Atlantic and Gulf Coasts of the USA) and Central America (Belize) [28,126–128,137–140].

Exsudoporus floridanus (Singer) Vizzini, Simonini and Gelardi, Index Fungorum 183:1, 2014.

Figure 6A,B.

MYCOBANK MB 550711.

≡ *Boletus frostii* subsp. *floridanus* Singer, Mycologia 37(6):799, 1945. (Basionym).

≡ *Suillellus floridanus* (Singer) Murrill, Lloydia 11(1):29, 1948 (nom. inval., art. 41.5, basyonim not cited).

≡ *Boletus floridanus* (Singer) Murrill, Lloydia 11(1):23, 1948 (nom. inval., art. 41.5, basyonim not cited).

≡ *Boletus floridanus* (Singer) Singer, Sydowia 30(1-6):255, 1977.

≡ *Butyriboletus floridanus* (Singer) G. Wu, Kuan Zhao and Zhu L. Yang, Fungal Diversity 81(1):72, 2016.

Holotype: USA, Florida, Alachua Co., Gainesville, June 1943, R. Singer, F 2428.

Epitype designated here: USA, Florida, Alachua Co., Gainesville, University of Florida Campus, Fifield Hall, under *Quercus virginiana*, 29°38′19″ N, 82°21′41″ W, 16 September 2014, legit. C. Ferguson, det. M. E. Smith, FLAS-F-59069, Index Fungorum 559409, GenBank: OL960514 for ITS, OL960488 for nrLSU, OL960496 for *tef1-α* and OL960503 for *rpb2*.

Selected morphological descriptions and illustrations: Murrill ([141], as "*Suillellus luridus*"?; [137]), Singer [126,127], Both [128], Bessette et al. [28,129,139,142], dubitatively García-Jiménez and Garza-Ocañas [143], Ortiz-Santana et al. [138], García-Jiménez [144], García-Jiménez et al. [145] and Gonzáles-Chicas et al. [146].

Edibility: Edible after prolonged cooking although with a somewhat acidic taste [129].

Ecology and phenology: Solitary to most frequently gregarious, growing in open stands, clearings or shaded lawns on sandy soil associated with various *Quercus* spp. (including *Q. chapmanii*, *Q. laurifolia* and *Q. virginiana*) (*Fagaceae*) or mixed with *Pinus clausa* (*Pinaceae*) and *Carya* (*Juglandaceae*). Late spring to fall (April to December), uncommon to rare.

Known distribution: Reported from eastern and southeastern USA (Tennessee and the Coastal Plain of North Carolina south to Florida and west to Texas), reported as far north as Long Island (New York) (MyCoPortal); records from Costa Rica, Mexico, Belize and Guatemala putatively belong to another related species (see below).

Examined material: USA, Florida: Alachua Co., Gainesville, University of Florida Campus, Fifield Hall, under *Quercus virginiana*, 29°38′19″ N, 82°21′41″ W, 1 April 2015, legit. R. Kneal and M. E. Smith, det. M. E. Smith, FLAS-F-59142; same loc., under *Quercus virginiana*, 16 September 2014, legit. C. Ferguson, det. M. E. Smith, FLAS-F-59069 (epitype); Putnam Co., Ordway-Swisher Biological Station, south of Goose Lake, along gravel road, under *Quercus hemisphaerica*, 29°41′50″ N, 81°58′53″ W, 22 June 2017, legit. G. LaPierre, det. M. E. Smith, FLAS-F-61008; same loc., NE of point C21, *Q. virginiana* dominated hardwood forest, with some *Carya*, *Pinus* and *Serenoa repens*, 29°42′00″ N, 81°59′26″ W, 1 August 2017, legit. D. Borland, L. Kaminsky, A. E. Bessette and A. R. Bessette, det. A. E. Bessette, FLAS-F-61189.

Figure 6. Basidiomes of *E. floridanus*: (**A**) FLAS-F-61008; (**B**) unnumbered collection. Basidiomes of *E. frostii s. l.*: (**C**) TO AVBB11; (**D**) TO AVBB10; (**E**) iNat 35326745 (TO AVBB12); (**F**) iNat 30897161 (TO AVBB13); (**G**) FLAS-F-60742. Basidiome of *E. ruber*: (**H**) TNS-F-37407, holotype collection of *Boletus kermesinus*. Photos by: (**A,G**) L. Kaminsky; (**B**) A. Farid; (**C**) L. Craig; (**D**) R. Abbott; (**E,F**) R. K. Antibus; (**H**) Y. Taneyama.

Comments: This species has been sufficiently described and illustrated. The exact collecting date of the original material is not clear and was not given in the publication or on the packet, but authentic samples from the description were collected in north Florida during June and September. Curiously, Both indicated a different collection (paratype F2204) as the type of *E. floridanus* [128]. This is likely because there were some confusing annotations on the specimens at FH and other herbaria and because F2428 does not appear to be present at either FH or F. An attempt was made to sequence the ITS region from two paratypes of *E. floridanus* from FH (F2204 and F650), but due to the age and condition of the specimens, this was unsuccessful. We consider it necessary to designate a new specimen of *E. floridanus* from the vicinity of the type locality as a modern epitype. Phylogenetic analysis shows that this entity might be a species complex, and *E. floridanus s. l.* appears to encompass two different species. One is the true *E. floridanus*, which is distributed in eastern and southeastern North America, including our specimens from Florida that were collected nearby the type locality. A second, morphologically similar species is recorded here from Central America, based on collections from Belize and Costa Rica. It seems likely that other collections from Belize [138] and Guatemala [147] might have been misidentified as *E. floridanus s. str.* but actually belong to this second species. Records from Mexico that were identified as *E. floridanus s. str.* [143–146] should be examined and/or have their DNA sequenced to determine which of the two species they represent. However, since many taxa from Florida extend into Mexico, it seems likely that both species may be present in Mexico.

Similar to *E. permagnificus*, the amyloid reaction of the hyphae of the stipe context with Melzer's reagent varies from weakly dextrinoid (FLAS-F-61008) and dextrinoid (FLAS-F-59069) to mixed amyloid and dextrinoid (FLAS-F-59142), as well as weakly amyloid (FLAS-F-61189). Farid et al. also noticed a dextrinoid reaction in the studied specimen (Farid 499) [148].

Exsudoporus frostii (J.L. Russell) Vizzini, Simonini and Gelardi, Index Fungorum 183:1, 2014.

Figure 6C–G.

MYCOBANK MB 550710.

≡ *Boletus frostii* J.L. Russell, Bulletin of the Buffalo Society of Natural Sciences 2:102, 1874. (Basionym).

≡ *Suillus frostii* (J.L. Russell) Kuntze, Revisio Generum Plantarum 3(2):535, 1898.

≡ *Suillellus frostii* (J.L. Russell) Murrill, Mycologia 1(1):17, 1909.

≡ *Tubiporus frostii* (J.L. Russell) S. Imai, Transactions of the Mycological Society of Japan 8(3):113, 1968.

≡ *Butyriboletus frostii* (J.L. Russell) G. Wu, Kuan Zhao and Zhu L. Yang, Fungal Diversity 81(1):72, 2016.

= *Boletus alveolatus* Berkeley and M.A. Curtis apud Frost, Bulletin of the Buffalo Society of Natural Sciences 2:102, 1874.

Holotype: USA, Vermont, Brattleboro, C.C. Frost. Preserved in VT and selected by Halling as lectotype (VT3156) [149]. According to Halling, several additional specimens currently housed in VT, FH and NYBG can be considered "original material" because these are labeled in Frost's handwriting and were apparently identified by Frost [149].

Selected morphological descriptions and illustrations: Curtis ([150], as "*Boletus purpureus*"), Frost ([130], also as "*Boletus alveolatus*"), Peck [151,152], Murrill [141,153], Farlow and Burt [154], Coker and Beers [155], Singer [127,156], Snell and Dick [157], Smith and Thiers [7], Miller [158], Smith and Weber [159], Lincoff [160], Halling [149], Imler [161], Weber and Smith [162], Arora [163], Bessette and Sundberg [164], McKnight and McKnight [165], Phillips [166], Metzler and Metzler [167], Both [128], Krisai-Greilhuber [131], Bessette et al. [28,129,139,168], Kibby [169]; Elliott and Stephenson [170].

Edibility: Edible after prolonged cooking [129,152,157] but not recommended [7] because this taxon sometimes causes gastrointestinal upset [28].

Ecology: Solitary to scattered or gregarious, growing primarily in forests but sometimes also in open grassy places, in association with *Quercus* (*Fagaceae*) or in mixed stands with *Fagus* (*Fagaceae*), *Tsuga* (*Pinaceae*) and *Arbutus menziesii* (*Ericaceae*). Summer to autumn (June to October), occasional to fairly common.

Known distribution: Reported from eastern North America, Canada (Quebec), and USA (New England south to Florida and west to Wisconsin and Arizona), repeatedly recorded in Mexico, Guatemala and Costa Rica, but identity not confirmed to date based on molecular data. Southern limits yet to be established.

Examined material: USA, Ohio: Swanton, Kitty Todd Nature Preserve, on sandy soil in oak savannah under *Quercus palustris*, 41°37′45″ N, 83°47′29″ W, 206 m, 15 August 2019, legit and det. R. K. Antibus, iNat 30897161 (TO AVBB13); same loc., 26 August 2019, legit. and det. R. K. Antibus, iNat 35326745 (TO AVBB12); Portage Co. Aurora, Aurora Sanctuary State Nature Preserve, in a mixed hardwood forest under *Quercus* sp., *Pinus* sp., *Fagus* sp. and *Acer* sp., 41°37′41″ N, 83°43′10″ W, 206 m, 29 August 2018, legit. and det. L. Craig, TO AVBB11; Michigan: Oakland Co., Commerce Township Preserve, Nike Missile Base Park, in a mixed hardwood forest under *Quercus* sp. and *Acer* sp., 42°35′56″ N, 83°28′05″ W, 285 m, 29 July 2019, legit. and det. R. Abbott, TO AVBB10; Florida: Putnam Co., Ordway-Swisher Biological Station, hammock near campground, with *Quercus* spp., 29°42′24″ N, 81°58′0.94″ W, 5 June 2017, legit. G. LaPierre, det. M. E. Smith, FLAS-F-60742.

Comments: This species has been sufficiently described and illustrated in several monographic treatments and popular field guides over the past century (see above). Similar to the case for *E. floridanus* (see above), there is cryptic diversity within *E. frostii s. l.*, including at least four distinct clades. Our molecular evidence indicates that specimens from the Eastern USA are found in three (Figure 1, *E.* cf. *frostii* 1, 3 and 4) of the four clades, so it is premature to epitypify *E. frostii* until further studies are carried out in New England to determine which taxon is likely to represent the type. *Exsudoporus* cf. *frostii* 1 (from New Hampshire, Ohio, Michigan, Tennessee) or *E.* cf. *frostii* 4 (from Ohio, Michigan and Florida) probably represent the original species as described by J.L. Russell from Vermont and New York [130,149].

One of the clades (Figure 1, *E.* cf. *frostii* 2) in the *E. frostii* complex includes specimens from Mexico and Costa Rica that likely represent a new species. It is clear that records of *E. frostii s. l.* from Mexico [144,145,163,171–176], Guatemala [147] and Costa Rica [163,177,178] are in need of urgent taxonomic revision because they likely represent an undescribed taxon. The occurrence of *Boletus frostii* in India as reported by Sharma et al. and Verma and Pandro are doubtful and likely represent a different taxon too [179,180].

Exsudoporus ruber (M. Zang) Gelardi, Biketova and Vizzini, **comb. nov.**
Figure 6H.
MYCOBANK MB 842306.

≡ *Leccinum rubrum* M. Zang, Acta Botanica Yunnanica 8(1):11, 1986. (Basionym).

≡ *Boletus ruber* M. Zang (as "*rubrus*") in Yuan MS and Sun PQ, the pictorial book of mushrooms of China: 194, 2007 (nom. inval., art. 41.5, basionym not cited).

≡ *Butyriboletus ruber* (M. Zang) K. Wu, G. Wu and Zhu L. Yang (as "*rubrus*"), Acta Edulis Fungi 27(2):96, 2020.

= *Boletus kermesinus* Har. Takah., Taneyama and Koyama, Mycoscience 52(6):419, 2011.

Holotype: China, Xizang Autonomous Region, Chayu Co., Linzhi City, Ridong, under *Abies* sp., 21 September 1982, D. Zhang, KUN-HKAS 17055.

Selected morphological descriptions and illustrations: Zang [132,133], Yoneyama ([181], named as "Miyama Aka Iguchi"), Yuan and Sun ([182], as "*Boletus rubrus*"), Takahashi et al. ([134], as "*Boletus kermesinus*"), Zang et al. [135], Mikšík ([107], as "*Boletus kermesinus*").

Edibility: Assumed to be edible by Wu et al. [136], edible after cooking according to Yoneyama [181].

Ecology: Solitary to scattered, growing in subalpine coniferous forests (1800–4200 m alt.) dominated by *Abies* spp. (including *A. chayuensis*, *A. georgei*, *A. mariesii* and *A. veitchii*), *Pinus densata* and *Tsuga diversifolia* (*Pinaceae*). Summer to autumn (July to October), uncommon.

Known distribution: Reported from southwestern China (Xizang, Tibet, Sichuan and Yunnan Provinces) and Japan (Nagano).

Examined material: JAPAN, Nagano Prefecture: Minamisaku-gun, Sakuho-cho, under *Abies mariesii*, 29 August 2009, legit. M. Taneyama, TNS-F-37407 (holotype of *B. kermesinus*); same loc., under *Tsuga diversifolia* and *Abies veitchii*, 5 September 2008, legit. A. Koyama, TNS-F-36802; same loc., under *Abies mariesii*, 6 September 2009, legit. T. Arano, TNS-F-37408; same loc., under *Abies mariesii*, 29 August 2009, legit. K. Kitahara, TNS-F-37409; same loc., under *Tsuga diversifolia* and *Abies veitchii*, 25 July 2010, legit. A. Koyama, TNS-F-36804; same loc., under *Tsuga diversifolia* and *Abies veitchii*, 3 August 2010, legit. A. Koyama, TNS-F-36805; same loc., under *Tsuga diversifolia* and *Abies veitchii*, 3 October 2010, legit. A. Koyama, TNS-F-36806; Azumino-shi, under *Abies mariesii*, 6 September 2008, legit. K. Itahana, TNS-F-37404; Nagano-shi, under *Abies mariesii*, 16 July 2009, legit. T. Fujisawa, TNS-F-37405; Simotakai-gun, Yamanouchi-cho, under *Abies mariesii*, 8 August 2009, legit. Y. Taneyama, TNS-F-37406; Suzaka-shi, under *Abies mariesii*, 29 August 2009, legit. T. Fujisawa, TNS-F-37410; Ina-shi, under *Tsuga diversifolia* and *Abies veitchii*, 18 September 2006, legit. A. Koyama, TNS-F-36801; Suwa-gun, Hara-Mura, 30 August 2012, under *Abies mariesii* and *Tsuga diversifolia*, legit. M. Koike, BTS-031J; Minamisaku-gun, Minamimaki-mura, 30 August 2011, under *Abies mariesii* and *Tsuga diversifolia*, legit. Y. Imai, BTS-031i.

Comments: *Exsudoporus ruber* was originally described from Yunnan Province (southwestern China) by Zang [132] and was also examined in his additional publications [133,135]. This species was recently re-described based on fresh collections, taxonomically revisited in the light of molecular inference and subsequently transferred to the genus *Butyriboletus* [136]. Two combinations of *B. rubrus* nom. inval. and *Bu. rubrus* were published with a grammatical mistake. The epithet was intended as the masculine form of the neuter basionym epithet *"rubrum"*, so the correct one should be *"ruber"*. A more comprehensive phylogenetic analysis carried out in the present study indicates that this species is more appropriately placed in *Exsudoporus*. This placement is supported by the close phylogenetic relationship between *E. ruber* and *E. frostii s. l.* but is also strengthened by the evident morphological resemblance with other species of *Exsudoporus*. *Exsudoporus ruber* was incorrectly identified from Japan as *B. frostii* by the Mycological Society of Japan during a past investigation of the mycoflora of Mount Fuji (Y. Taneyama, personal observation). Later, it was fully characterized by Takahashi et al. under the name *Boletus kermesinus* Har. Takah., Taneyama and Koyama, based on collections from Nagano Prefecture [134]. There was no mention in the original diagnosis or protologue of *B. kermesinus* about the yellow droplets exuding from the hymenophore, since at the time of publication it could not be determined with confidence whether it was a diagnostic trait for this species. However, these droplets are clearly visible in photos of this species, including from the holotype material (Figure 6H). Similarly, hymenophoral droplets have not been reported in the literature on *E. ruber*. According to the original diagnosis, *E. ruber* differs from *B. kermesinus* by the squamulose-punctate stipe surface, reddish tubes and lack of cheilocystidia [132,134]. These discrepancies, however, turned out to be unreliable [136]. Molecular phylogenetic inference confirmed that the Japanese *Boletus kermesinus* is conspecific with *E. ruber*, a heterotypic synonymy previously conjectured by Wu et al. [136] based solely on morphology.

In the phylogenetic tree (Figure 1), a visible distance between our two specimens from Japan (TNS-F-37407, BTS031J) and Chinese ones (KUN-HKAS 106891, KUN-HKAS 103513 and KUN-HKAS 103122) is an artefact, caused by a disproportion of missing data in the analysis: Japanese specimens have ITS and nrLSU sequences, while Chinese specimens have nrLSU, *tef1-α* and *rpb2*. BLASTn analysis of the only overlapping locus (nrLSU) between BTS031J and Chinese specimens shows 99.8% of similarity (the longest overlapping region is with the sequence of KUN-HKAS 103122: identities = 867/869 (99.77%), 0 gaps (0%)) as well as between TNS-F-37407 and Chinese specimens—100% of similarity (the longest overlapping region is with the sequence of KUN-HKAS 103122: identities = 588/588 (100%), 0 gaps (0%)).

Unfortunately, it has not been possible to examine or sequence DNA from the holotype material of *L. rubrum* (KUN-HKAS 17055) or from additional samples (KUN-HKAS 33928, KUN-HKAS 32431, KUN-HKAS 36578) preserved at the Herbarium of Cryptogams, Kunming Institute of Botany. The geographic range of this eastern Asian taxon is much wider than initially thought, spanning from the eastern Himalayas and the Hengduan Mountains of southwestern China [136,183–185] east to Japan [134], although with a possible disjunct distribution.

Extralimital Taxa

Amoenoboletus G. Wu, E. Horak and Zhu L. Yang 2021.
MYCOBANK MB 838620.
Generic type: *Boletus granulopunctatus* Hongo 1967.
Amoenoboletus weberi (Singer) Biketova, M.E. Sm. and Gelardi, **comb. nov.**
MYCOBANK MB 842315.
≡ *Boletus weberi* Singer, Mycologia 37(6): 797, 1945. (Basionym).
≡ *Suillellus weberi* (Singer) Murrill, Lloydia 11: 29, 1948.
= *Boletus pseudofrostii* B. Ortiz, Fungal Diversity 27(2): 322, 2007.
Holotype: USA, Florida, Alachua Co., Gainesville, Campus, 28 July 1943, C. Weber, F 3036.
Selected morphological descriptions and illustrations: Singer [126,127], Murrill [137], Both [128], Ortiz-Santana et al. ([138], as "*Boletus pseudofrostii*); Bessette et al. [28,139].
Edibility: Unknown.
Ecology and phenology: Solitary to gregarious, growing in tropical coniferous forests under *Pinus palustris, P. rigida, P. elliottii* and *P. caribaea* (*Pinaceae*) or mixed with *Quercus incana, Q. laurifolia, Fagus* sp. (*Fagaceae*) and *Acer* sp. (*Sapindaceae*). Fruiting in summer (June, July and August) and autumn (September and October), uncommon to rare [127,138,140].
Known distribution: Reported from Atlantic and Gulf Coasts of the USA (New Jersey, Florida west to Texas) and Belize. Almost certainly occurring in Mexico as well.
Examined material: USA, Florida: Putnam Co., Ordway-Swisher Biological Station, Ashley Lake boat ramp, under *Quercus* sp. and *Pinus* sp., 29°42′29″ N 81°59′06″ W, 28 August 2017, legit. L. Kaminsky and D. Borland, det. M. E. Smith, FLAS-F-61525; Alachua Co., Gainesville, University of Florida campus, Natural Area Teaching Laboratory, under *Quercus laurifolia* and *Pinus elliottii*, 29°38′4.81″ N, 82°22′2.67″ W, 18 September 2020, leg. P. M. Ramos Perez, det. M. E. Smith, FLAS-F-68076 (topotype).
Amoenoboletus mcrobbii (McNabb) G. Wu, E. Horak and Zhu L. Yang 2021.
MYCOBANK MB 838624.
≡ *Xerocomus mcrobbii* McNabb, New Zealand Journal of Botany 6(2): 147, 1968. (Basionym).
≡ *Boletus mcrobbii* (McNabb) G. Stev., Field guide to fungi: 91, 1982.
Holotype: New Zealand, Buller District, Maruia, Jackson's Creek, under *Fuscospora fusca* and *Lophozonia menziesii*, 23 March 1966, R.F.R. McNabb, PDD 25242.
Selected morphological descriptions and illustrations: McNabb [186], Stevenson [187], Wu et al. [42].
Edibility: Unknown.
Ecology and phenology: Gregarious or occasionally caespitose under *Fuscospora fusca, F. cliffortioides, Lophozonia menziesii* (*Nothofagaceae*), sometimes mixed with *Dacrydium cupressinum* (*Podocarpaceae*) and *Libocedrus bidwillii* (*Cupressaceae*), from summer to autumn (December to May), likely uncommon to rare [42,140,186,187].
Known distribution: Reported from New Zealand and Australia (Eastern coast of Queensland).
Examined material: NEW ZEALAND, West Coast: Buller District, Maruia, under *Fuscospora fusca* and *Lophozonia menziesii*, 23 March 1966, legit. J.A. McRobb, det. R.F.R. McNabb, K-M000025241 (paratype).

Additional material examined: Amoenoboletus cf. *mcrobbii*: AUSTRALIA, Queensland: Gold Coast, Springbrook National Park, Purling Brook Falls, under *Eucalyptus* sp. and *Eucalyptus grandis*, 28°11'22" S 153°16'08" E, 612 m, 30 April 2014, legit. R. E. Halling and N. Fechner, NY 2686023 (Halling 9916).

Notes on the genus Amoenoboletus: McNabb placed *X. mcrobbii* in *Xerocomus* sect. *Pseudogyrodontes* Singer [186]. *Boletus weberi* was tentatively placed by Singer in *Boletus* sect. *Subpruinosi* Fr., whereas there was no mention of *X. mcrobbii* [4]. Ortiz-Santana et al. recognized taxa that they considered in the "*Boletus weberi* complex" based on morphological characters [138]. Their discussion of this complex included six different species: *B. weberi* (North America), *X. mcrobbii* (Australia), *B. granulopunctatus* Hongo (Japan), *Boletus morrisii* Peck (North America), *B. rubropictus* Snell and A.H. Sm. (North America) and *B. guatemalensis* R. Flores and Simonini (Central America). The genus *Amoenoboletus* was described by Wu et al. and included four species: *Amoenoboletus granulopunctatus* (Hongo) G. Wu, E. Horak and Zhu L. Yang, *A. mcrobbii*, *A. miraculosus* E. Horak and G. Wu, and *A. phoeniculus* (Corner) G. Wu and Zhu L. Yang [42].

Our current study shows that the genus *Amoenoboletus* contains at least three additional species: *A. weberi*, *A.* cf. *mcrobbii* and *A.* cf. *granulopunctatus*. One of the studied collections, FLAS-F-68076, is a topotype of *A. weberi*, but unfortunately, we do not have a good quality photo of fresh basidiomes in order to designate it as an epitype.

The current phylogenetic analysis shows that samples from New Zealand (PDD97418) and Australia (NY 2686023), identified as *X. mcrobbii*, belong to two different species of the genus *Amoenoboletus*. Therefore, the Australian *A.* cf. *mcrobbii* represents a new, undescribed species. There is another sample PDD94435 identified as *X. mcrobbii*, whose ITS (JQ924297) and nrLSU (JQ924323) sequences are available in GenBank, but it has likely been misidentified and belongs to another genus. Unfortunately, our attempt to generate a sequence from the paratype collection of *A. mcrobbii* (K-M000025241) was unsuccessful. It should be mentioned that the type locality of *A. mcrobbii* in Maruia was wrongly attributed to Nelson Province by McNabb [186]. However, this province was abolished back in 1876, along with all other provinces of New Zealand [188].

Interestingly, our analysis also shows that *A. granulopunctatus* is a species complex. Wu et al. studied the type of *A. granulopunctatus* (Z-ZT 3103, isotype); however, they did not generate sequences from this collection and merged samples of two different species under the same name in the description and phylogenetic tree [42]. A specimen KUN-HKAS 56280 represents a distinct species from three other collections (MHHNU 9490, KUN-HKAS 80250 and KUN-HKAS 86007) included in the phylogenetic analysis. All of these collections were found in different regions of China, and fresh specimens from the type locality in Japan were not studied.

Boletus rubropictus and *B. guatemalensis* can be additional putative members of *Amoenoboletus*. Unfortunately, their sequences are unavailable, so the phylogenetic disposition of these taxa remains unknown. *Boletus morrisii*, another species of the "*Boletus weberi* complex" is rather distantly related to *Amoenoboletus*. Based on BLASTn analysis of ITS sequences in GenBank of the specimen Mushroom Observer 325450 [189], *B. morrisii* likely belongs in the subfamily *Xerocomoideae* and therefore outside of the "*Pulveroboletus* group". Further type studies of all actual and potential species of the genus *Amoenoboletus* are required.

Members of *Amoenoboletus* are generally small-sized boletes (pileus up to 6.5 cm diam., stipe up to 7 × 1.7 cm) displaying a fibrillose-squamulose to areolate pileal surface, yellow tubes with orange-reddish or yellow pores, reddish squamulose to scaly or floccose stipe surface, yellow context and non-bluing tissues, hymenophoral trama of the "*Pylloporus*-type" or the "*Boletus*-type", pileipellis composed of entangled hyphae and smooth basidiospores without a distinctive suprahilar depression [42,127,186]. This genus is morphologically similar to *Exsudoporus* but differs by generally larger basidiomes (pileus up to 15 cm diam.), bluing tissues, a non-squamulose and non-cracked pileus surface, a pronounced reticulum on the stipe and longer basidiospores (12–20 × 3.7–7.0 μm; Q_m = 2.47–3.35) [28,38,127,129,136]. As for the Melzer's reaction in the stipe base context in *Amoenoboletus* species, there are no data in the literature. Present studies of two specimens of *A. weberi* (FLAS-F-61525 and FLAS-F-68076) and one specimen of *A. mcrobbii* (K-M000025241) show an inamyloid reaction.

Hymenophoral droplets of *Amoenoboletus* spp. have not been reported in the published sources. However, pale greenish-yellow droplets are clearly visible in the picture of *A. weberi* 102459 on the Mushroom Observer website, and I.G. Safonov noted this feature in the description of the specimen [189].

4. Discussion

Over the past several years, numerous publications have treated *Exsudoporus* species as members of the genus *Butyriboletus* [18,136,190–193]. However, our integrated morphological and molecular analyses indicate that *Exsudoporus* is the unequivocally resolved sister group of *Butyriboletus*, as previously hypothesized by Vizzini [26] and confirmed by Gelardi et al. [25], Bozok et al. [38], Loizides et al. [39] and Farid et al. [148]. Since there are distinct morphological features associated with each of the two lineages and they are unambiguously autonomous evolutionary lineages, we recognize these as two distinct genera.

Boletus subsplendidus W.F. Chiu is the most closely related taxon to *Exsudoporus* and forms a monotypic generic clade (BS = 100%, PP = 1.00). A branch that unites these two sister clades has weak statistical support (BS = 52%, PP = 0.89). *Boletus subsplendidus* was also placed in *Butyriboletus* by Wu et al. but with the recognition of the genus *Exsudoporus* it will need to be transferred to a separate genus [18]. *Boletus subsplendidus* is a small to medium sized bolete with reddish-brown to dark brown pileus, 2.5–6 cm diam., stipe 5–9 × 0.7–2 cm, yellow on the upper part and pinkish to reddish towards the base, with a reddish reticulum covering nearly the entire length of the stipe or at least the upper half, orange-red to yellow hymenophore, tissues turning blue when exposed or bruised and basidiospores 9–12 × 3.5–4.0 µm, Q_m = 2.7 [18,194,195].

Along with *Exsudoporus*, Liang et al. also decided to include *Butyriboletus hainanensis* N.K. Zeng, Zhi Q. Liang and Dong Y. [190] in the genus *Butyriboletus*, despite the fact that the topology of these three well-supported clades in their phylogram is similar to the current phylogenetic reconstruction. They indicated that *Bu. hainanensis* is phylogenetically distinct from the described *Butyriboletus* taxa. There are also clear morphological features that distinguish *Bu. hainanensis* from the other known species in *Butyriboletus*. These characters include a blue-red-black color change of the hymenophore and context, a thick pileal context and a thin hymenophore (similar to that in *Baorangia*) and a stipe with a faint reticulation. *Butyriboletus hainanensis* and two unknown allied species from China constitute a well-supported generic clade (BS = 99%, PP = 1.00) and further research is needed to describe this group in more detail.

In this study, we also analyzed an additional generic lineage, *Amoenoboletus*, which has some morphological and genetical similarity with *Exsudoporus*. This group forms a strongly supported clade (BS = 100%, PP = 1.00) and includes at least six different species, but some other taxa (e.g., *B. rubropictus* and *B. guatemalensis*) should be evaluated in the future based on molecular data to see whether they are phylogenetically related or not.

In the present study we have performed the first comprehensive molecular phylogenetic reconstruction of the genus *Exsudoporus* with the inclusion of several new sequences of multiple genetic markers. Our analysis includes accessions across the Northern Hemisphere and has verified the separation of this group from the genus *Butyriboletus*. Phylogenetic results clearly support the separation of these two genera, and their independence is also reinforced by consistent morphological differences. Species of *Exsudoporus* exhibit several critical characters that discriminate them from *Butyriboletus* species, including: (1) an overall reddish color of the basidiomes, (2) red pores, (3) non-stuffed pores (although Smith and Thiers [7], quoting Coker, indicated stuffed pores for *E. frostii*), (4) hymenophore exuding golden-yellow droplets in fresh, young specimens, (5) stipe surface strongly and coarsely reticulate to reticulate-alveolate or with scaly patches, (6) generally stronger bluing reaction on bruising (with the exception of *E. ruber*) and (7) hyphae of the stipe base context usually weakly to strongly amyloid, although the iodine test may also result in a negative or even "pseudoamyloid" (dextrinoid) reaction. In sharp contrast, *Butyriboletus* is characterized by: (1) a different combination of colors with predominantly yellow tints, (2) yellow pores, (3) stuffed pores in early developmental stages, (4) a hymenophore that never exudes droplets, (5) a very shallowly and densely reticulate stipe surface, (6) generally weaker bluing reaction and no bluing at all in the context of the mid or lower stipe and (7) consistently inamyloid hyphae of the stipe base context [16,28,32,196,197].

Based on phylogenetic evidence, the closely related eastern North American species *E. frostii* and *E. floridanus* have been recognized as congeneric to the type species, the European *E. permagnificus*. Moreover, the Asian *Leccinum ruber* is an additional representative of the genus and accordingly is transferred to *Exsudoporus* based on morphological affinities and molecular evidence. There also appear to be at least four additional taxa from the USA, Mexico and Central America (Belize, Guatemala, Costa Rica) that were previously misidentified as either *B. frostii* or *B. floridanus* [138,143–147,163,171,173–178,198]. However, the taxonomic and geographic limits of these tentative new species are in need of further investigation.

SEM images of the basidiospores of *E. permagnificus* were previously published by Assyov and clearly show a smooth spore wall. Although we did not take SEM pictures of the basidiospores of *E. frostii*, *E. floridanus* and *E. ruber*, it is likely that they also possess a smooth spore wall [120].

Krisai-Greilhuber and Takahashi et al. reported an inamyloid reaction of the hyphae for *E. frostii* and *E. ruber*, respectively [131,134]. Conversely, we observed an amyloid reaction in several specimens of *E. permagnificus* as well as in the herbarium material of *E. frostii*, *E. floridanus* and *E. ruber*. A fleeting amyloid reaction was also reported previously for *E. frostii* by Smith and Thiers [7]. Moreover, a dextrinoid reaction was also observed in *E. floridanus* and *E. permagnificus*.

Results obtained in this study highlight a Holarctic distribution of *Exsudoporus*. Species in the genus are widespread across temperate to subtropical latitudes in both the Western Hemisphere (*E. frostii* and *E. floridanus* in eastern North America) and the Eastern Hemisphere (*E. ruber* in East Asia). The genus also extends into warm climatic regions, including two undescribed neotropical species reported from Central America and *E. permagnificus*, which occurs in dry and warm Mediterranean environments in Europe and the Middle East. It is worth noting that, with the exception of *E. ruber*, all other *Exsudoporus* species appear to be associated with angiosperms, especially with diverse species of oaks. Conversely, *E. ruber* seems to be restricted to the subalpine belt in association with coniferous trees such as *Abies*, *Pinus* and *Tsuga*. However, the evolutionary paleogeographic dynamics that led to the present distribution and ecological preferences of *Exsudoporus* species remain unknown.

Further research will be required to better assess the ecological requirements and distribution limits of the known species to formally describe cryptic species detected in this study and also to ascertain whether or not there are other overlooked species of *Exsudoporus* that remain undescribed from other parts of the world.

Key to the Described Species of Exsudoporus

1. Species occurring in East Asia in subalpine coniferous forests associated with *Pinaceae* (mainly *Abies*) at high altitudes (1800–4200 m a.s.l.), stipe surface typically disrupting into scaly patches at maturity . . . . **E. ruber**

1. Species occurring in Europe, West Asia or the Americas, in temperate, tropical or Mediterranean broadleaved forests, associated mostly with *Fagaceae* (especially *Quercus*) at lower altitudes, stipe surface reticulate to alveolate-reticulate 2

2. Stipe conspicuously reticulate-alveolate throughout with longitudinally stretched meshes, occurring in eastern North America . . . . **E. frostii**

2. Stipe finely to coarsely reticulate but predominantly in the upper half, occurring in eastern and southeastern North America or in Europe and West Asia 3
3. Basidiospores slightly narrower, (13.0) 13.2–16.7 (18.0) × (4.0) 4.5–5.0 (5.3) µm, occurring in eastern and southeastern North America . . . . *E. floridanus*
3. Basidiospores slightly broader, (12.7) 13.7–15.1 (15.4) × (5.5) 5.7–6.1 µm, occurring in Europe and West Asia . . . . *E. permagnificus*

Supplementary Materials: The following supplementary materials are available online at https://www.mdpi.com/article/10.3390/jof8020101/s1, File S1: aligned and concatenated dataset of *Exsudoporus* and closely related genera (ITS, nrLSU, *tef1-α*, *rpb2*). File S2: partition file for RAxML analysis.

Author Contributions: Conceptualization, M.G., A.Yu.B., G.S. and A.V.; methodology, A.Yu.B., M.G., G.S., A.V. and L.G.N.; sampling and morpho-anatomical investigation, G.S., M.G., A.Yu.B., M.E.S., Y.T., G.V., B.A.B. and A.V.; molecular experiments and phylogenetic data analysis, A.Yu.B., M.E.S., R.A.H., Á.K. and A.V.; resources, S.P.W., E.N., G.S., A.V., L.G.N., M.E.S. and U.P.; writing—original draft preparation, M.G., A.Yu.B. and G.S.; writing—review and editing, A.Yu.B., M.G., M.E.S., G.S., R.A.H., A.V., Y.T., U.P., L.G.N., B.A.B. and S.P.W.; visualization, A.Yu.B., M.G., A.V. and G.S.; project administration, M.G., A.Yu.B. and A.V.; funding acquisition, A.Yu.B., S.P.W., E.N., M.E.S., L.G.N., G.S. and A.V. All authors have read and agreed to the published version of the manuscript.

Funding: This work was supported in part by the University of Haifa (to A.Yu.B., E.N. and S.P.W.), the Mycological Society of Israel (to A.Yu.B.), US National Science Foundation grant DEB-2106130 (to M.E.S.), the Institute of Food and Agricultural Sciences (IFAS) at the University of Florida, the Florida Agricultural Experiment Station's Ordway-Swisher research grants and the USDA NIFA (Hatch 1001991, McIntire-Stennis 1011527) (to M.E.S. and R.A.H.).

Institutional Review Board Statement: Not applicable.

Informed Consent Statement: Not applicable.

Data Availability Statement: Data is contained within the article and Supplementary Materials. Also some data can be found in publicly available datasets: https://www.ncbi.nlm.nih.gov/; http://www.mycobank.org/; http://www.indexfungorum.org/, accessed on 24 December 2021.

Acknowledgments: The authors would like to express their sincere gratitude to our friends and colleagues who contributed collections and color pictures: Regina Kuhnert (Austria); Shani Alter, Yonah Cherniavsky, Roman Kuznetsov, Yaniv Segal and Zohar Shafranov (Israel); Claudia Aita, Ulderico Bonazzi, Federica Costanzo, Giuseppe Gelardi, Olivia Gelardi, Carmine Lavorato, Francesco Mondello, Andrea Pappalardo, Roberto Polverini, Salvatore Silviani and Maria Tullii‡ (Italy); Motokazu Koike and Yoshiyuki Imai (Japan); Robert K. Antibus (Bluffton University, Ohio, USA); Arian Farid (University of South Florida, USA); David Borland, Christopher Ferguson, Laurel Kaminsky, Gage LaPierre and Paula M. Ramos Perez (University of Florida, USA); Renee Abbott, Alan E. Bessette, Arleen R. Bessette, L. Craig and Richard Kneal (USA). We are very grateful to Yula Vilozni (Mycological Society of Israel, Pardesiya, Israel) and Zohar Shafranov for organizing collection trips to Mt. Carmel National Park and Odem Forest Reserve, Israel. Tatiana Yu. Svetasheva (Tula State Lev Tolstoy Pedagogical University, Tula; and Komarov Botanical Institute of Russian Academy of Sciences, St. Petersburg, Russian Federation) and Ivan V. Zmitrovich (Komarov Botanical Institute RAS), who kindly provided us anatomical and macrochemical features of the specimens LE 17906 and data about findings of *E. permagnificus* in Russia. We wish to express our gratitude to Raffaella Trabucco (Natural History Museum of Venice, Italy; MCVE), Alexander E. Kovalenko‡ and Olga V. Morozova (Komarov Botanical Institute RAS; LE), Begoña Aguirre-Hudson and Angela Bond (Fungarium, Royal Botanic Gardens, Kew, UK; K-M) and Roy E. Halling and Laura Briscoe (Cryptogamic Herbarium, New York Botanical Garden, New York, USA; NY) for providing collections on loan and permission for molecular study. We are very grateful to Federica Costanzo (Anguillara Sabazia, Italy) who first noted the salty taste of the droplets in *E. permagnificus*; to Naoki Yoshimura (Tokyo, Japan) who observed golden yellow droplets exuding from the hymenophore of *E. ruber* and provided a color picture; and to Konstanze Bensch (Mycobank; and Botanische Staatssammlung München, Munich, Germany) who pointed out a grammar mistake of the species epithet *"rubrus"*. We would like to acknowledge Philipp Dresch (Austria) for help with the molecular work on some specimens from the mycological collection of the University of Innsbruck (IB). Many thanks to Arian

Farid and Gang Wu (Kunming Institute of Botany, Chinese Academy of Sciences, Kunming, China) for providing sequences from their articles before they were released in GenBank. A special thanks goes to Marco Contu (Tempio Pausania, Sardinia, Italy) for the invaluable discussions on taxonomic and nomenclatural issues. We would also like to thank Ekaterina F. Malysheva and Vera F. Malysheva (Komarov Botanical Institute RAS) for their critical comments on phylogenetic analysis. Finally, thanks are also given to the anonymous reviewers for their valuable comments and suggestions.

Conflicts of Interest: The authors declare no conflict of interest. The funders had no role in the design of the study, in the collection, analyses or interpretation of data, in the writing of the manuscript or in the decision to publish the results.

References

1. Fries, E.M. *Epicrisis Systematis Mycologici Seu Synopsis Hymenomycetum*; Typographia Academica: Uppsala, Sweden, 1838.
2. Horak, E. Synopsis generum Agaricalium (Die Gattungstypen der Agaricales). In *Beiträge zur Kryptogamenflora der Schweiz*; Waber: Bern, Germany, 1968; Volume 13.
3. Moser, M.M. Die Röhrlinge und Blätterpilze (*Polyporales, Boletales, Agaricales, Russulales*). In *Kleine Kryptogamenflora IIb/2, Basidiomyceten, XIII*, 5th ed.; Gams, H., Ed.; Gustav Fischer: Stuttgart, Germany, 1983.
4. Singer, R. *The Agaricales in Modern Taxonomy*, 4th ed.; Koeltz Scientific Books: Koenigstein, Germany, 1986.
5. Lannoy, G.; Estadès, A. Flore Mycologique d'Europe 6–Les Bolets. In *Doc. Mycol. Mém. Hors Série*; Association d'Écologie et de Mycologie: Lille, France, 2001; Volume 6.
6. Watling, R. A manual and source book on the boletes and their allies. In *Synopsis Fungorum*; Fungiflora: Oslo, Norway, 2008; Volume 24.
7. Smith, A.H.; Thiers, H.D. *The Boletes of Michigan*; University of Michigan Press: Ann Arbor, MI, USA, 1971.
8. Bertéa, P.; Estadès, A. *Boletus luteocupreus*. Bull. Féd. Micol. Dauphiné-Savoie **1990**, *118*, 25–31.
9. Cazzoli, P. Contributo allo studio dei boleti del gruppo purpureus-torosus. In *"Il Fungo", Atti III Seminario Internazionale "Russulales e Boletales", Castelnovo ne' Monti, Reggio nell'Emilia, Italy, 6-9 September 1990*; Associazione Micologica Bresadola Gruppo "R. Franchi": Reggio Emilia, Italy, 1990 ("1991"); pp. 29–36.
10. Andary, C.; Cosson, L.; Bourrier, M.J.; Wylde, R.; Heitz, A. Chimiotaxonomie des bolets de la section *Luridi*. *Cryptogam. Mycol.* **1992**, *13*, 103–114.
11. Redeuilh, G. Contribution à l'étude des bolets II. Étude critique de *Boletus torosus* et *Boletus xanthocyaneus*. *Bull. Soc. Mycol. Fr.* **1992**, *108*, 155–172.
12. Watling, R.; Hills, A.E. Boletes and their allies—*Boletaceae, Strobilomycetaceae, Gyroporaceae, Paxillaceae, Coniophoraceae, Gomphidiaceae* (revised and enlarged edition). In *British Fungus Flora, Agarics and Boleti*; Henderson, D.M., Watling, R., Eds.; HMSO: Edinburgh, UK, 2005; Volume 1.
13. Zang, M. An annotated check-list of the genus *Boletus* and its sections from China. *Fungal Sci.* **1999**, *14*, 79–87.
14. Galli, R. *I Boleti—Atlante Pratico-Monografico per la Determinazione dei Boleti*, 3rd ed.; Dalla Natura: Milano, Italy, 2007.
15. Binder, M.; Hibbett, D.S. Molecular systematics and biological diversification of *Boletales*. *Mycologia* **2006**, *98*, 971–981. [CrossRef]
16. Nuhn, M.E.; Binder, M.; Taylor, A.F.S.; Halling, R.E.; Hibbett, D.S. Phylogenetic overview of the *Boletineae*. *Fungal Biol.* **2013**, *117*, 479–511. [CrossRef]
17. Wu, G.; Feng, B.; Xu, J.P.; Zhu, X.T.; Li, Y.C.; Zeng, N.K.; Hosen, I.; Yang, Z.L. Molecular phylogenetic analyses redefine seven major clades and reveal 22 new generic clades in the fungal family *Boletaceae*. *Fungal Divers.* **2014**, *69*, 93–115. [CrossRef]
18. Wu, G.; Li, Y.C.; Zhu, X.T.; Zhao, K.; Han, L.H.; Cui, Y.Y.; Li, F.; Xu, J.P.; Yang, Z.L. One hundred noteworthy boletes from China. *Fungal Divers.* **2016**, *81*, 25–188. [CrossRef]
19. Zeng, N.K.; Wu, G.; Li, Y.C.; Liang, Z.Q.; Yang, Z.L. *Crocinoboletus*, a new genus of Boletaceae (Boletales) with unusual boletocrocin polyene pigments. *Phytotaxa* **2014**, *175*, 133–140. [CrossRef]
20. Vizzini, A. Nomenclatural Novelties. Index Fungorum 2014a. No. 146. Available online: http://www.indexfungorum.org/Names/IndexFungorumPublicationsListing.asp (accessed on 10 December 2020).
21. Vizzini, A. Nomenclatural Novelties. Index Fungorum 2014c. No. 188. Available online: http://www.indexfungorum.org/Names/IndexFungorumPublicationsListing.asp (accessed on 10 December 2020).
22. Vizzini, A. Nomenclatural Novelties. Index Fungorum 2014d. No. 192. Available online: http://www.indexfungorum.org/Names/IndexFungorumPublicationsListing.asp (accessed on 10 December 2020).
23. Zhao, K.; Wu, G.; Yang, Z.L. A new genus, *Rubroboletus*, to accommodate *Boletus sinicus* and its allies. *Phytotaxa* **2014**, *188*, 61–77. [CrossRef]
24. Assyov, B.; Bellanger, J.M.; Bertéa, P.; Courtecuisse, R.; Koller, G.; Loizides, M.; Marques, G.; Muñoz, J.A.; Oppicelli, N.; Puddu, D.; et al. Nomenclatural Novelties. Index Fungorum 2015. No. 243. Available online: http://www.indexfungorum.org/Names/IndexFungorumPublicationsListing.asp (accessed on 10 December 2020).
25. Gelardi, M.; Simonini, G.; Ercole, E.; Davoli, P.; Vizzini, A. *Cupreoboletus* (*Boletaceae, Boletineae*), a new monotypic genus segregated from *Boletus* sect. *Luridi* to reassign the Mediterranean species *B. poikilochromus*. *Mycologia* **2015**, *107*, 1254–1269. [CrossRef] [PubMed]

26. Vizzini, A. Nomenclatural Novelties. Index Fungorum 2014b. No. 183. Available online: http://www.indexfungorum.org/Names/IndexFungorumPublicationsListing.asp (accessed on 10 December 2020).
27. Simonini, G.; Vizzini, A. *Boletus mendax*, una specie recentemente descritta in Italia ed i nuovi orientamenti sulla sistematica della sez. *Luridi* del genere *Boletus*. In Proceedings of the XL Mostra Reggiana del Fungo, Reggio Emilia, Italy, 10–11 October 2015; Associazione Micologica Bresadola: Reggio Emilia, Italy, 2015; pp. 3–24.
28. Bessette, A.E.; Roody, W.C.; Bessette, A.R. *Boletes of Eastern North America*; Syracuse University Press: Syracuse, NY, USA, 2016.
29. Dentinger, B.T.M.; Ammirati, J.F.; Both, E.E.; Desjardin, D.E.; Halling, R.E.; Henkel, T.W.; Moreau, P.-A.; Nagasawa, E.; Soytong, K.; Taylor, A.F.S.; et al. Molecular phylogenetics of porcini mushrooms (*Boletus* section *Boletus*). *Mol. Phy. Evol.* **2010**, *57*, 1276–1292. [CrossRef] [PubMed]
30. Feng, B.; Xu, J.; Wu, G.; Zeng, N.K.; Li, Y.C.; Tolgor, B.; Kost, G.W.; Yang, Z.L. DNA sequence analyses reveal abundant diversity, endemism and evidence for Asian origin of the porcini mushrooms. *PLoS ONE* **2012**, *7*, e37567. [CrossRef]
31. Zeng, N.K.; Cai, Q.; Yang, Z.L. *Corneroboletus*, a new genus to accommodate the southeast Asian *Boletus indecorus*. *Mycologia* **2012**, *104*, 1420–1432. [CrossRef] [PubMed]
32. Arora, D.; Frank, J.L. Clarifying the butter Boletes: A new genus, *Butyriboletus*, is established to accommodate *Boletus* sect. *Appendiculati*, and six new species are described. *Mycologia* **2014**, *106*, 464–480. [CrossRef] [PubMed]
33. Šutara, J.; Janda, V.; Kříž, M.; Graca, M.; Kolařík, M. Contribution to the study of genus *Boletus*, section *Appendiculati*: *Boletus roseogriseus* sp. nov. and neotypification of *Boletus fuscoroseus* Smotl. *Czech Mycol.* **2014**, *66*, 1–37. [CrossRef]
34. Zhao, K.; Wu, G.; Halling, R.E.; Yang, Z.L. Three new combinations of *Butyriboletus* (*Boletaceae*). *Phytotaxa* **2015**, *234*, 51–62. [CrossRef]
35. Smith, M.E.; Amses, K.R.; Elliott, T.F.; Obase, K.; Aime, M.C.; Henkel, T.W. New sequestrate fungi from Guyana: *Jimtrappea guyanensis* gen. sp. nov., *Castellanea pakaraimophila* gen. sp. nov., and *Costatisporus cyanescens* gen. sp. nov. (*Boletaceae, Boletales*). *IMA Fungus* **2015**, *6*, 297–317. [CrossRef]
36. Henkel, T.W.; Obase, K.; Husbands, D.; Uehling, J.K.; Bonito, G.; Aime, M.C.; Smith, M.E. New *Boletaceae* taxa from Guyana: *Binderoboletus segoi* gen. and sp. nov., *Guyanaporus albipodus* gen. and sp. nov., *Singerocomus rubriflavus* gen. and sp. nov., and a new combination for *Xerocomus inundabilis*. *Mycologia* **2016**, *108*, 157–173. [CrossRef]
37. Crous, P.W.; Luangsa-ard, J.J.; Wingfield, M.J.; Carnegie, A.J.; Hernández-Restrepo, M.; Lombard, L.; Roux, J.; Barreto, R.W.; Baseia, I.G.; Cano-Lira, J.F.; et al. Fungal Planet description sheets: 785–867 *Persoonia* **2018**, *41*, 238–417. [CrossRef]
38. Bozok, F.; Assyov, B.; Taşkin, H. First records of *Exsudoporus permagnificus* and *Pulchroboletus roseoalbidus* (*Boletales*) in association with non-native *Fagaceae*, with taxonomic remarks. *Phytol. Balc.* **2019**, *25*, 13–27.
39. Loizides, M.; Bellanger, J.-M.; Assyov, B.; Moreau, P.-A.; Richard, F. Present status and future of boletoid fungi (*Boletaceae*) on the island of Cyprus: Cryptic and threatened diversity unravelled by ten-year study. *Fungal Ecol.* **2019**, *41*, 65–81. [CrossRef]
40. He, M.Q.; Zhao, R.L.; Hyde, K.D.; Begerow, D.; Kemler, M.; Yurkov, A.; McKenzie, E.H.; Raspe, O.; Kakishima, M.; Sanchez-Ramirez, S.; et al. Notes, outline and divergence times of Basidiomycota. *Fungal Divers.* **2019**, *99*, 105–367. [CrossRef]
41. Wijayawardene, N.N.; Hyde, K.D.; Al-Ani, L.K.T.; Tedersoo, L.; Haelewaters, D.; Rajeshkumar, K.C.; Zhao, R.L.; Aptroot, A.; Leontyev, D.; Saxena, R.K. Outline of Fungi and fungus-like taxa. *Mycosphere* **2020**, *11*, 1060–1456. [CrossRef]
42. Wu, G.; Li, M.X.; Horak, E.; Yang, Z.L. Phylogenetic analysis reveals the new genus *Amoenoboletus* from Asia and New Zealand. *Mycosphere* **2021**, *12*, 1038–1076. [CrossRef]
43. Thiers, B. Index Herbariorum: A Global Directory of Public Herbaria and Associated Staff. New York Botanical Garden's Virtual Herbarium. Available online: http://sweetgum.nybg.org/science/ih/ (accessed on 14 December 2021).
44. Index Fungorum. Available online: http://www.indexfungorum.org/ (accessed on 24 December 2021).
45. Mycobank. Available online: https://www.mycobank.org/ (accessed on 24 December 2021).
46. MyCoPortal. Available online: https://www.mycoportal.org/ (accessed on 10 October 2021).
47. Ridgway, R. *Color Standards and Color Nomenclature*; Privately Published: Washington, DC, USA, 1912.
48. White, T.; Bruns, T.; Lee, S.; Taylor, J.W. Amplification and direct sequencing of fungal ribosomal RNA genes for phylogenetics. In *PCR Protocols: A Guide to Methods and Applications*; Innis, M.A., Gelfand, D.H., Snisky, J.J., White, T.J., Eds.; Academic Press: San Diego, CA, USA, 1990; pp. 315–322. [CrossRef]
49. Gardes, M.; Bruns, T.D. ITS primers with enhanced specificity for basidiomycetes. Application to the identification of mycorrhizae and rusts. *Mol. Ecol.* **1993**, *2*, 113–118. [CrossRef]
50. Vilgalys, R.; Hester, M. Rapid genetic identification and mapping of enzymatically amplified ribosomal DNA from several *Cryptococcus* species. *J. Bacteriol.* **1990**, *172*, 4238–4246. [CrossRef] [PubMed]
51. Cubeta, M.A.; Echandi, E.; Abernethy, L.; Vilgalys, R. Characterization of anastomosis groups of binucleate *Rhizoctonia* species using restriction analysis of an amplified ribosomal RNA gene. *Phytopathology* **1991**, *81*, 1395–1400. [CrossRef]
52. Rehner, S.A.; Buckley, E. A *Beauveria* phylogeny inferred from nuclear ITS and EF1-α sequences: Evidence for cryptic diversification and links to *Cordyceps* teleomorphs. *Mycologia* **2005**, *97*, 84–89. [CrossRef]
53. Matheny, P.B. Improving phylogenetic inference of mushrooms with RPB1 and RPB2 nucleotide sequences (*Inocybe*; *Agaricales*). *Mol. Phylogenet. Evol.* **2005**, *35*, 1–20. [CrossRef] [PubMed]
54. Katoh, K.; Rozewicki, J.; Yamada, K.D. MAFFT online service: Multiple sequence alignment, interactive sequence choice and visualization. *Brief. Bioinform.* **2019**, *20*, 1160–1166. [CrossRef] [PubMed]

55. Tamura, K.; Stecher, G.; Peterson, D.; Filipski, A.; Kumar, S. MEGA6: Molecular evolutionary genetic analysis version 6.0. *Mol. Biol. Evol.* **2013**, *30*, 2725–2729. [CrossRef] [PubMed]

56. Capella-Gutiérrez, S.; Silla-Martínez, J.M.; Gabaldón, T. trimAl: A tool for automated alignment trimming in large-scale phylogenetic analyses. *Bioinformatics* **2009**, *25*, 1972–1973. [CrossRef] [PubMed]

57. Edler, D.; Klein, J.; Antonelli, A.; Silvestro, D. RaxmlGUI 2.0: A graphical interface and toolkit for phylogenetic analyses using RAxML. *Meth. Ecol. Evol.* **2020**, *12*, 373–377. [CrossRef]

58. Stamatakis, A. RAxML version 8: A tool for phylogenetic analysis and post-analysis of large phylogenies. *Bioinformatics* **2014**, *30*, 1312–1313. [CrossRef]

59. Ronquist, F.; Teslenko, M.; van der Mark, P.; Ayres, D.L.; Darling, A.; Höhna, S.; Larget, B.; Liu, L.; Suchard, M.A.; Huelsenbeck, J.P. MrBayes 3.2: Efficient bayesian phylogenetic inference and model choice across a large model space. *Syst. Biol.* **2012**, *61*, 539–542. [CrossRef]

60. Rambaut, A.; Suchard, M.; Xie, D.; Drummond, A. Tracer v. 1.6. Institute of Evolutionary Biology, University of Edinburgh, Scotland, 2014. Available online: http://beast.bio.ed.ac.uk/Tracer (accessed on 16 December 2021).

61. Alessio, C.L. *Boletus frostii* Russell anche in Italia? *Micol. Ital.* **1980**, *9*, 15–20.

62. Angarano, M. *Boletus flammans.* *Boll. Gr. Micol. Bres.* **1980**, *23*. Front cover.

63. Cazzoli, P.; Consiglio, G. Micologia di Base—Approccio al Genere *Boletus*—II. *Riv. Micol.* **2001**, *44*, 195–213.

64. Alessio, C.L. Ancora sul presunto *Boletus frostii* Russel. *Micol. Ital.* **1981**, *10*, 39–41.

65. Alessio, C.L. E' *Boletus siculus* Inzenga il già presunto *B. frostii* Russel rinvenuto in Italia. *Micol. Ital.* **1981**, *10*, 40–42.

66. Inzenga, G. *Funghi Siciliani*; Centuria 2. F Lao: Palermo, Italy, 1869.

67. Pöder, R. *Boletus permagnificus* sp. nov., un appariscente boleto della Sez. *Luridi* Fr. associato alle querce. *Boll. Gr. Micol. Bres.* **1983**, *26*, 82–89.

68. Bellù, F. Precisazioni e commenti. La questione flammans. *Boll. Gr. Micol. Bres.* **1981**, *24*, 50.

69. Pöder, R. *Boletus permagnificus* sp. nov., ein Auffallender Röhrling der Sekt. *Luridi* Fr. assoziiert mit Eichen. *Sydowia* **1981** ("1982"), *34*, 149–156.

70. Alessio, C.L. *Boletus permagnificus* Pöder è sinonimo di *B. siculus* Inzenga. *Micol. Ital.* **1982**, *11*, 34.

71. Alessio, C.L. *Boletus siculus* Inz. e *Boletus permagnificus* Pöder. *Micol. Ital.* **1984**, *13*, 63–68.

72. Alessio, C.L. *Boletus* Dill. ex L. In *Fungi Europaei*, 1st ed.; Giovanna Biella: Saronno, Italy, 1985; Volume 2.

73. Alessio, C.L. Revisione dei miei lavori comparsi nei numeri 26–50 di Micologia Italiana. *Micol. Ital.* **1990**, *19*, 67–73.

74. Alessio, C.L. *Boletus* Dill. ex L. (Supplemento). In *Fungi Europaei*; Giovanna Biella: Saronno, Italy, 1991; Volume 2.

75. Bellù, F. Precisazioni e commenti. Ancora su *Boletus permagnificus* e *Boletus siculus*. *Boll. Gr. Micol. Bres.* **1986**, *29*, 75–79.

76. Lavorato, C. Chiave analitica e note bibliografiche della micoflora del cisto. *Boll. AMER* **1991**, *24*, 16–45.

77. Alessio, C.L. Un boleto non ancora noto. *Xerocomus ichnusanus* Alessio, Galli et Littini sp. nov. *Boll. Gr. Micol. Bres.* **1984**, *27*, 166–170.

78. Gelardi, M.; Simonini, G.; Ercole, E.; Vizzini, A. *Alessioporus* and *Pulchroboletus* (*Boletaceae*, *Boletineae*), two novel genera for *Xerocomus ichnusanus* and *X. roseoalbidus* from the European Mediterranean basin: Molecular and morphological evidence. *Mycologia* **2014**, *106*, 1168–1187. [CrossRef]

79. Blanco-Dios, J.B. Nomenclatural Novelties. Index Fungorum. 2015. No. 211. Available online: http://www.indexfungorum.org/Names/IndexFungorumPublicationsListing.asp (accessed on 10 December 2020).

80. Icard, C.; Hurtado, C. *Boletus permagnificus* Pöder au pied des Alpes. *Bull. Féd. Ass. Mycol. Médit.* **1997**, *12*, 5–10.

81. Muñoz Sánchez, J.A.; Cadiñanos Aguirre, J.A. Algunos Boletales interesantes de la Península Ibérica. *Belarra* **2001**, *17–18*, 55–64.

82. Muñoz, J.A. *Boletus* s.l. (excl. *Xerocomus*). In *Fungi Europaei*; Edizioni Candusso: Alassio, Italy, 2005; Volume 2.

83. Horak, E. *Röhrlinge und Blatterpilze in Europa*; Elsevier: Munich, Germany, 2005; Volume 6.

84. Assyov, B. New and rare Bulgarian boletes. *Mycol. Balc.* **2005**, *2*, 75–81.

85. Galli, R. *I Boleti—Atlante Pratico-Monografico per la Determinazione dei Boleti*, 4th ed.; Micologica: Vergiate, Italy, 2013.

86. Raya López, L.; Moreno Arroyo, B. *Flora Micológica de Andalucía*; Consejería de Medio Ambiente y Ordenación del Territorio: Sevilla, Spain, 2018.

87. Davoli, P.; Weber, R.W.S. Analisi cromatografica del profilo pigmentario di *Boletus permagnificus* mediante HPLC-MS. *Micol. Ital.* **2001**, *30*, 89–95.

88. Davoli, P.; Weber, R.W.S. Simple method for reversed-phase high-performance liquid chromatographic analysis of fungal pigments in fruit-bodies of *Boletales* (Fungi). *J. Chromatogr. A* **2002**, *964*, 129–135. [CrossRef]

89. Vasquez, G. Indagini Micologiche sulle *Boletales* Epigee del Territorio Siciliano—Mappatura e Censimento delle Specie. Ph.D. Thesis, Università degli Studi di Catania, Catania, Italy, 2014.

90. Di Rita, F.; Atzeni, M.; Tudino, F. The history of conifers in central Italy supports long-term persistence and adaptation of mesophilous conifer fungi in *Arbutus*-dominated shrublands. *Rev. Palaeobot. Palyn.* **2020**, *282*, 104300. [CrossRef]

91. Cetto, B. *I Funghi dal Vero*; Saturnia: Trento, Italy, 1983; Volume 4.

92. Engel, H.; Krieglsteiner, G.; Dermek, A.; Watling, R. *Dickröhrlinge. Die Gattung Boletus in Europa*; Verlag Heinz Engel: Weidhausen bei Coburg, Germany, 1983.

93. Lavorato, C. Chiave per la determinazione delle *Boletaceae* delle foreste della Calabria. *Pag. Micol.* **1996**, *5*, 2–27.

94. Migliozzi, V.; Coccia, M. Segnalazione per il territorio laziale di Boletacee interessanti e descrizione di *Boletus poikilochromus* Pöder, Cetto & Zuccherelli. *Boll. AMER* **1991**, *24*, 9–15.

95. Brotzu, R. *Guida ai Funghi della Sardegna—Parte Prima*; Editrice Archivio Fotografico Sardo: Nuoro, Italy, 1993.

96. Foiera, F.; Lazzarini, E.; Snabl, M.; Tani, O. *Funghi boleti*; Edizioni Edagricole: Bologna, Italy, 1993.

97. Galli, R. *I Boleti—Atlante Pratico-Monografico per la Determinazione dei Boleti*; Edinatura: Milano, Italy, 1998.

98. Papetti, C.; Consiglio, G.; Simonini, G. *Atlante Fotografico dei Funghi d'Italia*; AMB Centro Studi Micologici: Vicenza, Italy, 2001.

99. Lunghini, D.; Perrone, L. Contributo allo studio e al monitoraggio delle Boletacee del litorale laziale. 2. *Boll. AMER* **2002**, *54–55*, 39–60.

100. Migliozzi, V.; Camboni, M. La micoflora del litorale romano. 11° contributo. Descrizione di *Helvella juniperi*, *Boletus comptus*, *Boletus luridus*, *Boletus permagnificus* e *Xerocomus roseoalbidus*. *Boll. Gr. Micol. Bres.* **2002**, *45*, 7–28.

101. Napoli, M.; Signorello, P. Contributo alla conoscenza della flora macromicetica del siracusano. *Boll. AMER* **2004**, *62–63*, 14–32.

102. Corea, E. Il genere *Boletus* L. in Calabria—2a parte. *Riv. Micol.* **2006**, *49*, 99–113.

103. Boccardo, F.; Traverso, M.; Vizzini, A.; Zotti, M. *Funghi d'Italia*; Zanichelli: Bologna, Italy, 2008.

104. Brotzu, R.; Colomo, S. *I Funghi della Sardegna. Basidiomiceti: Cortinariaceae, Bolbitiaceae, Boletaceae, Gomphidiaceae, Paxillaceae*; Editrice Archivio Fotografico Sardo: Nuoro, Italy, 2009; Volume 6.

105. Oppicelli, N. *Funghi in Italia*; Erredi Grafiche Editoriali: Genova, Italy, 2020.

106. Šutara, J.; Mikšík, M.; Janda, V. *Hřibovité Houby*; Academia: Praha, Czech Republic, 2009.

107. Mikšík, M. *Hřibovité Houby Evropy*; Vydalo Nakladatelství Svojtka & Co.: Praha, Czech Republic, 2017.

108. Moreno, G.; Esteve-Raventós, F. *Boletus aemilii* Barbier, *B. permagnificus* Pöder and *Xerocomus truncatus* Singer, Snell et Dick, in Spain. *Lazaroa* **1988**, *10*, 253–258.

109. Moreno, G.; Esteve-Raventós, F.; Illana, C. Estudios micologicos en el parque natural de Monfragüe y otras zonas de Extremadura (España), 4. Agaricales. *Bol. Soc. Micol. Madr.* **1989**, *14*, 115–141.

110. Mendaza, R.; Diaz, G. *Las Setas en la Naturaleza*; Iberdrola: Bilbao, Spain, 1996; Tomo 1.

111. Gelpi, C.; De Castro, J. *Boletus permagnificus* (Pöeder) Citado por Alessio (1985) como *B. siculus* (Inzenga 1869). *Bol. Soc. Micol. Extrem.* **2003**, *14*, 44.

112. Moreno Arroyo, B. *Inventario Micològico Básico de Andalucía*; Consejería de Medio Ambiente: Córdoba, Spain, 2004.

113. Pardo, F.M.V.; Maqueda, S.M.; Pimienta, A.B.L.; Pacheco, D.P. Aproximación al Catálogo de las especies del Orden *Boletales* (*Basidiomycetes, Fungi*) en Extremadura (España). *Rev. Estud. Extrem.* **2004**, *40*, 1255–1291.

114. Calzada Domínguez, A.C. *Guía de los boletos de España y Portugal*; Náyade Editorial: Medina del Campo, Spain, 2007.

115. Becerra Parra, M.; Robles Domínguez, E.; Díaz Romera, J.A.; Astete Sánchez, G.; Olivera Amaya, M.; López Pastora, A.; Gaona Ríos, J.M.; Peña Máquez, M.I. Nuevas aportaciones al conocimiento de los *Boletales* Andaluces. *Lactarius* **2015**, *22*, 87–99.

116. Chevassut, G.; Bertéa, P. La poussée fongique de l'automne 1991 en Languedoc. *Bull. Féd. Ass. Mycol. Médit.* **1992**, *2*, 11–16.

117. Joset, H. *Les Bolets à pores rouges en Corse*; Soc. Mycol. Ajaccio: Ajaccio, France, 1992.

118. Roth, A. *Boletus permagnificus* Pöder, *Xerocomus roseoalbidus* Alessio & Littini. *Bull. Soc. Mycol. Fr.* **1994**, *110*, 292–294.

119. Estadès, A.; Lannoy, G. Les bolets européens. *Bull. Féd. Mycol. Dauphinée-Savoie* **2004**, *174*, 3–79.

120. Assyov, B. Scanning electron microscopic study of the European members of *Baorangia*, *Exsudoporus* and *Lanmaoa* (*Boletales*, fungi). *Comprend. Acad. Bulg. Sci.* **2017**, *70*, 657–662.

121. Denchev, C.M.; Assyov, B. Checklist of the larger basidiomycetes in Bulgaria. *Mycotaxon* **2010**, *111*, 279–282. [CrossRef]

122. Jurc, D.; Ogris, N.; Piltaver, A.; Dolenc, A. *Seznam vrst in Razširjenost Makromicet v Sloveniji z Analizo Stopnje Ogroženosti: Ciljni Raziskovalni Program "Konkurenčnost Slovenije 2001–2006"*; Projekt št. V4-0703; Gozdarski Inštitit Slovenije: Ljubljana, Slovenia, 2004.

123. Constantinidis, G. *Mushrooms, a Photographic Guide for Collectors*; Privately Published: Athens, Greece, 2009.

124. Wagensommer, R.P.; Flores, G.A.; Arcangeli, A.; Bistocchi, G.; Maneli, F.; Materozzi, G.; Perini, C.; Venanzoni, R.; Angelini, P. Application of IUCN red listing criteria at the regional level: A case study with Boletales across the Appennine province ecoregion and EU-habitats of central Italy. *Plant Biosyst.* **2021**. [CrossRef]

125. Bedenko, E.P. Makromitsety Srednerusskoy Vozvyshennosti [Macromycetes of the Central Russian Upland]. Ph.D. Thesis, Lomonosov Moscow State University, Moscow, Russia, 1989. (In Russian).

126. Singer, R. New Boletaceae from Florida (a preliminary communication). *Mycologia* **1945**, *37*, 797–799. [CrossRef]

127. Singer, R. The Boletoideae of Florida. The Boletineae of Florida with notes on extralimital species III. *Am. Mid. Nat.* **1947**, *37*, 1–135. [CrossRef]

128. Both, E.E. *The Boletes of North America. A Compendium*; Buffalo Museum of Science: Buffalo, NY, USA, 1993.

129. Bessette, A.E.; Roody, W.C.; Bessette, A.R. *North American Boletes. A Color Guide to the Fleshy Pored Mushrooms*; Syracuse University Press: Syracuse, NY, USA, 2000.

130. Frost, C.C. Catalogue of boleti of New England, with descriptions of new species. *Bull. Buffalo Soc. Nat. Sci.* **1874**, *2*, 100–105.

131. Krisai-Greilhuber, I. *Boletus frostii*, Frosts Dickröhrling—Ein prachtvoller nordamerikanischer Röhrling [*Boletus frostii*, Frost's Bolete—A splendid North American bolete]. *Beih. Zeit. Mykol.* **1999**, *9*, 9–13.

132. Zang, M. Notes on the *Boletales* from eastern Himalayas and adjacent areas of China. *Acta Bot. Yunnanica* **1986**, *8*, 1–22. (In Chinese)

133. Zang, M. *Fungi of the Hengduan Mountains—The Series of the Scientific Expedition to the Hengduan Mountains, Qinghai-Xizang Plateau*; Science Press: Beijing, China, 1996. (In Chinese)

134. Takahashi, H.; Taneyama, Y.; Koyama, A. *Boletus kermesinus*, a new species of *Boletus* section *Luridi* from central Honshu, Japan. *Mycoscience* **2011**, *52*, 419–424. [CrossRef]
135. Zang, M.; Li, X.J.; He, Y.S. *Boletaceae* (II). In *Flora Fungorum Sinicorum*; Science Press: Beijing, China, 2013; Volume 44. (In Chinese)
136. Wu, K.; Wu, G.; Yang, Z.L. A taxonomic revision of *Leccinum rubrum* in subalpine coniferous forests, southwestern China. *Acta Edulis Fungi* **2020**, *27*, 92–100. [CrossRef]
137. Murrill, W.A. Florida boletes. *Lloydia* **1948**, *11*, 21–35.
138. Ortiz-Santana, B.; Lodge, D.J.; Baroni, T.J.; Both, E.E. Boletes from Belize and the Dominican Republic. *Fungal Div.* **2007**, *27*, 247–416.
139. Bessette, A.E.; Bessette, A.R.; Lewis, D.P. *Mushrooms of the Gulf Coast States: A Field Guide to Texas, Louisiana, Missisipi, Alabama, and Florida*; University of Texas Press: Austin, TX, USA, 2019.
140. GBIF. Available online: https://www.gbif.org/ (accessed on 24 May 2021).
141. Murrill, W.A. *American Boletes*; Privately Published: New York, NY, USA, 1914.
142. Bessette, A.E.; Roody, W.C.; Bessette, A.R.; Dunaway, D.L. *Mushrooms of the Southeastern United States*; Syracuse University Press: Syracuse, NY, USA, 2007.
143. García-Jiménez, J.; Garza-Ocañas, F. Conocimiento de los hongos de la familia Boletaceae de México. *Ciencia UANL* **2001**, *4*, 336–343.
144. García-Jiménez, J. Diversidad de Macromicetos en el Estado de Tamaulipas, México. Ph.D. Thesis, Universidad Autónoma de Nuevo León, San Nicolás de los Garza, Mexico, 2013.
145. García-Jiménez, J.; Singer, R.; Estrada, E.; Garza-Ocañas, F.; Valenzuela, R. Dos especies nuevas del género *Boletus* (*Boletales: Agaricomycetes*) en México. *Rev. Mex. Bioxbio* **2013**, *84*, 152–162. [CrossRef]
146. Gonzáles-Chicas, E.; Cappello, S.; Cifuentes, J.; Torres-De la Cruz, M. New records of *Boletales* (*Basidiomycota*) in a tropical oak forest from Mexican Southeast. *Bot. Sci.* **2019**, *97*, 423–432. [CrossRef]
147. Flores Arzù, R. Diversity and importance of edible ectomycorrhizal fungi in Guatemala. In *Mushrooms, Humans and Nature in a Changing World—Perspectives from Ecological, Agricultural and Social Sciences*; Pérez-Moreno, J., Guerin-Laguette, A., Flores Arzù, R., Yu, F.Q., Eds.; Springer Nature: Cham, Switzerlnd, 2020; Chapter 4; pp. 101–139.
148. Farid, A.; Bessette, A.E.; Bessette, A.R.; Bolin, J.A.; Kudzma, L.V.; Franck, A.R.; Garey, J.R. Investigations in the boletes (*Boletaceae*) of southeastern USA: Four novel species, and three novel combinations. *Mycosphere* **2021**, *12*, 1038–1076. [CrossRef]
149. Halling, R.E. Boletes described by Charles, C. Frost. *Mycologia* **1983**, *75*, 70–92. [CrossRef]
150. Curtis, M.A. Contributions to the mycology of North America. *Am. J. Sci. 2 Ser.* **1848**, *6*, 349–353.
151. Peck, C.H. Boleti of the United States. *Ann. Rep. N. Y. State Mus. Nat. Hist.* **1889**, *2*, 73–166.
152. Peck, C.H. Report of the State Botanist, 1906. *Bull. N. Y. State Mus.* **1907**, *116*, 1–117.
153. Murrill, W.A. The *Boletaceae* of North America—I. *Mycologia* **1909**, *1*, 4–18. [CrossRef]
154. Farlow, W.G.; Burt, E.A. *Icones Farlowianae: Illustrations of the Larger Fungi of Eastern North America*; The Farlow Library and Herbarium of Harvard University: Cambridge, MA, USA, 1929.
155. Coker, W.C.; Beers, A.H. *The Boletaceae of North Carolina*; University of North Carolina Press: Chapel Hill, NC, USA, 1943.
156. Singer, R. Die Röhrlinge. Teil II. Die Boletoideaea und Strobilomycetaceae. In *Die Pilze Mitteleuropas*; Verlag Julius Klinkhardt: Bad Heilbrunn, Germany, 1967; Band 6.
157. Snell, W.H.; Dick, E.A. *The Boleti of Northeastern North America*; Cramer: Lehre, Germany, 1970.
158. Miller, O.K., Jr. *Mushrooms of North America*; E.P. Dutton: New York, NY, USA, 1972.
159. Smith, A.H.; Weber, N.S. *The Mushroom Hunter's Field Guide*, 3rd ed.; University of Michigan Press: Ann Arbor, MI, USA, 1980.
160. Lincoff, G.H. *The Audubon Society Field Guide to North American Mushrooms*; Alfred, A. Knopf: New York, NY, USA, 1981.
161. Imler, L. *Icones Mycologicae*; Jardin Botanique National de Belgique: Meise, Belgium, 1985; p. 96.
162. Weber, N.S.; Smith, A.H. *A Field Guide to Southern Mushrooms*; University of Michigan Press: Ann Arbor, MI, USA, 1985.
163. Arora, D. *Mushrooms Demystified*, 2nd ed.; Ten Speed Press: Berkeley, CA, USA, 1986.
164. Bessette, A.E.; Sundberg, W.J. *Mushrooms: A Quick Reference Guide to Mushrooms of North America*; McMillian Field Guides: New York, NY, USA, 1987.
165. McKnight, K.H.; McKnight, B.B. *A Field Guide to Mushrooms of North America: Peterson Field Guides*; Houghton Mifflin: Boston, MA, USA, 1987; Volume 34.
166. Phillips, R. *Mushrooms of North America*; Little Brown Company: Boston, MA, USA, 1991.
167. Metzler, S.; Metzler, V. *Texas Mushrooms: A Field Guide*; University of Texas Press: Austin, TX, USA, 1992.
168. Bessette, A.E.; Bessette, A.R.; Fischer, E.W. *Mushrooms of Northeastern North America*; Syracuse University Press: Syracuse, NY, USA, 1997.
169. Kibby, G.G. Editorial. *Field Mycol.* **2005**, *6*, 38. [CrossRef]
170. Elliott, T.F.; Stephenson, S.L. *Mushrooms of the Southeast*; Timber Press: Portland, OR, USA, 2018.
171. Ibarra, L.V. Etnomicología de la Etnia Wirrárixa (Huichol), Jalisco, México. Master's Thesis, Universidad de Guadaljara, Guadalajara, Mexico, 1999.
172. Morris, M.H.; Perez-Perez, M.A.; Smith, M.E.; Bledsoe, C.S. Influence of host species on ectomycorrhizal communities associated with two co-occurring oaks (*Quercus* spp.) in a tropical cloud forest. *FEMS Microbiol. Ecol.* **2009**, *69*, 274–287. [CrossRef]
173. García-Jiménez, J. Estudio Sobre la Taxonomía, Ecología y Distributión de Algunos Hongos de la Familia *Boletaceae* (*Agaricales, Basidiomycetes*) de México. Master's Thesis, Universidad Autónoma de Nuevo León, San Nicolás de los Garza, Mexico, 1999.

174. Gándara, E.; Guzmán-Dávalos, L.; Guzmán, G.; Rodríguez, O. Inventario micobiótico de la región de Tapalpa, Jalisco, México. *Acta Bot. Mex.* **2014**, *107*, 165–185. [CrossRef]
175. Robles-García, D.; Suzán-Azpiri, H.; Montoya-Esquivel, A.; García-Jiménez, J.; Esquivel-Naranjo, E.U.; Yahia, E.; Landeros-Jaime, F. Ethnomycological knowledge in three communities in Amealco, Querétaro, México. *J. Ethnobiol. Ethnomed.* **2018**, *14*, 7. [CrossRef] [PubMed]
176. Saldivar, Á.E.; García Jiménez, J.; Herrera Fonseca, M.J.; Rodríguez Alcántar, O. Listado actualizado y nuevos registros de *Boletaceae (Fungi, Basidiomycota, Boletales)* en Jalisco, México. *Polibotánica* **2021**, *52*, 25–49. [CrossRef]
177. Halling, R.E.; Mueller, G.M. Agarics and boletes of neotropical oakwoods. In *Tropical Mycology. Macromycetes*; Watling, R., Frankland, J.C., Ainsworth, A.M., Isaac, S., Robinson, C.H., Eds.; CABInternational: Wallingford, UK, 2002; Volume 1, pp. 1–10.
178. Halling, R.E.; Mueller, G.M. Common mushrooms of the Talamanca mountains, Costa Rica. In *Memoirs of the New York Botanical Garden*; New York Botanical Garden Press: Bronx, NY, USA, 2005; Volume 90.
179. Sharma, A.D.; Jandaik, C.L.; Munjal, R.L.; Seth, P.K. Some fleshy fungi from Himachal Pradesh–I. *Indian J. Mush.* **1978**, *4*, 1–4.
180. Verma, B.; Pandro, V. Distribution of boletaceous mushrooms in India, some new records from sal forest of central India. *Int. J. Curr. Microbiol. Appl. Sci.* **2018**, *7*, 1694–1713. [CrossRef]
181. Yoneyama, T. *Mushrooms of Mt. Fuji*; Gotenba Food Hygiene Association: Shizuoka, Japan, 2007. (In Japanese)
182. Yuan, M.S.; Sun, P.Q. *The Pictorial Book of Mushrooms of China*; Sichuan Science and Technology Press: Chengdu, China, 2007. (In Chinese)
183. Li, T.H.; Song, B. Chinese boletes: A comparison of boreal and tropical elements. In *Tropical Mycology 2000*; The Millenium Meeting on Tropical Mycology (Main Meeting 2000); Walley, A.J.S., Ed.; Liverpool John Moores University: Liverpool, UK, 2000; pp. 1–10.
184. Fu, S.Z.; Wang, Q.B.; Yao, Y.J. An annotated checklist of *Leccinum* in China. *Mycotaxon* **2006**, *96*, 47–50.
185. Zhang, Y.; Zhou, D.Q.; Zhou, T.S.; Ou, X.K. New records and distribution of macrofungi in Laojun Mountain, northwest Yunnan, China. *Mycosystema* **2012**, *31*, 196–212.
186. McNabb, R.F.R. The boletaceae of New Zealand. *N. Z. J. Bot.* **1968**, *6*, 137–176. [CrossRef]
187. Stevenson, G. *Field Guide to Fungi*; University of Canterbury: Canterbury, UK, 1982.
188. New Zealand Provinces 1848–77. Available online: http://rulers.org/newzprov.html (accessed on 17 December 2021).
189. Mushroom Observer. Available online: https://mushroomobserver.org/ (accessed on 10 September 2021).
190. Liang, Z.Q.; An, D.Y.; Jiang, S.; Su, M.S.; Zeng, N.K. *Butyriboletus hainanensis (Boletaceae, Boletales)*, a new species from tropical China. *Phytotaxa* **2016**, *267*, 256–262. [CrossRef]
191. Zhang, M.; Li, T.H. *Erythrophylloporus* (Boletaceae, Boletales), a new genus inferred from morphological and molecular data from subtropical and tropical China. *Mycosystema* **2018**, *37*, 1111–1126. [CrossRef]
192. Janda, V.; Kříž, M.; Kolařík, M. *Butyriboletus regius* and *Butyriboletus fechtneri*: Typification of two well-known species. *Czech Mycol.* **2019**, *71*, 1–32. [CrossRef]
193. Kuo, M.; Ortiz-Santana, B. Revision of leccinoid fungi, with emphasis on North American taxa, based on molecular and morphological data. *Mycologia* **2020**, *112*, 197–211. [CrossRef]
194. Chiu, W.F. The Boletes of Yunnan. *Mycologia* **1948**, *40*, 199–231. [CrossRef]
195. Chiu, W.F. *Atlas of the Yunnan Bolets*; Science Press: Beijing, China, 1957. (In Chinese)
196. Šutara, J. Anatomical structure of pores in European species of genera *Boletus* s.str. and *Butyriboletus (Boletaceae)*. *Czech Mycol.* **2014**, *66*, 157–170. [CrossRef]
197. Janda, V.; Kříž, M. Evropské druhy hřibů rodu *Butyriboletus* [European representatives of genus *Butyriboletus*]. *Mykol. Listy* **2016**, *135*, 11–51.
198. Morris, M.H.; Perez-Perez, M.A.; Smith, M.E.; Bledsoe, C.S. Multiple species of ectomycorrhizal fungi are frequently detected on individual oak root tips in a tropical cloud forest. *Mycorrhiza* **2008**, *18*, 375–383. [CrossRef]

Journal of
Fungi

Article

Distribution and Origin of Major, Trace and Rare Earth Elements in Wild Edible Mushrooms: Urban vs. Forest Areas

Maja Ivanić, Martina Furdek Turk, Zdenko Tkalčec, Željka Fiket * and Armin Mešić

Ruđer Bošković Institute, Division for Marine and Environmental Research, Bijenička 54, 10000 Zagreb, Croatia; mivanic@irb.hr (M.I.); mfurdek@irb.hr (M.F.T.); ztkalcec@irb.hr (Z.T.); armin.mesic@irb.hr (A.M.)
* Correspondence: zeljka.fiket@irb.hr

Abstract: This paper investigates the composition of major, trace, and rare earth elements in 15 different species of wild edible mushrooms and the possible effect of urban pollution on elemental uptake. The collected mushrooms include different species from the green areas of the city, exposed to urban pollution, and from the forests, with limited anthropogenic influence. Through a comprehensive approach that included the analysis of 46 elements, an attempt was made to expand knowledge about element uptake by mushroom fruiting bodies. The results showed a wide variability in the composition of mushrooms, suggesting a number of factors influencing their element uptake capacity. The data obtained do not indicate significant exposure to anthropogenic influences, regardless of sampling location. While major elements' levels appear to be influenced more by species-specific affinities, this is not true for trace elements, whose levels presumably reflect the geochemical characteristics of the sampling site. However, the risk assessment showed that consumption of excessive amounts of the mushrooms studied, both from urban areas and from forests, may have adverse health effects.

Keywords: mushrooms; trace elements; rare earth elements; urban soils; forest

Citation: Ivanić, M.; Furdek Turk, M.; Tkalčec, Z.; Fiket, Ž.; Mešić, A. Distribution and Origin of Major, Trace and Rare Earth Elements in Wild Edible Mushrooms: Urban vs. Forest Areas. *J. Fungi* **2021**, *7*, 1068. https://doi.org/10.3390/jof7121068

Academic Editors: Anush Kosakyan, Rodica Catana, Alona Biketova and Laurent Dufossé

Received: 8 October 2021
Accepted: 8 December 2021
Published: 12 December 2021

Publisher's Note: MDPI stays neutral with regard to jurisdictional claims in published maps and institutional affiliations.

1. Introduction

Mushrooms have been a part of the human diet for centuries. Nowadays, they are not only readily consumed, but they are also very popular in the category of healthy foods and delicacies. Beyond diets, mushrooms feature in some types of traditional medicine, but also as fractions in various medical supplements [1,2]. In Europe, they are also defined as "novel foods", that is, foods that were not consumed to any significant extent before 1997.

Since food consumption is generally considered the most likely route of human exposure to heavy metals, the levels of individual metal species in mushrooms became of interest not only to the scientific community, but also to many regulatory agencies. Due to the increased intake of heavy metals and their persistence in the environment, these pollutants are among the most serious environmental problems. Their accumulation in soils as well as slow degradation processes enable their efficient transfer into living organisms, and consequently through the food chain to humans [3]. For this reason, the content of metals and metalloids in mushrooms, both wild and cultivated, should be continuously monitored.

Mushrooms are known to readily accumulate high concentrations of metals and metalloids, such as mercury, cadmium, lead, copper, arsenic, and radionuclides [4–7]. They take up elements from a substrate through an extensive mycelium, and their total content is influenced by both fungal and environmental factors [8–10]. In addition to species-specific affinity, the morphological part of the fruiting body, developmental stages, age of the mycelium, biochemical composition, interval between fructifications, and element accumulation in mushrooms is also influenced by environmental factors, such as the amount of organic matter, pH, temperature, water content, and element concentrations in the soil [9,11,12]. In urban areas, soils are additionally burdened by the influence of traffic and industrial pollution, which may lead to the accumulation of heavy metals and metalloids that are of particular importance for human health due to their toxic effects even at low concentrations, such as As, Cd, and Pb [3,13–15]. Mushrooms growing near pollution sources are generally avoided for consumption, and those from forests are considered safe, while the geochemical composition of the underlying soil is usually neglected as a potentially significant source of elements. This is particularly important when considering mushrooms growing in areas with naturally elevated concentrations of metals in the soil [12]. Consequently, elevated levels of metals and metalloids in mushrooms have been reported not only near industrial areas [16,17] and intensive traffic [9,10,13,16,18,19], but also in areas with naturally elevated metal concentrations due to the geological composition of the bedrock [12].

To date, numerous studies have been carried out on the content of individual elements in fruiting bodies from different areas [20], such as France [21], the Czech Republic [22], Poland [4,23–25], Slovakia [6], Spain [9,10,26], Portugal [16], Turkey [27–30], USA [31], China [32,33], Greece [18], Serbia [34,35], and Italy [36–39]. However, published results often differ significantly and usually include a narrow range of elements and/or mushroom species studied [20]. Studies on the metal content in mushrooms from Croatia are also sparse and mostly limited to several measured elements [40–42].

In this work, we report on the concentration of 46 elements, including macroelements, trace elements, and rare earth elements with Y (REY), determined in 15 edible wild mushroom species from urban and forest areas in northwestern Croatia. As this region is characterized by naturally elevated levels of aluminum (Al), bismuth (Bi), cobalt (Co), chromium (Cr), cesium (Cs), iron (Fe), molybdenum (Mo), nickel (Ni), lead (Pb), antimony (Sb), scandium (Sc), titanium (Ti), thallium (Tl), uranium (U), vanadium (V), zinc (Zn), and rare earth elements, including yttrium (REY) [12,43,44], the distribution of elements in edible mushroom species from the study area will provide new insights into metal uptake by mushroom fruiting bodies.

This comprehensive study aims to increase the knowledge on element distribution in different edible mushrooms, and particularly to investigate the importance of site-specific properties (soil geochemistry) versus the species-specific uptake of elements in the investigated mushrooms and the influence of urban pollution on element uptake by mushrooms. This was accomplished by (i) studying the distribution of elements in different mushrooms from specific locations, and (ii) comparing element content in mushrooms from urban and forest locations. To assess the health risks associated with the consumption of the wild mushrooms studied, the daily intake and hazard index for elements that may have a negative impact on human health were also calculated.

2. Materials and Methods

2.1. Study Area

During the sampling campaign in October 2016, an attempt was made to collect different edible wild mushroom species from the urban and forest areas (Table 1); mushrooms of a particular species were collected from different locations, and different species were collected from each sampling location. Sampling was conducted in northwestern Croatia (Figure 1), in green areas within the capital city of Zagreb, and in rural parts of Karlovac County. The city of Zagreb has a population of approximately 800,000 people inhabiting an area of 641 km^2. Green areas of the city where the mushrooms were collected include forested areas of the city parks: Dotrščina (2 km^2; samples 1/7272–1/7275), Jelenovac (54 ha; samples 1/7258–1/7263), and Maksimir (316 ha; samples 1/7267–1/7269), as well as smaller green areas in residential areas of the city (samples 1/7256, 1/7293, and 1/7294). These mushrooms, although exposed to urban pollution due to their location in the capital, were not sampled in the immediate vicinity of roads, but in the central areas of city parks. Only samples 1/7293 and 1/7294 were collected in the immediate vicinity of a road with low traffic volume. The city of Zagreb, despite being a capital city with industries and dense traffic, had good air quality with low PM10 content at the time of sampling, as shown by the data collected by the Croatian Environmental Agency. The samples from Karlovac County (1/7284, 1/7288, 1/7291, and 1/7292) included various forest sites distributed throughout the district, far from roads and other pollution sources.

A total of 19 samples, representing 15 edible mushroom species from the phylum Basidiomycota, were collected and analyzed for total element concentration (Table 1). Photographs of the identified mushroom species and details of collection, morphological and molecular identification, and screening of bioactive compounds can be found elsewhere [45]. Basic information on the mushroom species and sampling is also presented here (Table 1) to be easily accessible for interpretation of the data from this study. All mushrooms used in this study are stored in the Croatian National Fungarium (CNF).

Table 1. Investigated species of mushrooms, and their lifestyle and location information. Trophic mode ([a] saprotroph, [b] pathotroph/saprotroph or [c] symbiotroph).

	Taxon/Family	Locality	Habitat	Substrate
Urban area, Zagreb				
1/7256 [b]	*Pleurotus dryinus* Pleurotaceae	Črnomerec	forest of *Populus nigra, Acer campestre, Sambucus nigra*	wood
1/7258 [a]	*Infundibulicybe gibba* Tricholomataceae	Jelenovac	forest of *Quercus robur, Carpinus betulus, Fagus sylvatica*	soil
1/7259 [a]	*Lycoperdon excipuliforme* Agaricaceae	Jelenovac	forest of *Quercus robur, Carpinus betulus, Fagus sylvatica*	soil
1/7260 [a]	*Lycoperdon perlatum* Agaricaceae	Jelenovac	forest of *Quercus robur, Carpinus betulus, Fagus sylvatica*	soil
1/7262 [a]	*Paralepista flaccida* Tricholomataceae	Jelenovac	forest of *Quercus robur, Carpinus betulus, Fagus sylvatica*	soil
1/7263 [a]	*Psathyrella multipedata* Psathyrellaceae	Jelenovac	forest of *Quercus robur, Carpinus betulus, Fagus sylvatica*	soil
1/7267 [a]	*Lycoperdon perlatum* Agaricaceae	Maksimir Park	forest of *Quercus robur, Q. petraea, Carpinus betulus*	soil
1/7268 [a]	*Psathyrella piluliformis* Psathyrellaceae	Maksimir Park	forest of *Quercus robur, Q. petraea, Carpinus betulus*	wood
1/7269 [a]	*Macrolepiota procera* Agaricaceae	Maksimir Park	forest of *Quercus robur, Q. petraea, Carpinus betulus*	soil
1/7272 [a]	*Infundibulicybe gibba* Tricholomataceae	Dotrščina Park	forest of *Quercus petraea, Carpinus betulus, Fagus sylvatica*	soil
1/7273 [a]	*Psathyrella piluliformis* Psathyrellaceae	Dotrščina Park	forest of *Quercus* sp., *Carpinus betulus*	wood
1/7274 [a]	*Coprinus comatus* Agaricaceae	Dotrščina Park	forest of *Quercus* sp., *Carpinus betulus*	soil
1/7275 [a]	*Hymenopellis radicata* Physalacriaceae	Dotrščina Park	forest of *Quercus* sp., *Carpinus betulus*	wood
1/7293 [a]	*Leucoagaricus leucothites* Agaricaceae	Ruđer Bošković Institute	grassland	soil
1/7294 [b]	*Agrocybe cylindracea* Strophariaceae	Ruđer Bošković Institute	grassland, on old living tree of *Populus nigra*	wood
Forest area, Karlovac County				
1/7284 [c]	*Lactarius deterrimus* Russulaceae	Dvorišće Ozaljsko	grassland, near planted *Picea abies*	soil
1/7288 [c]	*Lactarius deterrimus* Russulaceae	Vukova Gorica	grassland, near planted *Picea abies*	soil
1/7291 [a]	*Lepista nuda* Tricholomataceae	Vukova Gorica	forest of *Quercus robur, Corylus avellana*	soil
1/7292 [c]	*Craterellus cornucopioides* Cantharellaceae	Novaki Ozaljski	forest of *Fagus sylvatica*	soil

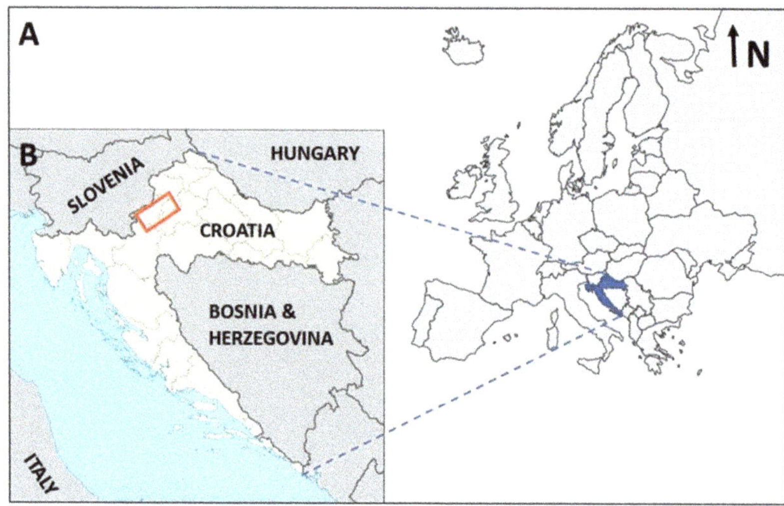

Figure 1. Map of the study area, its geographical position (**A**), and sampling area (**B**).

2.2. Sample Collection and Preparation

In the laboratory, soil particles were carefully removed from the fruiting bodies. Approximately 5 g of raw fruiting bodies from each sample was stored in a separate plastic bag at $-82\ °C$. Prior to analyses, the frozen samples were freeze-dried (Finn-Aqua Lyovac GT 2) and pulverized with liquid nitrogen.

2.3. Multi-Element Analysis

Prior to multi-element analysis, subsamples (0.05 g) of the previously freeze-dried samples were subjected to digestion in the microwave oven (Multiwave ECO, Anton Paar, Graz, Austria) with 7 mL of HNO_3 (65%, *supra pur*) and 0.1 mL of HF (48%, *pro analysi*) [46]. Digests were acidified with 2% (*v/v*) HNO_3 (65%, *supra pur*), without further dilution, and indium (In, $1\ \mu g\ L^{-1}$) was added as an internal standard.

The multi-element analysis was performed by High-Resolution Inductively Coupled Plasma Mass Spectrometry (HR-ICP-MS) using an Element 2 instrument (Thermo, Bremen, Germany). The instrument conditions and measurement parameters used in this work are described in the ref. [47]. Standards for multi-element analysis were prepared by appropriate dilution of a multi-element reference standard (Analytika, Prague, Czech Republic) containing Al, As, Be, Cd, Co, Cr, Cu, Fe, Li, Mn, Ni, Pb, Rb, Sr, and Ti, to which single-element standard solutions of Sn (Analytika, Prague, Czech Republic) and Sb (Analytika, Prague, Czech Republic) were added. For the analysis of REY, separate standards were prepared by appropriate dilution of a multi-elemental reference standard (Analytika, Prague, Czech Republic) containing Ce, Dy, Er, Eu, Gd, Ho, La, Lu, Nd, Pr, Sm, Tb, Tm, Y, and Yb.

The samples were analyzed for the total concentration of 46 elements (Al, As, Ba, Be, Bi, Ca, Cd, Ce, Co, Cr, Cs, Cu, Dy, Er, Eu, Fe, Gd, Ho, K, La, Li, Lu, Mg, Mn, Mo, Na, Nd, Ni, Pb, Pr, Rb, Sb, Sc, Se, Sn, Sm, Sr, Tb, Ti, Tl, Tm, U, V, Y, Yb, and Zn). All concentrations refer to dry matter.

2.4. Quality Control

Limits of detection (LOD) and limits of quantification (LOQ) were calculated as 3 and 10 times, respectively, the standard deviation of 10 consecutive measurements of analyte concentration in the procedural blank. Quality control of the analytical procedure was performed by simultaneous analysis of the blank sample and the certified reference material for Citrus leave (NCS ZC73018, China National Analysis Center for Iron and Steel). Good agreement was obtained between the analyzed and certified concentrations for all the elements measured, with recoveries ranging from 85% to 106%. Measurement precision was determined from five consecutive measurements in two Citrus leave CRM samples, and averaged 5%.

2.5. Health Risk Assessment

The assessment of non-carcinogenic health risks of heavy metals from mushroom consumption was performed according to the USEPA procedure described in the ref. [48]. The risk assessment included metals known to be harmful to human health, such as As, Ba, Be, Cd, Cu, Fe, Mn, Ni, Pb, Sb, Sr, You, and Zn. First, the intake factor, that is, the estimated daily intake, was calculated according to Equation (1):

$$EDI \left(\mu g\ kg^{-1}\ day^{-1} \right) = \frac{C_{metal} \times IR \times ED \times EF}{BW \times AT}, \tag{1}$$

where C_{metal} is the metal concentration in dried mushrooms ($\mu g\ kg^{-1}$), IR is the intake rate (in kg per person per day) corresponding to 0.03 kg of dried mushrooms [11], ED is the exposure duration (30 years for adults), EF is the exposure frequency (350 days per $year^{-1}$), BW is the average body weight (70 kg for adults), and AT is the average exposure duration (10,500 days, calculated C_{meta} as EF × ED).

Then, the target hazard quotient (*THQ*) of a given metal (Equation (2)) was calculated as the ratio between the estimated daily intake (*EDI*) and the reference dose (*RfD*). The *RfD* value is an estimate of the daily dose that is unlikely to pose a significant risk to human health over a lifetime [49]. This value is specific to the element being assessed and is defined in Integrated Risk Information System assessments (IRIS) under the USEPA programme.

$$THQ = \frac{EDI \left(\mu g\ kg^{-1}\ day^{-1} \right)}{RfD \left(\mu g\ kg^{-1}\ day^{-1} \right)} \tag{2}$$

Finally, the risk assessment related to the cumulative effect of several heavy metals present in the mushrooms was performed by calculating the overall hazard index as the sum of exposures scaled by the toxicity of each metal (Equation (3)),

$$HI = \sum_{i=1}^{n} THQ_i, \tag{3}$$

where THQ_i is the target hazard quotient for a single metal.

2.6. Statistical Analysis

Data were statistically analyzed using STATISTICA 7.0 (StatSoft Inc., Tulsa, OK, USA). The Spearman rank correlation coefficient was used for correlation of investigated elements in studied mushrooms. Multivariate principal component analysis (PCA) was performed on the data matrix consisting of element concentrations. The significance level was set at $p < 0.05$.

3. Results

The concentrations of macro- and trace elements determined in the studied mushrooms, along with calculated LODs and LOQs, are presented in Table 2. Elements that showed the widest range of concentrations among different mushrooms were As, Co, Cs, Mo, Rb, and Tl, while Be, Cr, Fe, K, Li, Mg, Mn, Ni, Sn, and Zn were present in a narrow range. The measured elements appeared in the following order of abundance: K > Mg > Na > Ca > Rb > Zn > Fe > Al > Cu > Mn > Ti > Cd > Pb > As > Cs > Ba > Ni > Se > Sr > Mo > $\sum REY$ > Cr > Co > V > Sn > Tl > Li > Sb > U > Bi > Be.

Table 2. Concentrations (in mg kg^{-1}, dry matter) of macro- and trace elements in the investigated mushrooms, including calculated LODs and LOQs (in mg kg^{-1}), minimum (min), maximum (max), and average (avg) values.

	Al	As	Ba	Be	Bi	Ca	Cd	Co	Cr	Cs	Cu
LOD	2	0.003	0.05	0.001	0.001	15	0.002	0.002	0.03	0.002	0.03
LOQ	6	0.01	0.15	0.003	0.003	45	0.006	0.006	0.1	0.006	0.1
1/7256	7.83	0.019	0.096	<DL	<DL	26	0.681	0.006	0.313	0.016	2.19
1/7258	30.9	0.318	1.599	<DL	0.001	171	0.379	0.026	0.265	0.124	46.6
1/7259	19.4	0.394	0.518	<DL	0.004	187	0.356	0.125	0.419	0.002	38.6
1/7260	17.0	0.542	0.484	<DL	0.003	164	0.313	0.171	0.378	0.003	84.3
1/7262	33.2	0.482	1.367	<DL	0.001	273	0.144	0.031	0.246	0.015	28.8
1/7263	23.5	3.684	1.126	0.001	0.003	374	0.365	0.100	0.221	0.124	78.0
1/7267	24.4	0.549	0.528	<DL	0.010	146	0.271	0.092	0.198	0.017	67.2
1/7268	99.3	0.087	1.725	0.001	0.002	216	0.396	0.084	0.305	0.333	11.3
1/7269	12.6	0.219	0.307	<DL	0.002	188	0.384	0.033	0.279	0.008	56.6
1/7272	27.6	0.661	0.979	0.002	0.001	262	0.216	0.018	0.306	0.037	52.4
1/7273	43.9	0.151	0.591	0.001	0.001	231	0.490	0.139	0.343	0.604	20.0
1/7274	22.9	0.489	0.580	<DL	0.001	545	0.486	0.058	0.608	0.006	42.3
1/7275	40.7	0.008	0.810	<DL	<DL	223	0.116	0.021	0.159	0.035	9.44
1/7284	91.0	0.639	1.453	<DL	<DL	276	0.593	0.069	0.268	0.167	16.5
1/7288	34.4	0.260	0.554	<DL	0.002	257	0.550	0.048	0.337	0.581	8.85
1/7291	26.1	0.593	0.512	<DL	0.001	255	0.177	0.125	0.223	0.043	41.0
1/7292	77.2	0.012	0.793	0.002	0.001	256	0.549	0.701	0.761	1.956	32.9
1/7293	13.0	0.310	0.294	<DL	<DL	144	4.547	0.043	0.278	0.482	26.2
1/7294	16.0	<DL	0.227	<DL	<DL	128	3.956	0.011	0.230	0.419	44.8
avg	34.8	0.52	0.77	0.001	0.002	227	0.79	0.10	0.32	0.26	37.3
min	7.83	0.01	0.10	0.001	0.001	25.6	0.12	0.01	0.16	0.002	2.19
max	99.3	3.68	1.73	0.002	0.010	545	4.55	0.70	0.76	1.96	84.3

	Fe	K	Li	Mg	Mn	Mo	Na	Ni	Pb	Rb
LOD	3	15	0.002	6	0.25	0.005	3	0.03	0.03	0.03
LOQ	10	45	0.006	18	0.75	0.015	10	0.1	0.1	0.1
1/7256	11.0	42327	0.019	1706	5.58	0.018	77.2	0.325	0.072	14.7
1/7258	95.7	25183	0.031	1086	21.5	0.557	60.4	0.559	0.135	10.3
1/7259	52.5	17100	0.015	1194	13.6	0.332	70.3	0.611	1.11	5.48
1/7260	103	22924	0.012	1597	27.7	0.355	36.1	0.742	0.509	2.59
1/7262	49.4	41252	0.026	1395	31.4	0.812	53.5	0.363	0.228	2.69
1/7263	49.0	65291	0.025	2014	11.6	0.488	829	0.547	0.352	24.6
1/7267	79.7	17410	0.013	1520	38.2	0.375	41.6	0.472	0.868	26.5
1/7268	87.9	66745	0.078	2067	22.9	0.015	932	0.327	0.261	289
1/7269	60.6	33029	0.017	1300	11.5	0.409	49.5	0.358	0.437	10.1
1/7272	76.1	27369	0.016	959	36.4	0.606	47.9	0.663	0.128	40.1
1/7273	67.4	80836	0.037	2093	25.6	0.016	653	0.247	0.406	185
1/7274	57.0	49527	0.013	1639	14.5	0.164	835	0.181	0.161	27.9
1/7275	32.9	38559	0.027	1060	21.3	0.045	410	0.187	0.122	27.3
1/7284	62.7	34991	0.079	1248	8.13	0.007	60.6	0.508	0.180	144
1/7288	29.7	25886	0.021	1153	4.93	0.010	35.9	0.505	0.187	191
1/7291	48.0	31622	0.010	1031	37.6	1.145	72.5	0.432	0.535	37.6
1/7292	77.3	46135	0.074	939	25.1	0.103	74.1	1.595	4.019	299
1/7293	23.9	60912	0.025	1876	7.26	0.235	1011	0.162	0.299	46.8
1/7294	22.1	43277	0.012	1694	5.87	0.050	25.2	0.259	0.234	133
avg	57.1	40546	0.029	1451	19.5	0.302	283	0.476	0.539	79.8
min	11.0	17100	0.010	939	4.93	0.007	25.2	0.162	0.072	2.59
max	103	80836	0.079	2093	38.2	1.145	1011	1.595	4.019	299

	Sb	Sc	Se	Sn	Sr	Ti	Tl	U	V	Zn
LOD	0.002	0.002	0.02	0.003	0.5	0.5	0.001	0.001	0.002	1.5
LOQ	0.006	0.006	0.06	0.01	1.5	1.5	0.003	0.003	0.006	4.5
1/7256	0.002	<DL	<DL	0.095	0.052	0.52	<DL	<DL	0.011	17.3
1/7258	0.015	<DL	0.144	0.094	0.386	2.89	0.001	<DL	0.153	74.2

Table 2. *Cont.*

	Sb	Sc	Se	Sn	Sr	Ti	Tl	U	V	Zn
1/7259	0.006	0.099	0.725	0.195	0.394	1.15	<DL	0.010	0.094	143
1/7260	0.006	0.047	0.788	0.140	0.391	0.65	0.001	0.005	0.081	109
1/7262	0.013	0.031	0.842	0.156	0.734	2.16	0.005	0.004	0.131	100
1/7263	0.068	<DL	1.593	0.118	1.198	2.29	0.003	0.004	0.083	97.6
1/7267	0.004	<DL	0.679	0.086	0.282	1.20	<DL	<DL	0.068	127
1/7268	0.012	0.012	<DL	0.085	0.966	9.70	0.006	0.002	0.174	89.4
1/7269	0.005	0.026	0.387	0.107	0.309	1.91	0.003	<DL	0.052	72.8
1/7272	0.007	0.011	0.254	0.116	0.609	1.69	0.001	0.001	0.111	66.9
1/7273	0.003	0.012	<DL	0.117	0.497	3.51	0.011	0.001	0.083	69.8
1/7274	0.011	0.004	0.386	0.083	0.756	1.57	0.003	0.003	0.054	58.3
1/7275	0.004	<DL	<DL	0.102	0.428	3.58	0.001	<DL	0.067	45.8
1/7284	0.002	0.024	0.414	0.068	0.734	8.65	0.052	0.008	0.138	133
1/7288	0.003	0.013	0.318	0.122	0.479	2.73	0.129	0.003	0.038	103
1/7291	0.004	<DL	0.205	0.057	0.318	1.18	0.004	<DL	0.136	58.3
1/7292	0.004	0.094	<DL	0.114	0.543	7.76	0.002	0.005	0.150	64.2
1/7293	0.035	<DL	1.49	0.066	0.277	0.93	0.007	0.001	0.018	75.3
1/7294	0.003	<DL	<DL	0.069	0.488	1.51	0.007	0.001	0.025	91.5
avg	0.011	0.034	0.633	0.105	0.518	2.93	0.012	0.003	0.088	84.1
min	0.002	0.004	0.144	0.057	0.052	0.52	0.001	0.001	0.011	17.3
max	0.068	0.099	1.593	0.195	1.198	9.70	0.129	0.010	0.174	143

DL—detection limits.

The results of the measurement of REY (La, Ce, Pr, Nd, Sm, Eu, Gd, Tb, Dy, Ho, Er, Tm, Yb, Lu, and Y), along with calculated LODs and LOQs, are shown in Table 3. The description of REY distribution in the samples was based on geochemical classification into light (LREE—La, Ce, Pr, Nd, Sm, Eu, Gd) and heavy (HREE—Tb, Dy, Ho, Er, Tm, Yb, Lu, and Y). The concentrations of REY in all the analyzed samples ranged over three orders of magnitude, from below the detection limit (Eu, Gd, Tb, Ho, Er, Tm, and Lu in some samples) to 0.137 mg kg^{-1} (Ce), with ΣREY ranging from 0.031 mg kg^{-1} to 0.452 mg kg^{-1} (Table 3). Among them, Ce was present in the highest concentrations in all samples and accounted for between 19% and 36% of the total REY, while Eu, Gd, Tb, Ho, Tm, and Lu had the lowest values. In general, the mushrooms showed great variability in terms of REY concentrations, with an RSD of up to 120% and ΣREY variability of up to one order of magnitude. The highest REY values were measured in *L. deterrimus* (1/7284, 0.452 mg kg^{-1}), while the lowest were observed in *P. dryinus* (1/7256, 0.031 mg kg^{-1}). LREEs were found to be more abundant in the majority of the samples studied, with LREE/HREE ratios ranging from 0.7 to 3.8 and the average being 1.9. The content of LREEs accounted for 41.9–79.3% of the total REYs in the mushrooms studied.

Table 3. Concentration of REY (in mg kg^{-1}, dry matter) in the investigated mushrooms, including calculated LODs and LOQs (in mg kg^{-1}), minimum (min), maximum (max), and average (avg) values.

	La	Ce	Pr	Nd	Sm	Eu	Gd	Tb	Dy	Ho	Er	Tm	Yb	Lu	Y	ΣREY
LOD	0.001	0.001	0.001	0.001	0.001	0.001	0.001	0.001	0.001	0.001	0.001	0.001	0.001	0.001	0.001	
LOQ	0.003	0.003	0.003	0.003	0.003	0.003	0.003	0.003	0.003	0.003	0.003	0.003	0.003	0.003	0.003	
1/7256	0.003	0.006	0.001	0.002	0.001	<DL	<DL	<DL	0.009	<DL	<DL	<DL	0.007	<DL	0.002	0.031
1/7258	0.012	0.028	0.003	0.012	0.003	0.001	<DL	<DL	0.011	0.001	0.001	<DL	0.008	<DL	0.012	0.092
1/7259	0.005	0.009	0.001	0.004	0.001	0.000	<DL	<DL	0.010	<DL	0.001	<DL	0.008	<DL	0.004	0.043
1/7260	0.006	0.010	0.001	0.005	0.001	0.001	<DL	<DL	0.011	<DL	0.001	<DL	0.008	<DL	0.007	0.051
1/7262	0.016	0.031	0.004	0.016	0.003	0.001	<DL	<DL	0.011	0.001	0.001	<DL	0.009	<DL	0.013	0.094
1/7263	0.021	0.039	0.006	0.022	0.006	0.001	0.002	0.001	0.013	0.001	0.002	<DL	0.008	<DL	0.024	0.146
1/7267	0.009	0.017	0.002	0.007	0.001	0.001	<DL	<DL	0.011	<DL	<DL	<DL	0.008	<DL	0.004	0.060
1/7268	0.053	0.115	0.013	0.057	0.010	0.002	<DL	0.001	0.016	0.002	0.003	0.001	0.010	0.001	0.032	0.316
1/7269	0.005	0.012	0.001	0.003	0.001	<DL	<DL	<DL	0.010	<DL	<DL	<DL	0.007	<DL	0.006	0.045
1/7272	0.015	0.026	0.003	0.012	0.002	0.001	<DL	<DL	0.011	<DL	0.001	<DL	0.008	<DL	0.009	0.088
1/7273	0.019	0.035	0.004	0.014	0.003	0.001	<DL	<DL	0.011	0.001	0.001	<DL	0.008	<DL	0.012	0.109
1/7274	0.010	0.019	0.003	0.010	0.003	0.001	<DL	<DL	0.010	<DL	0.001	<DL	0.007	<DL	0.006	0.070
1/7275	0.024	0.051	0.006	0.026	0.005	0.001	<DL	<DL	0.011	0.000	0.001	<DL	0.008	<DL	0.011	0.144
1/7284	0.093	0.137	0.017	0.063	0.011	0.002	<DL	0.002	0.020	0.003	0.006	0.001	0.014	0.001	0.082	0.452
1/7288	0.028	0.031	0.004	0.014	0.003	0.001	<DL	0.001	0.013	0.001	0.003	<DL	0.007	<DL	0.040	0.146
1/7291	0.016	0.024	0.003	0.014	0.003	0.001	<DL	<DL	0.008	<DL	0.001	<DL	0.006	<DL	0.009	0.085
1/7292	0.050	0.080	0.008	0.034	0.007	0.001	<DL	0.001	0.013	0.001	0.004	0.001	0.009	0.001	0.029	0.239

Table 3. *Cont.*

	La	Ce	Pr	Nd	Sm	Eu	Gd	Tb	Dy	Ho	Er	Tm	Yb	Lu	Y	∑REY
1/7293	0.008	0.013	0.001	0.006	0.002	<DL	<DL	<DL	0.010	<DL	<DL	<DL	0.007	<DL	0.004	0.051
1/7294	0.010	0.019	0.002	0.008	0.002	<DL	<DL	<DL	0.010	<DL	0.001	<DL	0.007	<DL	0.005	0.064
avg	0.021	0.037	0.004	0.017	0.004	0.001	0.002	0.001	0.012	0.001	0.002	0.001	0.008	0.001	0.016	0.122
min	0.003	0.006	0.001	0.002	0.001	<DL	<DL	<DL	0.008	<DL	<DL	<DL	0.006	<DL	0.002	0.031
max	0.093	0.137	0.017	0.063	0.011	0.002	0.002	0.002	0.02	0.003	0.006	0.001	0.014	0.001	0.082	0.452

DL—detection limits.

4. Discussion

4.1. Toxic Elements (Pb, Cd, As)

The non-essential metals Pb, Cd, and As show adverse effects on human health even at low concentrations. They readily accumulate in living organisms and interfere with metabolic processes and biological functions. In the investigated mushrooms, with a few exceptions, they were mostly present in low concentrations. A comparison of measured element concentrations with those from other studies in the region is shown in Table 4.

Table 4. Concentration of macro- and trace elements in different mushrooms from the region (expressed in mg/kg or * g/kg). If not otherwise specified, the results refer to the fruiting body.

	Croatia			Serbia			Italy	
Element	Edible (Caps and Stipes) [40–42]	*Agaricus* sp., *Trichaptum biforme* [12]	*M. procera* (Caps and Stipes) [34,35]	Edible [50]	Boletaceae (Edible) [51]	Different Fungal Species [38]	*Morchella* Group [37]
Al		623–925	29–5664	54.2–396	27.7–608		0.05–0.33
As		0.19–3.64	0.01–3.40		0.03–1.66	0.10–11.6	<DL–0.23
Ba		19.2–72.4	0.20–46	1.20–8.7			0.33–4.06
Be		0.02–0.41					<DL
Bi		0.04–0.08			0.02–1.40		<DL
Cd	0.60–3.23	0.02–0.18	0.04–43.5	0.27–2.93	0.68–2.94	0.16–101	0.18–21.5
Co		0.17–2.92	<DL–12.0		0.07–0.72		0.06–0.28
Cr	0.77–3.85	1.32–13.19	0.20–13.8	3.25–10.8	0.16–1.34		0.26–2.18
Cs		0.15–1.18					0.05–0.48
Cu	7.41–78.2	6.67–32.4	29–304	12.2–73.6	4.66–34.3		10.1–63
Fe	49.3–154	365–4905	30–4018	45.9–319	24.4–515		35–517
Li		0.49–7.57		0.21–1.27			0.03–0.67
Mn		18.9–248	7.6–367	9.0–35.5	2.91–23.8		7.5–46.3
Mo		0.16–0.43		1.14–2.3			0.2–0.45
Ni	2.02–4.10	1.02–7.16	0.09–11.7	2.24–5.04	0.40–1.69		0.6–2.33
Pb	0.48–1.91	1.29–5.73	<DL–14.3	6.32–9.8	0.29–10.6	0.58–10.6	<DL–1.09
Rb		2.28–12.6			23.74–500		2.55–36.6
Sb		0.05–0.28		4.9–26.1			0.01–0.05
Sc							
Se			0.17–3.3		0.04–2.32	0.2–94.4	<DL–0.49
Sn		0.14–0.61					<DL
Sr		7.98–21.7	0.08–29.9	1.15–4.34	0.50–4.71		0.8–5.8
Ti		59.02–647	1.2–156				6.31–18.8
Tl		0.01–0.17					0.002–0.13
You		0.04–0.46					0.001–0.02
V		1.34–15.7					0.06–1.26
Zn	41.99–90.56	21.3–59.1	27.3–535	38.7–117	17.6–301.6		99.7–259
Ca *			0.02–4.43	0.55–2.12	0.11–2.33		0.28–3.51
K *			11–120	11.3–41.8	10.1–21.2		25.6–> 43
Mg *			0.69–3.4	0.29–1.12			0.79–2.64
Na *			11–1900	0.41–0.93	39.8–916		0.1–0.83

The content of lead in the studied mushrooms ranged from 0.072 mg kg^{-1} in *P. dryinus* (1/7256) to 1.11 mg kg^{-1} in *L. excipuliforme* (1/7259) (Table 2). The exception was *C. cornucopioides* (1/7292) with the determined concentration of 4.02 mg kg^{-1}. These values are within the usual range of concentrations of Pb in wild mushrooms from un-contaminated sites (<5 mg kg^{-1}; [52]). Pb concentrations in certain species (*A. cylindracea*, *M. procera*, *L. deterrimus*, *C. comatus*, *L. nuda*, *L. leucothites*, and *L. perlatum*) were sometimes significantly lower compared to available data from other studies in the region [35,40–42] (Table 4) and in other areas [10,19,40,53–55].

Uptake from contaminated soil, where it is mostly deposited by atmospheric processes, is considered the main source of Pb in wild mushrooms [19]. Although its use in gasoline has been banned, traffic is still the main source of Pb in urban soils due to its accumulated amounts [15], and recent studies have demonstrated a direct influence of traffic on Pb concentrations [16,19].

The narrow range of Pb concentrations in this study suggests a limited influence of traffic on Pb content in the studied mushrooms. This is also supported by the low content in the species *C. comatus* (1/7274), *L. perlatum* (1/7260, 1/7267), and *L. nuda* (1/7291), which readily accumulate Pb and are considered good indicators of traffic pollution [9,52]. However, it should be noted that mushrooms were mostly sampled from larger green areas within the city, while in studies where traffic pollution caused elevated Pb levels in mushrooms, samples were collected closer to roads (up to 50 m), which may have contributed to pollution levels [9,13,16,56].

Schlecht and Säumel [19] found that although traffic pollution undoubtedly affected Pb levels in mushrooms, high Pb concentrations were also determined in mushrooms from areas where low urban pollution is expected, while low concentrations were found in mushrooms from potentially polluted areas [17]. Although the Pb concentrations determined in this study do not exceed the usual level in mushrooms, it can be observed that the highest concentration was determined in *C. cornucopioides* (1/7292) sampled in a forest far from any significant source of pollution, such as the impact of the industry or traffic (Table 2), implying that Pb uptake in this case is mainly determined by the geochemistry of the underlying soil. This is supported by the fact that this species also had elevated or maximal concentrations of lithogenic elements (Al, Be, Cr, Cs, Li, Ni, Rb, REY, Ti, and V; Table 2). However, genetic factors appeared to predominate in Pb uptake by *L. deterrimus*, as similar Pb concentrations were obtained in the two samples from distant sites (1/7284 and 1/7288).

Cadmium concentrations in the investigated mushrooms varied from 0.116 mg kg^{-1} in *H. radicata* (1/7275) to 0.681 mg kg^{-1} in *P. dryinus* (1/7256), with the exception of *L. leucothites* (1/7293) and *A. cylindracea* (1/7294), for which much higher values (4.55 mg kg^{-1} and 3.96 mg kg^{-1}, respectively) were obtained (Table 2). The values obtained are in agreement with previous studies from the region [40–42,50,51] (Table 4) and are within the accepted range of values for wild mushrooms (0.5–5 mg kg^{-1}; [52]). The values obtained for certain species (*C. comatus*, *L. nuda*, *L. perlatum*, *M. procera*) are similar to those previously determined [40,53]. However, higher [33,54,57,58], as well as lower [27,53,55] concentrations have also been reported.

According to Melgar et al. [58], bioaccumulation of Cd by mushrooms is species-dependent, and the age of the mycelium and the period between fructifications are suggested as important factors in determining the metal content. However, higher Cd levels in plants, like Pb, are mostly associated with urban contamination [59]. Schlecht and Säumel [19] found that traffic pollution affected Cd levels only in some mushroom species, highlighting site-specific characteristics as a determining factor for uptake. This is supported by the results of our study, as the two mushroom species (*L. leucothites* (1/7293) and *A. cylindracea* (1/7294)) with higher Cd levels (Table 2) were sampled at the same site, near a road in the capital city of Zagreb (Table 1), and thus were exposed to the influence of nearby traffic. Considering that these two species have different lifestyles, that is, grow on different substrates (*L. leucothites* in soil and *A. cylindracea* on wood), but both have significantly elevated Cd levels, it seems that the sampling location, such as soil properties and/or urban pollution, had a predominant influence on the uptake of Cd by these mushrooms. A positive correlation between Cd and typical lithophilic elements, such as Cs and Rb (r = 0.53 and 0.51, respectively, $p < 0.05$; Table S1), confirms that soil geochemistry is an important factor in Cd uptake.

Arsenic concentrations were below 0.7 mg kg^{-1} (Table 2), with the exception of *P. multipedata* (1/7263), in which 3.68 mg kg^{-1} was determined. Other mushrooms from the same sampling area (Jelenovac; 1/7258, 1/7259, 1/7260, 1/7262) or the same family

(*Psathyrellaceae*; 1/7268 and 1/7273) had significantly lower values, indicating species-specific affinities as the main source of the elevated As content. However, *P. multipedata* was not recognized as an As accumulator in previous studies, and this maximum is outside the accepted range for As concentration for mushrooms from uncontaminated areas (<1 mg kg^{-1}, [52]).

Vetter [60] found that the level of As in a given species is the result of several factors, both environmental and genetic, with some species having an accumulative affinity for As, regardless of their habitat. In this study, lower As levels were found in mushrooms growing on wood (1/7256, 1/7268, 1/7273, 1/7275, 1/7294), supporting the work of Falandysz and Rizal [61], who propose geogenic sources as the most important factor for As uptake.

The established maximum level for Cd and Pb in edible mushrooms is regulated by EU regulations [62,63] and is 0.3 mg kg^{-1} for Pb and 1 mg kg^{-1} for Cd, wet weight (3 and 10 mg kg^{-1} dry weight at 90% moisture content). The maximum amount of As in fresh mushrooms is regulated by Croatian legislation to 0.3 mg kg^{-1} (3 mg kg^{-1} dry weight) [64]. According to these regulations, *P. multipedata* (1/7263) and *C. cornucopioides* (1/7292) collected in the forest area far from direct anthropogenic influences are not considered safe for consumption due to elevated As and Pb content, respectively. Cd levels were well below the maximum allowable limit in dry products. These results support the conclusions of Fu et al. [32] who advise caution in the consumption of wild mushrooms due to possible accumulation of these potentially toxic elements.

4.2. Macronutrients (Ca, Mg, K, Na)

In general, the highest concentrations of macroelements were determined in mushrooms belonging to the *Psathyrellaceae* family, while *C. comatus* (1/7274) was rich in Ca and Na.

The determined concentrations of Ca, Mg, and K were in the usual range for wild mushrooms (100–500 mg kg^{-1}, 800–1800 mg kg^{-1}, and 20–40 g kg^{-1}, respectively; [50]), although the values for individual mushroom species (*L. perlatum*, *M. procera*, *C. cornucopioides*, *L. nuda*, *A. cylindracea*, *C. comatus*) did not always agree with those determined by other researchers [23,52,53,56].

Concentrations of Ca ranged from 128 to 545 mg kg^{-1} (Table 2), except in *P. dryinus* (1/7256), where an exceptionally low concentration (26 mg kg^{-1}) was determined. The concentrations of Mg and K varied from 0.939 to 1.88 g kg^{-1} and 17.1 to 49.5 g kg^{-1}, respectively, except for mushrooms from the family *Psathyrellaceae* (*P. piluliformis* (1/7268, 1/7273) and *P. multipedata* (1/7263)), where slightly higher values of Mg (around 2 g kg^{-1}) and significantly higher values of K (65.3–80.8 g kg^{-1}) were determined. The determined Na concentrations were outside the range suggested by Kalač [52] (100–400 mg kg^{-1}); in the majority of samples they were below 100 mg kg^{-1}, while in mushrooms from the family *Psathyrellaceae* (*P. piluliformis* (1/7268, 1/7273) and *P. multipedata* (1/7263)) and *L. leucothites* (1/7293) and *C. comatus* (1/7274) from the family *Agaricaeae* were in the range of 0.653 to 1.01 g kg^{-1} (Table 2).

While the uptake of K was correlated with that of Mg and Na (r = 0.6 and 0.7, respectively, $p < 0.05$; Table S1), taking into account all studied mushrooms, the uptake of Na was significantly increased in species of the family *Psathyrellaceae* from different sites, but was also site-specific, as samples 1/7273, 1/7274, and 1/7275 from Dotrščina Park had significantly increased values compared to other mushrooms. However, the fact that mushrooms of the same species collected from different sites (*L. perlatum* (1/7260 and 1/7267), *I. gibba* (1/7258 and 1/7272) and *P. piluliformis* (1/7268 and 1/7273)) had similar concentrations of macroelements supports the finding that macronutrient uptake is mostly species-dependent. *L. deterrimus* (1/7284 and 1/7288), a mycorrhizal mushroom, showed different concentrations of K and Na, which could indicate a possible importance of the individual host tree in the uptake of major elements.

4.3. Micronutrients (Cu, Fe, Mn, Mo, Se, Zn)

The obtained values for micronutrients were within the accepted range for wild mushrooms (20–100 mg kg^{-1} for Cu, 30–150 mg kg^{-1} for Fe, 10–60 mg kg^{-1} for Mn, <2 mg kg^{-1} for Se, 25–200 mg kg^{-1} for Zn; [52]). Only for Mo, several samples exceeded the values found in Kalač [52] (<0.6 mg kg^{-1}).

The concentrations of Cu and Fe ranged from 2.19 to 84.3 mg kg^{-1}, and 11.0 to 105 mg kg^{-1}, respectively (Table 2). The lowest values were found in *P. dryinus* (1/7256), and the highest in *L. perlatum* (1/7260), and no site-, species-, or lifestyle-related trends were observed. A comparison with data from related studies for specific mushrooms (*L. perlatum*, *C. comatus*, *M. procera*, *L. nuda*, *C. cornucopioides*, *L. leucothites*, *L. deterrimus*, *A. cylindracea*) revealed similar, but also divergent values [23,40,53–57,65].

The values for Mn ranged from 4.9 mg kg^{-1} to 38.2 mg kg^{-1}, with the lowest concentrations obtained in *P. dryinus* (1/7256), *L. deterrimus* (1/7288), and *A. cylindracea* (1/7294), and the highest in *L. perlatum* (1/7267), *I. gibba* (1/7272), and *L. nuda* (1/7291). In general, the values obtained for certain species (*L. perlatum*, *M. procera*, *C. comatus*, *L. nuda*, *C. cornucopioides*) were lower than those reported in the literature [23,28,33,57,65], although higher values were found for *L. nuda* [29] and similar values for *A. cylindracea* and *C. comatus* [53], *L. leucothites* [53], and *M. procera* [23,30].

A positive correlation between Mn and Fe (r = 0.62, *p* < 0.05) and Ni and Fe (r = 0.53, *p* < 0.05; Table S1) in the investigated mushrooms suggests their interdependent uptake, as previously reported by Kokkoris et al. [18]. The co-occurrence of Mn and Fe results from the Fe-Mn oxides and oxyhydroxides associated with clay minerals in the soil. Moreover, a strong correlation was observed between V and Fe (r = 0.66, *p* < 0.05; Table S1) due to its association with Fe oxides [3]. This agrees with previously reported findings that Fe, Mn, and partially Ni in topsoils from the Zagreb area are mostly of geological/pedological origin [66].

The determined concentrations of Mo ranged from 0.007 mg kg^{-1} to 1.15 mg kg^{-1}. The lowest Mo content was determined in *L. deterrimus* (1/7284, 1/7288), *P. dryinus* (1/7256), and *P. piluliformis* (1/7268, 1/7273), while *P. flaccida* (1/7262) and *L. nuda* (1/7291) exceeded the usual content for mushrooms from uncontaminated areas [52]. In general, it was observed that mushrooms growing on wood had lower Mo concentrations.

Selenium is a micronutrient normally present in mushrooms at 2 mg kg^{-1} [52]. This is in agreement with the results obtained; the concentrations obtained were mostly <1 mg kg^{-1}, except for the two samples with higher values; 1.59 mg kg^{-1} in *P. multipedata* (1/7263) and 1.49 mg kg^{-1} in *L. leucothites* (1/7293). The important role of Se in biological functions is supported by its positive correlation with macronutrients K and Mg (r = 0.62 and 0.5, respectively; *p* < 0.05; Table S1).

Concentrations of Zn ranged from 17.3 mg kg^{-1} in *P. dryinus* (1/7256) to 143 mg kg^{-1} in *L. excipuliforme* (1/7259). *L. perlatum* (1/7260 and 1/7267), as an accumulator of Zn [52], showed higher values; however, they were not the highest recorded. Species from the *Psathyrellaceae* family (*P. multipedata* (1/7263) and *P. piluliformis* (1/7268 and 1/7273)) and Russulaceae (*L. deterrimus*, 1/7284 and 1/7288), which were from different sites, showed similar Zn values, suggesting that species-specific affinities rather than soil geochemistry are important in Zn uptake. Considering specific mushroom species, similar values were previously reported for *C. comatus* [27,30], *M. procera* [23,30,40], *L. leucothites* [67], *L. deterrimus* [40], and *A. cylindracea* [53], but different concentrations can also be found [33,53,55,57,65].

4.4. Lithogenic Group of Elements

The group of lithogenic elements (Al, Ba, Li, Ti, Cs, and Rb) was positively correlated with each other (r = 0.50–0.87, *p* < 0.05; Table S1), indicating a common mechanism of their uptake. Their concentrations were in a narrow range of values, considering the differences between localities. Positive correlations of REY, Sr, and V with Al (Table S1) indicate their common origin from the soil.

Concentrations of Al varied from 7.83 mg kg^{-1} in *P. dryinus* (1/7256) to 99.3 mg kg^{-1} in *P. piluliformis* (1/7268). Similar values were reported for *C. comatus* [53] and *L. nuda* [28], lower for *A. cylindracea* [53] and *C. cornucopioides* [28], and higher for *L. perlatum* [28]. According to the data for a number of wild species [68], the Ba content in mushrooms is mostly below 1.6 mg kg^{-1}. This is in agreement with the present results; only *P. piluliformis* (1/7268) slightly exceeded this range with a value of 1.73 mg kg^{-1}. Similar concentrations of Ba were previously reported for *C. comatus*, and higher for *A. cylindracea* [53]. Lithium is not a commonly occurring element in mushrooms, and the results obtained are consistent with the usual reported range, such as below 0.19 mg kg^{-1} [69]. The content of Ti was up to 9.7 mg kg^{-1}, in agreement with the usual content in wild mushrooms (<10 mg kg^{-1}, [52]), although Niedzielski et al. [53] found lower content in *A. cylindracea* and *C. comatus*. The Cs content was mostly below 0.6 mg kg^{-1}, except in *C. cornucopioides* (1/7292), where the determined value was 1.96 mg kg^{-1}, confirming an increased input of lithogenic elements, as previously suggested. A strong positive correlation between Cs and Rb (r = 0.85, $p < 0.05$; Table S1) results from their similar geochemical behaviour [70]. According to data collected by Kalač [52], Rb is present in mushrooms in tens to hundreds of mg kg^{-1}; in this study, the content of Rb varied from 2.59 mg kg^{-1} in *L. perlatum* (1/7260) to 299 mg kg^{-1} in *C. cornucopioides* (1/7292).

The highest concentrations of Cr and Ni were determined to be 0.761 mg kg^{-1} and 1.595 mg kg^{-1}, respectively, in *C. cornucopioides* (1/7292) (Table 2). The concentrations of Ni and Cr were generally lower compared to other studies [33,41,42,55], while similar levels of Ni were determined in *A. cylindracea* [53]. The results showed equal Ni content in the two samples of *L. deterrimus* sampled at widely separated sites (1/7284 and 1/7288), while similar values were also observed in mushrooms belonging to the *Tricholomataceae* family (1/7258, 1/7262, 1/7272, 1/7291) (Table 2); this may indicate that the accumulation of Ni is species-specific. However, the positive correlation of Ni with Fe and V (r = 0.53, $p < 0.05$; Table S1) suggests its association with Fe oxides and oxyhydroxides from soil. In contrast, the lack of a significant correlation between Cr and other elements suggests that there are multiple factors controlling its bioaccumulation.

4.5. Rare Earth Elements

The REY concentrations obtained (Table 3) are in accordance with the values reported for saprobic macrofungi [71] and aboveground species [72], but are significantly higher compared to the values reported for *Suillus luteus* [70] and lower compared to the values reported by the ref. [73] for caps of *M. procera* (Table 5). Moreover, the mushrooms from northwestern Croatia investigated in this study have much lower REY values compared to the mushrooms of the genus *Agaricus* from the Prašnik area (eastern part of Croatia; [44]), although all of them were grown on a pedological substrate enriched with REEs [43]. The reason for the observed discrepancy is probably the fact that mushrooms of the genus *Agaricus* are known to easily accumulate metals [11].

Table 5. The comparison of the determined REY concentration with the literature values for mushrooms from different regions (all expressed in mg/kg). If not otherwise specified, the results refer to the fruiting body.

	Croatia	Poland			Czech Republic	Italy
Element	*Agaricus* sp. (Caps and Stipes) [44]	*Agaricus* sp., *Trichaptum biforme* [12]	*M. procera* (Caps and Stipes) [34,35]	**Edible [48]**	*Boletaceae* (Edible) [49]	**Different Fungal Species [38]**
La	1.373	0.083 ± 0.049	0.06	<0.01–0.08	0.023	0.015–0.488
Ce	4.137	0.18 ± 0.091	0.12		0.042	0.025–0.843
Pr	0.526	0.017 ± 0.009	0.04	0.02–1.8	0.0056	<DL–0.109
Nd	2.137	0.058 ± 0.003	0.17	0.11–0.45	0.020	0.011–0.446
Sm	0.467	0.012 ± 0.006	0.03		0.0041	0.0021–0.0822
Eu	0.099	0.0027 ± 0.018	<0.01		0.00068	<DL–0.019
Gd	0.305	0.011 ± 0.006	0.05	<0.01–0.05	0.0023	0.002–0.081
Tb	0.057	0.0018 ± 0.0009	0.03		0.00059	<DL–0.012

Table 5. *Cont.*

| | Croatia | | Poland | | Czech Republic | Italy |
Element	*Agaricus* sp. (Caps and Stipes) [44]	*Agaricus* sp., *Trichaptum biforme* [12]	*M. procera* (Caps and Stipes) [34,35]	Edible [48]	*Boletaceae* (Edible) [49]	Different Fungal Species [38]
Dy	0.285	0.010 ± 0.005	<0.01		0.0022	0.001–0.0724
Ho	0.062	0.0023 ± 0.0012	0.04	<0.01–0.30	0.00042	0.0003–0.015
Er	0.163	0.0070 ± 0.0035	0.72		0.0013	0.0005–0.0429
Tm	0.027	0.0011 ± 0.0006	0.03		0.00017	0.0001–0.006
Yb	0.150	0.0073 ± 0.0037	0.02		0.0013	<DL–0.041
Lu	0.027	0.0011 ± 0.0005	0.02		0.00013	<DL–0.0058
Y	0.693	0.074 ± 0.039	0.04			0.009–0.549
∑REY	9.797	0.481	0.23		0.104	

In general, the results indicate a large variability between the different species studied in terms of their ability for REY uptake and suggest that both the soil substrate and the mushroom species influence the accumulation of this group of elements in the mushroom tissue.

4.6. Factors Influencing the Element Uptake

In order to identify the role of different factors affecting the intake of elements, such as species, lifestyle, dietary habits, and geographical location, the relationship between the obtained element concentrations was investigated using principal component analysis (PCA). The eigenvalues of the first two principal components (PCs) were greater than 1, indicating their significance. The first two PCs explained 42.2% of the total variability among 32 variables; the first component (PC1) contributed 26.3%, while the second corresponded to 15.9% of the total variance of the data set. PCA results are presented on the PCA loading plot (Figure 2a) and PCA score plot (Figure 2b), which illustrate the alignment of variables (elements) and samples (mushrooms) with respect to principal components.

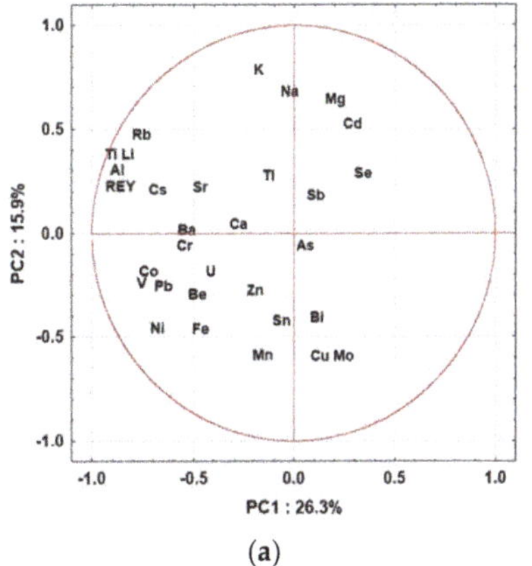

(a)

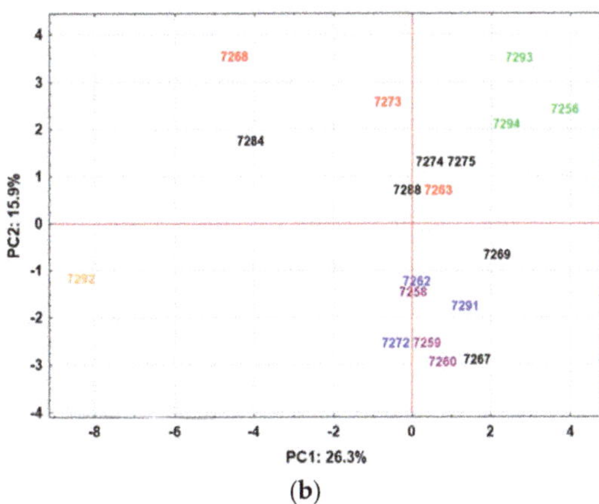

(b)

Figure 2. Principal component analysis: PC1–PC2 loading (**a**) and score plot (**b**).

Based on these parameters, the following can be observed:

(i) Mushrooms belonging to the *Psathyrellaceae* family (1/7263, 1/7268, and 1/7273; colored red in Figure 2b) showed high variability in element distribution, suggesting that soil geochemistry/properties are the main factor determining the uptake of elements. However, *P. piluliformis* (1/7268 and 1/7273) collected from widely separated sites showed similarities in terms of high uptake of macronutrients (K, Na, Mg) and low uptake of Mo and Cu, indicating that the uptake of these elements is species-specific.

(ii) *L. leucothites* (1/7293) and *A. cylindracea* (1/7294) sampled from the same site in the city of Zagreb, and *P. dryinus* (1/7256) (all colored green in Figure 2b) from another site within the urban area were characterized by high Cd concentrations; these mushrooms were collected near a road and, therefore, the elevated Cd levels could be explained by traffic pollution. However, elevated levels of other elements derived from traffic pollution, such as Pb, were not observed.

(iii) Mushrooms from the *Tricholomataceae* family (1/7262, 1/7272, and 1/7291; colored blue) were grouped independently of sampling sites, suggesting that species-specific affinities in element uptake predominate. These mushrooms were described by slightly higher levels of Cu, Mn, and Mo and lower levels of Cd and macronutrients.

(iv) Significant similarity in the content of elements was observed in different mushrooms sampled within the green area at the Jelenovac site: *L. excipuliforme* (1/7259), *I. gibba* (1/7258), *L. perlatum* (1/7260) (all colored purple in Figure 2b) and *P. flaccida* (1/7262) (colored blue in Figure 2b). This strongly suggests that geochemistry and soil properties are the main factors controlling elemental uptake, as these samples are characterized by higher levels of Cu, Mn, and Mo and lower levels of K, Mg, and Na.

(v) PCA analysis confirmed the unique elemental composition of *C. cornucopioides* (1/7292; colored orange in Figure 2b), which is distinguished from all other samples by its significantly elevated values of Co, Ni, Pb, and REY.

The results presented showed that mushrooms belonging to the same species or family collected in different areas had considerably different elemental contents, suggesting different soil geochemical composition and/or anthropogenic influences as the main factors determining elemental content. Since the main distribution of trace elements in the studied areas did not indicate an anthropogenic influence, it can be assumed that soil geochemical properties are the most important factor determining these variations in chemical composition. The influence of different soil properties and geochemistry on the uptake of elements by mushrooms may be more evident in studies comparing soils with markedly different geological composition, as in the study by Nikkarinen and Mertanen [74] and Alaimo et al. [37], than when comparing mushrooms growing in areas where differences in soil geochemistry are less pronounced. However, several elements, such as As, Ni, Zn, and macronutrients, showed similar behavior in certain mushroom species or families, indicating an important influence of species-specific affinity for certain elements. Finally, both factors—soil geochemistry and species characteristics—have an important influence on the uptake of elements, as also shown by other studies [36,38,39,52], although the prevalence of each factor varies from species to species and depends on the element in question.

4.7. Health Risk Assessment

The data (THQ and HI values) obtained in the assessment of non-carcinogenic health risk due to metal ingestion from the consumption of the wild mushrooms studied are presented in Table 6. Toxicity was evaluated based on the calculated HI values. If the HI value for a mixture is less than 1, it is considered unlikely that assessed exposure from mushroom consumption will result in significant toxicity. On the other hand, there is great concern about potential toxicity when the HI value is greater than 1 [48].

Table 6. Non-carcinogenic risk data (expressed by health risk indexes THQ and HI) obtained in the health risk assessment of metals through the consumption of the edible wild mushrooms from Croatia. The RfD data are expressed as $\mu g\ kg^{-1}\ day^{-1}$.

	Target Hazard Quotient (THQ)													**Hazard Index (HI)**
	As	Ba	Be	Cd	Cu	Fe	Mn	Ni	Pb	Sb	Sr	U	Zn	
1/7256	0.026	0.000	0.000	0.280	0.023	0.006	0.016	0.007	0.008	0.002	0.000	0.000	0.024	0.392
1/7258	0.436	0.003	0.000	0.156	0.479	0.056	0.063	0.011	0.016	0.015	0.000	0.000	0.102	1.337
1/7259	0.540	0.001	0.000	0.146	0.397	0.031	0.040	0.013	0.130	0.006	0.000	0.001	0.196	1.501
1/7260	0.742	0.001	0.000	0.129	0.866	0.060	0.081	0.015	0.060	0.006	0.000	0.001	0.149	2.111
1/7262	0.660	0.003	0.000	0.059	0.296	0.029	0.092	0.007	0.027	0.013	0.001	0.001	0.137	1.325
1/7263	5.047	0.002	0.000	0.150	0.801	0.029	0.034	0.011	0.041	0.070	0.001	0.001	0.134	6.321
1/7267	0.752	0.001	0.000	0.111	0.690	0.047	0.112	0.010	0.102	0.004	0.000	0.000	0.174	2.004

Table 6. *Cont.*

	Target Hazard Quotient (THQ)													Hazard Index (HI)
	As	Ba	Be	Cd	Cu	Fe	Mn	Ni	Pb	Sb	Sr	U	Zn	
1/7268	0.119	0.004	0.000	0.163	0.116	0.052	0.067	0.007	0.031	0.012	0.001	0.000	0.122	0.694
1/7269	0.300	0.001	0.000	0.158	0.582	0.036	0.034	0.007	0.051	0.005	0.000	0.000	0.100	1.273
1/7272	0.905	0.002	0.000	0.089	0.538	0.045	0.107	0.014	0.015	0.007	0.000	0.000	0.092	1.815
1/7273	0.207	0.001	0.000	0.201	0.205	0.040	0.075	0.005	0.048	0.003	0.000	0.000	0.096	0.882
1/7274	0.670	0.001	0.000	0.200	0.435	0.033	0.043	0.004	0.019	0.011	0.001	0.000	0.080	1.496
1/7275	0.011	0.002	0.000	0.048	0.097	0.019	0.063	0.004	0.014	0.004	0.000	0.000	0.063	0.324
1/7284	0.875	0.003	0.000	0.244	0.170	0.037	0.024	0.010	0.021	0.002	0.001	0.001	0.182	1.570
1/7288	0.356	0.001	0.000	0.226	0.091	0.017	0.014	0.010	0.022	0.003	0.000	0.000	0.141	0.883
1/7291	0.812	0.001	0.000	0.073	0.421	0.028	0.110	0.009	0.063	0.004	0.000	0.000	0.080	1.602
1/7292	0.016	0.002	0.000	0.226	0.338	0.045	0.074	0.033	0.472	0.004	0.000	0.001	0.088	1.299
1/7293	0.425	0.001	0.000	1.869	0.269	0.014	0.021	0.003	0.035	0.036	0.000	0.000	0.103	2.776
1/7294	0.000	0.000	0.000	1.626	0.460	0.013	0.017	0.005	0.027	0.003	0.000	0.000	0.125	2.278
RfD	0.3 [a]	200 [a]	2 [a]	1 [a]	40 [b]	700 [b]	140 [a]	20 [a]	3.5 [b]	0.4 [a]	600 [a]	3 [a]	300 [a]	

[a]—IRIS assessments (USEPA) [49]. [b]—from Dowlati et al. [48].

The risk assessment revealed that 14 of the 19 mushrooms studied, from both urban and forest areas, had HI levels above 1; therefore, consumers of these mushrooms are exposed to a significant non-carcinogenic health risk due to the high metal content. As, Cd, and Cu, which have the highest THQ values, account for the largest proportion of total toxicity.

The mushrooms that do not pose a risk to human health are those growing on wood—1/7256, 1/7268, 1/7273, and 1/7275. The only mushroom growing in soil that is considered safe for consumption is *L. deterrimus* (1/7288), which was growing in grassland within the forest area.

Therefore, a diet that includes many wild mushrooms can be very risky. Dowlati et al. [48] conducted a comprehensive review, meta-analysis, and evaluation of non-carcinogenic health risk to humans based on the collected data on the concentrations of toxic metals in edible mushrooms from different countries of the world. The health risk assessment revealed that consumers from 8 out of 19 countries studied were exposed to a significant non-carcinogenic health risk.

5. Conclusions

In the studied mushrooms, the most abundant elements were essential nutrients, whose uptake seems to be more under a genetic influence and less influenced by site-specific characteristics.

The content of toxic elements was low, and no influence of traffic pollution could be detected, as similar concentrations of Pb were found in mushrooms from forest and urban sites. While the concentrations of metals did not show variations according to substrate (wood and soil), it was observed that mushrooms growing on wood generally had lower levels of As and Mo.

Nonetheless, the health risk assessment revealed a high non-carcinogenic risk of metals in 14 out of 19 mushrooms studied from both urban and forest areas, suggesting that these mushrooms should be consumed with caution.

Despite a limited data set, the results suggest that both the geochemical composition of substrate and the species-specific affinities influence the uptake of elements, with the predominance of one of these factors depending on both the mushroom species and the element of interest. The diversity of factors affecting the chemical composition of edible mushrooms should certainly be considered not only in future studies, but also in the evaluation of edible mushrooms as a source of certain metals in the diet.

Supplementary Materials: The following are available online at https://www.mdpi.com/article/10.3390/jof7121068/s1, Table S1: title Spearman correlation coefficients of the investigated elements in mushrooms. Correlation coefficients marked in red are significant at $p < 0.05$.

Author Contributions: Conceptualization, M.I. and Ž.F.; sampling: A.M. and Z.T.; methodology, M.F.T.; data curation: M.I., M.F.T. and Ž.F.; writing—original draft preparation, M.I. and Ž.F.; writing—review and editing, M.I., M.F.T. and Ž.F. All authors have read and agreed to the published version of the manuscript.

Funding: This study has been partially supported by Croatian Science Foundation under the project ForFungiDNA (IP-2018-01-1736).

Institutional Review Board Statement: Not applicable.

Informed Consent Statement: Not applicable.

Data Availability Statement: Not applicable.

Acknowledgments: A.M. and Z.T. are grateful to Dunja Šamec and Vedran Bahun for freeze-drying of samples for this study and to Margita Jadan for generating DNA barcode sequences (ITS rDNA) used for molecular identification of mushrooms.

Conflicts of Interest: The authors declare no conflict of interest.

References

1. Katz, A. Alternative Medicine for Prostate Cancer: Diet Vitamins Minerals and Supplements. In *Early Diagnosis and Treatment of Cancer Series: Prostate Cancer*; Su, L.-M., Ed.; Elsevier: Amsterdam, The Netherlands, 2010.
2. Marchand, L.R.A.; Stewart, J. Breast Cancer. In *Integrative Medicine*; Rakel, D., Ed.; Lancet: London, UK, 2018.
3. Kabata-Pendias, A. *Trace Elements in Soils and Plants*; CRC Press: Boca Raton, FL, USA, 2010.
4. Falandysz, J.; Kawano, M.; Swieczkowski, A.; Brzostowski, A.; Dadej, M. Total mercury in wild-grown higher mushrooms and underlying soil from Wdzydze Landscape Park Northern Poland. *Food Chem.* **2003**, *81*, 21–26. [CrossRef]
5. Kalač, P. A review of edible mushroom radioactivity. *Food Chem.* **2001**, *75*, 29–35. [CrossRef]
6. Svoboda, L.; Zimmermannova, K.; Kalač, P. Concentrations of mercury cadmium lead and copper in fruiting bodies of edible mushrooms in an emission area of a copper smelter and a mercury smelter. *Sci. Total Environ.* **2000**, *246*, 61–67. [CrossRef]
7. Falandysz, J.; Borovička, J. Macro and trace mineral constituents and radionuclides in mushrooms: Health benefits and risks. *Appl. Microbiol. Biotechnol.* **2012**, *97*, 477–501. [CrossRef]
8. Das, N. Heavy metal biosorption by mushrooms. *Nat. Prod. Radiance* **2005**, *4*, 454–459.
9. García, M.A.; Alonso, J.; Fernandez, M.I.; Melgar, M.J. Lead content in edible wild mushrooms in Northwest Spain as indicator of environmental contamination. *Arch. Environ. Contam. Toxicol.* **1998**, *34*, 330–335. [CrossRef]
10. García, M.A.; Alonso, J.; Melgar, M.J. Lead in edible mushrooms; Levels and bioaccumulation factors. *J. Hazard. Mater.* **2009**, *167*, 777–783. [CrossRef] [PubMed]
11. Kalač, P.; Svoboda, L. A review of trace element concentrations in edible mushrooms. *Food Chem.* **2000**, *69*, 273–281. [CrossRef]
12. Ivanić, M.; Fiket, Ž.; Medunić, G.; Furdek Turk, M.; Marović, G.; Senčar, J.; Kniewald, G. Multi-element composition of soil mosses and mushrooms and assessment of natural and artificial radioactivity of a pristine temperate rainforest system (Slavonia, Croatia). *Chemosphere* **2019**, *215*, 668–677. [CrossRef]
13. Mleczek, M.; Niedzielski, P.; Kalač, P.; Budka, A.; Siwulski, M.; Gąsecka, M.; Rzymski, P.; Magdziak, Z.; Sobieralski, K. Multielemental analysis of 20 mushroom species growing near a heavily trafficked road in Poland. *Environ. Sci. Pollut. Res.* **2016**, *23*, 16280–16295. [CrossRef] [PubMed]
14. Sager, M. Urban Soils and Road Dust—Civilization Effects and Metal Pollution—A Review. *Environments* **2020**, *7*, 98. [CrossRef]
15. Rodríguez-Seijo, A.; Arenas-Lago, D.; Andrade, M.L.; Vega, F.A. Identifying sources of Pb pollution in urban soils by means of MC-ICP-MS and TOF-SIMS. *Environ. Sci. Pollut. Res.* **2015**, *22*, 7859–7872. [CrossRef]
16. Carvalho, M.L.; Pimentel, A.C.; Fernandes, B. Study of heavy metals in wild edible mushrooms under different pollution conditions by X-ray fluorescence spectrometry. *Anal. Sci.* **2005**, *21*, 747–750. [CrossRef]
17. Petkovšek, S.A.S.; Pokorny, B. Lead and cadmium in mushrooms from the vicinity of two large emission sources in Slovenia. *Sci. Total Environ.* **2013**, *443*, 944–954. [CrossRef] [PubMed]
18. Kokkoris, V.; Massas, I.; Polemis, E.; Koutrotsios, G.; Zervakis, G.I. Accumulation of heavy metals by wild edible mushrooms with respect to soil substrates in the Athens metropolitan area (Greece). *Sci. Total Environ.* **2019**, *685*, 280–296. [CrossRef] [PubMed]
19. Schlecht, M.T.; Säumel, I. Wild growing mushrooms for the Edible City? Cadmium and lead content in edible mushrooms harvested within the urban agglomeration of Berlin. *Ger. Environ. Pollut.* **2015**, *204*, 298–305. [CrossRef] [PubMed]
20. Świsłowski, P.; Dołhańczuk-Śródka, A.; Rajfur, M. Bibliometric analysis of European publications between 2001 and 2016 on concentrations of selected elements in mushrooms. *Environ. Sci. Pollut. Res.* **2020**, *27*, 22235–22250. [CrossRef] [PubMed]
21. Michelot, D.; Siobud, E.; Dore, J.C.; Viel, C.; Poirier, F. Update of metal content profiles in mushrooms – toxicological implications and tentative approach to the mechanisms of bioaccumulation. *Toxicon* **1998**, *36*, 1997–2012. [CrossRef]
22. Svoboda, L.; Kalač, P.; Špička, J.; Janoušková, D. Leaching of cadmium lead and mercury from fresh and differently preserved edible mushroom *Xerocomus badius* during soaking and boiling. *Food Chem.* **2002**, *79*, 41–45. [CrossRef]

23. Falandysz, J.; Szymczyk, K.; Ichihashi, H.; Bielawski, L.; Gucia, M.; Frankowska, A.; Yamasaki, S.I. ICP/MS and ICP/AES elemental analysis (38 elements) of edible wild mushrooms growing in Poland. *Food Addit. Contam.* **2001**, *18*, 503–513. [CrossRef]
24. Malinowska, E.; Szefer, P.; Falandaysz, J. Metals bioaccumulation by bay bolete *Xerocomus badius* from selected sites in Poland. *Food Chem.* **2004**, *84*, 405–416. [CrossRef]
25. Nowakowski, P.; Markiewicz-Żukowska, R.; Soroczyńska, J.; Puścion-Jakubik, A.; Mielcarek, K.; Borawska, M.H.; Socha, K. Evaluation of toxic element content and health risk assessment of edible wild mushrooms. *J. Food Compos. Anal.* **2021**, *96*, 103698. [CrossRef]
26. Alonso, J.; García, M.; Pérez-López, M.; Melgar, M.J. The concentrations and bioconcentration factors of copper and zinc in edible mushrooms. *Arch. Environ. Contam. Toxicol.* **2003**, *44*, 180–188. [CrossRef] [PubMed]
27. Mendil, D.; Uluözlü, Ö.D.; Tüzen, M.; Hasdemir, E.; Sarı, H. Trace metal levels in mushroom samples from Ordu Turkey. *Food Chem.* **2005**, *91*, 463–467. [CrossRef]
28. Sesli, E.; Tuzen, M.; Soylak, M. Evaluation of trace metal contents of some wild edible mushrooms from Black sea region Turkey. *J. Hazard. Mater.* **2008**, *160*, 462–467. [CrossRef] [PubMed]
29. Turkekul, I.; Elmastas, M.; Tüzen, M. Determination of iron copper manganese zinc lead and cadmium in mushroom samples from Tokat Turkey. *Food Chem.* **2004**, *84*, 389–392. [CrossRef]
30. Yamaç, M.; Yıldız, D.; Sarıkürkcü, C.; Çelikkollu, M.; Solak, M.H. Heavy metals in some edible mushrooms from the Central Anatolia Turkey. *Food Chem.* **2007**, *103*, 263–267. [CrossRef]
31. Aruguete, D.M.; Aldstadt, J.H.; Mueller, G.M. Accumulation of several heavy metals and lanthanides in mushrooms (*Agaricales*) from the Chicago region. *Sci. Total Environ.* **1998**, *224*, 43–56. [CrossRef]
32. Fu, Z.; Liu, G.; Wang, L. Assessment of potential human health risk of trace element in wild edible mushroom species collected from Yunnan Province China. *Environ. Sci. Pollut. Res. Int.* **2020**, *27*, 29218–29227. [CrossRef]
33. Zhu, F.; Qu, L.; Fan, W.; Qiao, M.; Hao, H.; Wang, X. Assessment of heavy metals in some wild edible mushrooms collected from Yunnan Province China. *Environ. Monit. Assess.* **2011**, *179*, 191–199. [CrossRef]
34. Stefanović, V.; Trifković, J.; Djurdjić, S.; Vukojević, V.; Tešić, Ž.; Mutić, J. Study of silver selenium and arsenic concentration in wild edible mushroom *Macrolepiota procera* health benefit and risk. *Environ. Sci. Pollut. Res.* **2016**, *23*, 22084–22098. [CrossRef] [PubMed]
35. Stefanović, V.; Trifković, J.; Mutić, J.; Tešić, Ž. Metal accumulation capacity of parasol mushroom (*Macrolepiota procera*) from Rasina region (Serbia). *Environ. Sci. Pollut. Res.* **2016**, *23*, 13178–13190. [CrossRef] [PubMed]
36. Alaimo, M.G.; Dongarrà, G.; La Rosa, A.; Tamburo, E.; Vasquez, G.; Varrica, D. Major and trace elements in *Boletus aereus* and *Clitopilus prunulus* growing on volcanic and sedimentary soils of Sicily (Italy). *Ecotoxicol. Environ. Saf.* **2018**, *157*, 182–190. [CrossRef] [PubMed]
37. Alaimo, M.G.; Saitta, A.; Ambrosio, E. Bedrock and soil geochemistry influence the content of chemical elements in wild edible mushrooms (*Morchella* group) from South Italy (Sicily). *Acta Mycol.* **2019**, *54*, 1122. [CrossRef]
38. Cocchi, L.; Vescovi, L.; Petrini, L.; Petrin, E. Heavy metals in edible mushrooms in Italy. *Food Chem.* **2006**, *98*, 277–284. [CrossRef]
39. Giannaccini, G.; Betti, L.; Palego, L.; Mascia, G.; Schmid, L.; Lanza, M.; Mela, A.; Fabbrini, L.; Biondi, L.; Lucacchini, A. The trace element content of top-soil and wild edible mushroom samples collected in Tuscany Italy. *Environ. Monit. Assess.* **2012**, *184*, 7579–7595. [CrossRef]
40. Širić, I.; Kos, I.; Kasap, A.; Petković, F.Z.; Držaić, V. Heavy metals bioaccumulation by edible saprophytic mushrooms. *J. Cent. Eur. Agric.* **2016**, *17*, 884–900.
41. Širić, I.; Humar, M.; Kasap, A.; Kos, I.; Mioč, B.; Pohleven, F. Heavy metal bioaccumulation by wild edible saprophytic and ectomycorrhizal mushrooms. *Environ. Sci. Pollut. Res.* **2016**, *23*, 18239–18252. [CrossRef]
42. Širić, I.; Kasap, A.; Bedeković, D.; Falandysz, J. Lead, cadmium and mercury contents and bioaccumulation potential of wild edible saprophytic and ectomycorrhizal mushrooms Croatia. *J. Environ. Sci. Health B.* **2017**, *52*, 156–165. [CrossRef]
43. Salminen, R.; Batista, M.J.; Bidovec, M.; Demetriades, A.; De Vivo, B.; De Vos, W.; Duris, M.; Gilucis, A.; Gregorauskiene, V.; Halamić, J.; et al. FOREGS Geochemical Atlas of Europe Part 1: Background Information Methodology and Maps. Geological Survey of Finland Espoo. 2005, p. 526. Available online: http://weppi.gtk.fi/publ/foregsatlas/ (accessed on 17 March 2021).
44. Fiket, Ž.; Medunić, G.; Furdek Turk, M.; Ivanić, M.; Kniewald, G. Influence of soil characteristics on rare earth fingerprints in mosses and mushrooms: Example of a pristine temperate rainforest (Slavonia Croatia). *Chemosphere* **2017**, *179*, 92–100. [CrossRef]
45. Mešić, A.; Šamec, D.; Jadan, M.; Bahun, V.; Tkalčec, Z. Integrated morphological with molecular identification and bioactive compounds of 23 Croatian wild mushrooms samples. *Food Biosci.* **2020**, *37*, 100720. [CrossRef]
46. Filipović Marijić, V.; Raspor, B. Site-specific gastrointestinal metal variability in relation to the gut content and fish age of indigenous european chub from the Sava River. *Water Air Soil Pollut.* **2012**, *223*, 4769–4783. [CrossRef]
47. Fiket, Ž.; Mikac, N.; Kniewald, G. Mass fractions of forty-six major and trace elements including rare earth elements in sediment and soil reference materials used in environmental studies. *Geostand. Geoanal. Res.* **2017**, *41*, 123–135. [CrossRef]
48. Dowlati, M.; Sobhi, H.R.; Esrafili, A.; FarzadKia, M.; Yeganeh, M. Heavy metals content in edible mushrooms: A systematic review, meta-analysis and health risk assessment. *Trends Food. Sci. Technol.* **2021**, *109*, 527–535. [CrossRef]
49. USEPA. *Reference Dose (RfD): Description and Use in Health Risk Assessment*; USEPA: Washington, DC, USA, 1993.
50. Cvetković, J.S.; Mitić, V.D.; Stankov-Jovanović, V.P.; Dimitrijević, M.V.; Nikolić-Mandić, S.D. Elemental composition of wild edible mushrooms from Serbia. *Anal. Lett.* **2015**, *48*, 2107–2121. [CrossRef]

51. Dimitrijević, M.V.; Mitić, V.D.; Cvetković, J.S.; Jovanović, V.P.S.; Mutić, J.J.; Mandić, S.D.N. Update on element content profiles in eleven wild edible mushrooms from family *Boletaceae*. *Eur. Food Res. Technol.* **2016**, *242*, 1–10. [CrossRef]
52. Kalač, P. Trace element contents in European species of wild growing edible mushrooms: A review for the period 2000–2009. *Food Chem.* **2010**, *122*, 2–15. [CrossRef]
53. Niedzielski, P.; Mleczek, M.; Budka, A.; Rzymski, P.; Siwulski, M.; Jasińska, A.; Gąsecka, M.; Budzyńska, S. A screening study of elemental composition in 12 marketable mushroom species accessible in Poland. *Eur. Food Res. Technol.* **2017**, *243*, 1759–1771. [CrossRef]
54. Sarikurkcu, C.; Popović-Djordjević, J.; Solak, M.H. Wild edible mushrooms from Mediterranean region: Metal concentrations and health risk assessment. *Ecotoxicol. Environ. Saf.* **2020**, *190*, 110058. [CrossRef] [PubMed]
55. Sevindik, M.; Rasul, A.; Hussain, G.; Anwar, H.; Zahoor, M.K.; Sarfraz, I.; Kamran, K.S.; Akgul, H.; Akata, I.; Selamoglu, Z. Determination of anti-oxidative anti-microbial activity and heavy metal contents of *Leucoagaricus leucothites*. *Pak. J. Pharm. Sci.* **2018**, *31*, 2163–2168.
56. Jorhem, L.; Sundström, B. Levels of some trace elements in edible fungi. *Z. Lebensm. Unters. Forsch.* **1995**, *201*, 311–316. [CrossRef] [PubMed]
57. Brzezicha-Cirocka, J.; Mędyk, M.; Falandysz, J.; Szefer, P. Bio-and toxic elements in edible wild mushrooms from two regions of potentially different environmental conditions in eastern Poland. *Environ. Sci. Pollut. Res.* **2016**, *23*, 21517–21522. [CrossRef] [PubMed]
58. Melgar, M.J.; Alonso, J.; García, M.A. Cadmium in edible mushrooms from NW Spain: Bioconcentration factors and consumer health implications. *Food Chem. Toxicol.* **2016**, *88*, 13–20. [CrossRef]
59. Sardans, J.; Peñuelas, J. Trace element accumulation in the moss *Hypnum cupressiforme* Hedw.; and the trees *Quercus ilex* L.; and *Pinus halepensis* Mill.; in Catalonia. *Chemosphere* **2005**, *60*, 1293–1307. [CrossRef]
60. Vetter, J. Arsenic content of some edible mushroom species. *Eur. Food Res. Technol.* **2004**, *219*, 71–74. [CrossRef]
61. Falandysz, J.; Rizal, L.M. Arsenic and its compounds in mushrooms: A review. *J. Environ. Sci. Health C* **2016**, *34*, 217–232. [CrossRef]
62. Commission Regulation (EC). No 629/2008 of 2 July 2008 Amending Regulation. (EC) No 1881/2006 Setting Maximum Levels for Certain Contaminants in Foodstuffs. 2008; L 173/6–173/9.
63. Commission Regulation (EC). No 2015/1005 of 25 June 2015 Amending Regulation. (EC) No 1881/2006 as Regards Maximum Levels of Lead in Certain Foodstuffs. 2015; L 161/9–161/13.
64. Official Gazette. Regulations on Maximum Levels of Certain Contaminants in Foodstuffs; National Journal Zagreb Croatia 2008, 154/08. Available online: https://narodne-novine.nn.hr/clanci/sluzbeni/2008_12_154_4198.html (accessed on 9 December 2021).
65. Brzezicha-Cirocka, J.; Grembecka, M.; Grochowska, I.; Falandysz, J.; Szefer, P. Elemental composition of selected species of mushrooms based on a chemometric evaluation. *Ecotoxicol. Environ. Saf.* **2019**, *173*, 353–365. [CrossRef]
66. Romić, M.; Romić, D. Heavy metals distribution in agricultural topsoils in urban area. *Environ. Geol.* **2003**, *43*, 795–805. [CrossRef]
67. Borovička, J.; Řanda, Z. Distribution of iron cobalt zinc and selenium in macrofungi. *Mycol. Prog.* **2007**, *6*, 249. [CrossRef]
68. Demirbaş, A. Concentrations of 21 metals in 18 species of mushrooms growing in the East Black Sea region. *Food Chem.* **2001**, *75*, 453–457. [CrossRef]
69. Vetter, J. Lithium content of some common edible wild-growing mushrooms. *Food Chem.* **2005**, *90*, 31–37. [CrossRef]
70. Zocher, A.L.; Kraemer, D.; Merschel, G.; Bau, M. Distribution of major and trace elements in the bolete mushroom Suillus luteus and the bioavailability of rare earth elements. *Chem. Geol.* **2018**, *483*, 491–500. [CrossRef]
71. Borovička, J.; Kubrová, J.; Rohovec, J.; Řanda, Z.; Dunn, C.E. Uranium thorium and rare earth elements in macrofungi: What are the genuine concentrations? *Biometals* **2011**, *24*, 837–845. [CrossRef] [PubMed]
72. Mleczek, M.; Niedzielski, P.; Kalač, P.; Siwulski, M.; Rzymski, P.; Gąsecka, M. Levels of platinum group elements and rare-earth elements in wild mushroom species growing in Poland. *Food Addit. Contam.* **2015**, *33*, 86–94. [CrossRef] [PubMed]
73. Falandysz, J.; Chudzińska, M.; Barałkiewicz, D.; Drewnowska, M.; Hanć, A. Toxic elements and bio-metals in *Cantharellus* mushrooms from Poland and China. *Environ. Sci. Pollut. Res. Int.* **2017**, *24*, 11472–11482. [CrossRef]
74. Nikkarinen, M.; Mertanen, E. Impact of geological origin on trace element composition of edible mushrooms. *J. Food Compos. Anal.* **2004**, *17*, 301–310. [CrossRef]

Journal of
Fungi

Article

An Updated Global Species Diversity and Phylogeny in the Genus *Wickerhamomyces* with Addition of Two New Species from Thailand

Supakorn Nundaeng [1,2], Nakarin Suwannarach [2,3], Savitree Limtong [4,5], Surapong Khuna [2,3], Jaturong Kumla [2,3,*] and Saisamorn Lumyong [2,3,5,*]

[1] Master of Science Program in Applied Microbiology (International Program), Faculty of Science, Chiang Mai University, Chiang Mai 50200, Thailand; Supakorn.ning@gmail.com
[2] Department of Biology, Faculty of Science, Chiang Mai University, Chiang Mai 50200, Thailand; suwan.462@gmail.com (N.S.); Trio_9@hotmail.com (S.K.)
[3] Research Center of Microbial Diversity and Sustainable Utilization, Chiang Mai University, Chiang Mai 50200, Thailand
[4] Department of Microbiology, Faculty of Science, Kasetsart University, Bangkok 10900, Thailand; fscistl@ku.ac.th
[5] Academy of Science, The Royal Society of Thailand, Bangkok 10300, Thailand
* Correspondence: Jaturong_yai@hotmail.com (J.K.); scboi009@gmail.com (S.L.); Tel.: +66-81-881-3658 (S.L.)

Citation: Nundaeng, S.; Suwannarach, N.; Limtong, S.; Khuna, S.; Kumla, J.; Lumyong, S. An Updated Global Species Diversity and Phylogeny in the Genus *Wickerhamomyces* with Addition of Two New Species from Thailand. *J. Fungi* 2021, 7, 957. https://doi.org/10.3390/jof7110957

Academic Editors: Anush Kosakyan, Rodica Catana and Alona Biketova

Received: 14 October 2021
Accepted: 8 November 2021
Published: 11 November 2021

Publisher's Note: MDPI stays neutral with regard to jurisdictional claims in published maps and institutional affiliations.

Abstract: Ascomycetous yeast species in the genus *Wickerhamomyces* (Saccharomycetales, Wickerhamomycetaceae) are isolated from various habitats and distributed throughout the world. Prior to this study, 35 species had been validly published and accepted into this genus. Beneficially, *Wickerhamomyces* species have been used in a number of biotechnologically applications of environment, food, beverage industries, biofuel, medicine and agriculture. However, in some studies, *Wickerhamomyces* species have been identified as an opportunistic human pathogen. Through an overview of diversity, taxonomy and recently published literature, we have updated a brief review of *Wickerhamomyces*. Moreover, two new *Wickerhamomyces* species were isolated from the soil samples of Assam tea (*Camellia sinensis* var. *assamica*) that were collected from plantations in northern Thailand. Herein, we have identified these species as *W. lannaensis* and *W. nanensis*. The identification of these species was based on phenotypic (morphological, biochemical and physiological characteristics) and molecular analyses. Phylogenetic analyses of a combination of the internal transcribed spacer (ITS) region and the D1/D2 domains of the large subunit (LSU) of ribosomal DNA genes support that *W. lannaensis* and *W. nanensis* are distinct from other species within the genus *Wickerhamomyces*. A full description, illustrations and a phylogenetic tree showing the position of both new species have been provided. Accordingly, a new combination species, *W. myanmarensis* has been proposed based on the phylogenetic results. A new key for species identification is provided.

Keywords: ascomycetous yeast; distribution; new species; phylogeny; taxonomy; *Wickerhamomyces*

1. Introduction

The genus *Wickerhamomyces* was first proposed by Kurtzman et al. [1] in 2008 with *W. canadensis* (basionym *Hansenula canadensis*) as the type species. This genus belongs to the family Wickerhamomycetaceae of the order Saccharomycetales [1]. *Wickerhamomyces* species can reproduce both asexually and sexually. Through asexual reproduction, the species reproduce by budding and some species produce pseudohyphae and/or true hyphae. Alternatively, in sexual reproduction they produce hat-shaped or spherical ascospores with an equatorial ledge for sexual reproduction [1,2]. Most of the known *Wickerhamomyces* species can utilize various carbon sources, but not methanol or hexadecane. Nitrate utilization was observed in some species, while the diazonium blue B reaction was negative for all species. The predominant ubiquinone in the *Wickerhamomyces* species is CoQ-7 [1,3].

Most species of the genus *Wickerhamomyces* have been transferred from the genera *Candida, Hansenula, Pichia* and *Williopsis* based on phylogenetic analyses [1,2,4–6]. Currently, a total of 35 species have been accepted and recorded in the Index Fungorum [7]. A phylogenetic study of Arastehfar et al. [8] has suggested that *Pichai myanmarensis* [9] should be transferred to the genus *Wickerhamomyces*, but *W. myanmarensis* has remained invalidly published. Therefore, in this study, we have proposed the validation of this name for a new combination species. In this case, 35 type species of *Wickerhamomyces* and *P. myanmarensis* have been reported since 1891. It has been revealed that the highest number of the *Wickerhamomyces* species were discovered during the period of 2011–2020, followed by the period of 1951–1960 (6 species), the period of 1971–1980 (4 species) and the period of 2001–2010 (3 species) (Figure 1). An increasing trend with regard to the discovery of new species of *Wickerhamomyces* is expected to continue in the future. *Wickerhamomyces* species are widely distributed in tropical, subtropical, temperate and subpolar areas throughout the world (Figure 2). It has been reported that the highest number of *Wickerhamomyces* species were found in Asia (18 species), followed by Europe (12 species), South America (8 species), Africa (7 species), North America (3 species), Oceania (1 species) and Antartica (1 species) [7–82] (Table 1). Both *W. anomalus* and *W. onychis* are known to be from Asia, Africa, Europe and South America [13−35,54−59]. Moreover, *W. anomalus* and *W. rabaulenis* have been discovered in Antarctica (King George Island) [17] and Oceania (Papua New Guinea) [67], respectively. Recently, *W. psychrolipolyticus* has been discovered from Japan [65]. Consequently, *Wickerhamomyces* species have been successfully isolated from various habitats, as has been summarized in Table 1.

Many species of *Wickerhamomyces* have been used in a variety of industries including the medicinal, agricultural, biofuel, food and beverage industries, and a number of others [64]. Most previous studies have focused on different strains of *W. anomalus* for biotechnological applications. For example, the *W. anomalus* strains CBS261, HN006 and HN010 are capable of excessively producing ethyl acetate. As a result, this species has been used in the brewing of Baijiu (Chinese liquor) and in winemaking to improve the aroma and quality of the finished product [83–85].

Wickerhamomyces anomalus strains BS91 and DMKU-RP04 could effectively inhibit plant pathogenic fungi and have been used as a biological control agent in agriculture [86–88]. Notably, *W. anomalus* strains SDBR-CMU-S1-06 and Wa-32 have exhibited plant growth promotion potential by solubilizing insoluble minerals, producing indole-3-acitic acid (IAA) and siderophores, and by secreting various extracellular enzymes [16,89]. Moreover, most strains of *W. anomalus* are known to produce killer toxins that possess antimicrobial and larvicidal activities [90,91]. *Wickerhamomyces bovis* and *W. silvicola* have been observed to produce mycocin, which exhibited fungicidal activity [92,93]. In addition, *W. lynferdii* and *W. sydowiorum* have been recognized as relevant yeast species for the improvement of the fermentation processes for coffee cherries and cocoa, respectively [79,94]. Furthermore, *W. subpelliculosus* has been used as an alternative to baker's yeast [95], while *W. chambardii* could produce amylase and cellulase enzymes that could be used to produce bioethanol from corn straw [96,97]. Previous studies have found that the biosurfactants produced by *W. anomalus* and *W. edaphicus* [47,98,99], the saturated fatty acids produced by *W. siamensis* [100], xylitol produced by *W. rabaulensis* [101], cellulase enzymes produced by *W. psychrolipolyticus* [65] and the extracellular polysaccharide produced by *W. mucosus* [102] could be applied in the bioremediation, biotechnological and cosmetic industries. Furthermore, these substances could also be employed in the production of detergents, food and various pharmaceuticals, as well as in the process of oil recovery enhancement.

On the other hand, some *Wickerhamomyces* species (e.g., *W. anomalus* and *W. lynferdii*) have been responsible for the spoilage of beer and bakery products [103–106]. Some cases of human infection caused by *W. anomalus, W. myanmarensis* and *W. onychis* have also been reported, but only with patients with serious illness [8,107–111]. Based on this evidence,

W. anomalus has been labeled a biosafety level 1 organism by the European Food Safety Authority [112] and is considered safe for consumption by healthy individuals.

Currently, only eight *Wickerhamomyces* species, namely *W. anomalus*, *W. ciferrii*, *W. edaphicus*, *W. rabaulensis*, *W. siamensis*, *W. sydowiorum*, *W. tratensis* and *W. xylosicus*, have been reported in Thailand [4,8,16,17,19,20,46,70,78]. Accordingly, Thailand has been identified as a hotspot for unexpected novel species and the newly recorded discovery of many microorganisms [113,114]. In our previous investigation on yeasts in soil samples collected from Assam tea (*Camellia sinensis* var. *assamica*) plantations in northern Thailand [16], we obtained five yeast strains belonged to the genus *Wickerhamomyces* that represent potentially new species. In our present study, we have described them into two novel species. These two novel species are introduced based on their phenotypic (morphological, biochemical and physiological data) and molecular characteristics. To confirm their taxonomic status, phylogenetic relationship was determined by analysis of the combined sequence dataset of the D1/D2 domains of LSU and ITS sequences.

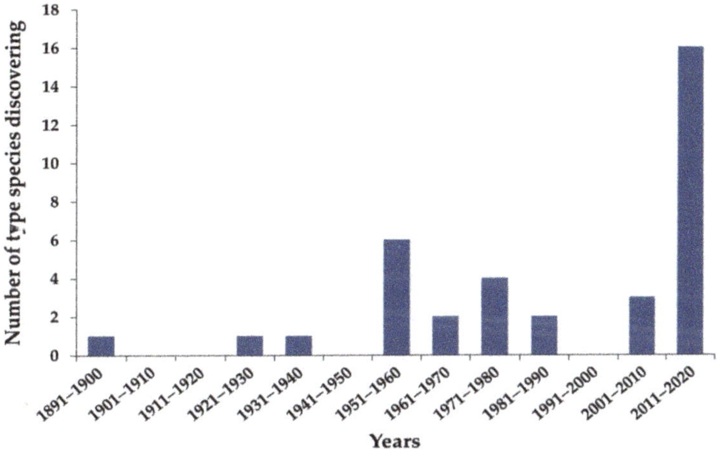

Figure 1. The discovering of *Wickerhamomyces* type species since 1891 till the present time.

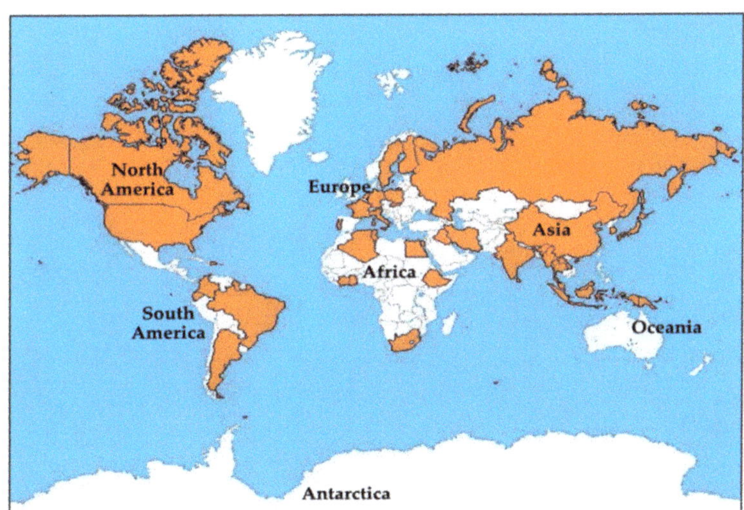

Figure 2. Global distribution of *Wickerhamomyces* species. The countries where *Wickerhamomyces* species have been discovered are indicated in orange color.

Table 1. Global distribution and isolation sources of *Wickerhamomyces* species.

No.	Spices	Known Distribution	Isolation Source	Reference
1	*Wickerhamomyces alni* (Phaff, M.W. Mill. and M. Miranda) Kurtzman, Robnett and Bas.-Powers	Canada	Exudate of *Alnus rubra*	[12]
2	*Wickerhamomyces anomalus* (E.C. Hansen) Kurtzman, Robnett and Bas.-Powers	Algeria, Brazil, China, Colombia, Ethiopia, India, Iraq, King George Island, Lao, Russia, Slovakia Sweden and Thailand	Process of beer and wine, phylloplane, soil, water, coral reefs, Thai traditional alcoholic starter, mangrove forest, fermented food, flowers, fruits, fermented grains, coffee processing, wastewater treatment plant, Colombian fermented beans and Brazilian spirit	[13–35]
3	*Wickerhamomyces arborarius* S.A. James, E.J. Carvajal, Barahona, T.C. Harr., C.F. Lee, C.J. Bond and I.N. Roberts	Ecuador	Flower	[36]
4	*Wickerhamomyces bisporus* (O. Beck) Kurtzman, Robnett and Bas.-Powers	Finland, France and USA	*Platypus compositus*, phoretic mites on *Ips typographus* and bark beetles (*Dendroctonus*)	[37–39]
5	*Wickerhamomyces bovis* (Uden and Carmo Souza) Kurtzman, Robnett and Bas.-Powers	Portugal	Caecum of feral cattle (*Bos taurus*)	[40]
6	*Wickerhamomyces canadensis* (Wick.) Kurtzman, Robnett and Bas.-Powers	Canada	Beetle frass from *Pinus resinosa*	[41]
7	*Wickerhamomyces chambardii* (C. Ramírez and Boidin) Kurtzman, Robnett and Bas.-Powers	France	Chestnut	[42]
8	*Wickerhamomyces chaumierensis* M. Groenew., V. Robert and M.T. Sm.	Guyana	Surface of flower	[43]
9	*Wickerhamomyces ciferrii* (Lodder) Kurtzman, Robnett and Bas.-Powers	Dominican Republic, USA and Thailand	Fruit of *Dipteryx odorata* and male olive fruit fly (*Bactrocera oleae*)	[44–46]
10	*Wickerhamomyces edaphicus* Limtong, Yongman., H. Kawas. and Fujiyama	India and Thailand	Forest and mangrove soils	[4,47]
11	*Wickerhamomyces hampshirensis* (Kurtzman) Kurtzman, Robnett and Bas.-Powers	USA	Frass of cut and dead of *Quercus* and beetle (*Xyloterinus politus*)	[48,49]
12	*Wickerhamomyces kurtzmanii* A.H. Li, Y.G. Zhou and Q.M. Wang	China	Crater lake water	[6]
13	*Wickerhamomyces lynferdii* (Van der Walt and Johannsen) Kurtzman, Robnett and Bas.-Powers	South Africa	Soil	[50]
14	*Wickerhamomyces menglaensis* F.L. Hui and L.N. Huang	China	Rotting wood	[51]
15	*Wickerhamomyces mori* F.L. Hui, Liang Chen, X.Y. Chu, Niu and T. Ke	China	Gut of larvae of wood-boring insect on trunk of *Morus alba*	[52]
16	*Wickerhamomyces mucosus* (Wick. and Kurtzman) Kurtzman, Robnett and Bas.-Powers	USA	Soil	[53]
17	*Wickerhamomyces myanmarensis* (Nagats., H. Kawas. and T. Seki) J. Kumla, N. Suwannarach and S. Lumyong	Iran and Myanmar	Palm sugar in rum distiller, and blood and central venous catheter of patients	[8,9]
18	*Wickerhamomyces ochangensis* K.S. Shin	South Korea	Soil of potato field	[11]
19	*Wickerhamomyces onychis* (Yarrow) Kurtzman, Robnett and Bas.-Powers	Brazil, Ethiopia, Iraq, Malaysia, Netherlands, Poland and Tunisia	Nail infection of *Homo sapiens*, fermented food, cocoa beans, grape and tomato during spontaneous fermentation, and soil	[23,54–59]
20	*Wickerhamomyces orientalis* Sipiczki, S. Nasr, H.D.T. Nguyen and Soudi	Iran and Sri Lanka	Fruits and rhizosphere soil	[27,60]
21	*Wickerhamomyces patagonicus* V. de García, Brizzio, C.A. Rosa, Libkind and Van Broock	Argentina	Sap exudate on cut branches of *Nothofagus dombeyi* and glacier meltwater river	[61]
22	*Wickerhamomyces pijperi* (Van der Walt & Tscheuschner) Kurtzman, Robnett and Bas.-Powers	Egypt, Ghana and South Africa	Buttermilk, cocoa fermentation and orange juice	[62–64]
23	*Wickerhamomyces psychrolipolyticus* Y. Shimizu and K. Konno	Japan	Soil	[65]
24	*Wickerhamomyces queroliae* C.A. Rosa, P.B. Morais, Lachance and Pimenta	Brazil	Larva of *Anastrepha mucronata* from fruit of *Peritassa campestris*	[66]
25	*Wickerhamomyces rabaulensis* (Soneda and S. Uchida) Kurtzman, Robnett and Bas.-Powers	Ethiopia, Papua New Guinea and Thailand	Excreta of snail, soils, decaying agricultural residues, decaying leaves and tree bark, and fermented food	[23,67,68]
26	*Wickerhamomyces scolytoplatypi* Ninomiya	Japan	Gallery of beetles (*Scolytoplatypus shogun*) in *Fagus crenata*	[69]
27	*Wickerhamomyces siamensis* Kaewwich., Yongman., H. Kawas. and Limtong	Thailand	Phylloplane of *Saccharum officinarum*	[70]

Table 1. *Cont.*

No.	Spices	Known Distribution	Isolation Source	Reference
28	*Wickerhamomyces silvicola* (Wick.) Kurtzman, Robnett and Bas.-Powers	Germany, South Korea and USA	Flowers, gum of *Prunus serotina* and *Prunus* wood	[41,71,72]
29	*Wickerhamomyces spegazzinii* Masiulionis and Pagnocca	Argentina	The fungus garden of an attine ant nest (*Acromyrmex lundii*)	[73]
30	*Wickerhamomyces strasburgensis* (C. Ramírez and Boïdin) Kurtzman, Robnett and Bas.-Powers	France	On leather tanned by vegetable means	[74]
31	*Wickerhamomyces subpelliculosus* (Kurtzman) Kurtzman, Robnett and Bas.-Powers	Egypt and USA	Fermenting cucumber brines, gut of honey bee and molasses	[75,76]
32	*Wickerhamomyces sydowiorum* (D.B. Scott and Van der Walt) Kurtzman, Robnett and Bas.-Powers	Brazil, Ivory Coast, South Africa and Thailand	Frass of *Sinoxylon ruficorne* in dead *Combretum apiculatum*, decayed plant leaf, fermented cocoa, honey, sand and water	[59,77–80]
33	*Wickerhamomyces sylviae* Moschetti and J.P. Samp.	Italy	Cloaca of migratory birds (*Sylvia communis*)	[81]
34	*Wickerhamomyces tratensis* Nakase, Jindam., Am-In, Ninomiya and H. Kawas.	Thailand	Flower of mangrove apple (*Sonneratia caseolaris*)	[82]
35	*Wickerhamomyces xylosicus* Limtong, Nitiyon, Kaewwich., Jindam., Am-In and Yongman	Thailand	Soil	[5]
36	*Wickerhamomyces xylosivorus* R. Kobay., A. Kanti and H. Kawas.	Indonesia	Decayed wood	[4]

2. Materials and Methods

2.1. Yeast Strain

Five yeasts strains (SDBR-CMU-S2-02, SDBR-CMU-S2-06, SDBR-CMU-S2-14, SDBR-CMU-S2-17 and CMU-S3-15) isolated from soils of Assam tea (*C. sinensis* var. *assamica*) plantations in Thep Sadej, Doi Saket District, Chiang Mai Province and Sri Na Pan, Muang District, Nan Province, northern Thailand [16] were selected for this present study. All strains were deposited in the culture collection of the Sustainable Development of Biological Resources, Faculty of Science, Chiang Mai University (SDBR-CMU), Chiang Mai Province and Thailand Bioresource Research Center (TBRC), Pathum Thani Province, Thailand.

2.2. Yeast Identification

2.2.1. Morphological Study

The morphological characteristics of yeast strains were determined according to established methods by Kurtzman et al. [2], Yarrow [3] and Limtong et al. [10]. Colony characters were observed on yeast extract-malt extract agar (YMA) after two days of incubation in darkness at 30 °C. Ascospore formation was investigated on YMA, 5% malt extract agar (MEA), potato dextrose agar (PDA) and V8 agar after incubation at 25 °C in the dark for four weeks. Micromorphological characteristics were examined under a light microscope (Nikon Eclipse Ni U, Tokyo, Japan). Size data of the anatomical structure (e.g., cells, pseudohyphae, asci and ascospores) were based on at least 50 measurements of each structure using the Tarosoft (R) Image Frame Work program.

2.2.2. Biochemical and Physiological Studies

Biochemical and physiological characterizations of yeast strains was followed the previous studies [2,3,115]. Fermentation of carbohydrates including glucose, galactose, maltose, sucrose, trehalose, melibiose, lactose, raffinose, and xylose were performed. Additionally, assimilation tests for carbon (D-glucose, D-galactose, L-sorbose, N-acetyl glucosamine, D-ribose, D-xylose, L-arabinose, D-arabinose, rhamnose, sucrose, maltose, α,α-trehalose, α-methyl-D-glucoside, cellobiose, salicin, melibiose, lactose, raffinose, melezitose, inulin, soluble starch, glycerol, erythritol, ribitol, D-glucitol, D-mannitol, galactitol, myo-inositol, D-glucono-1,5-lactone, 2-ketogluconic acid, 5-ketogluconic acid, D-gluconate, D-glucuronate, D-galacturonic acid, DL-lactate, succinate, citrate, methanol, ethanol, and xylitol) and nitrogen compounds (ammonium sulfate, potassium nitrate, sodium nitrite, ethylamine hydrochloride, L-lysine, cadaverine, and creatine) were determined. Moreover, the effects of temperature on growth were examined by cultivation on YMA at temperature ranging from 15–45 °C and diazonium blue B reactions were tested [116].

2.2.3. Molecular Study

Each yeast strain was grown in 5 mL of yeast extract-malt extract broth in 18 × 180 mm test tubes with shaking at 150 rpm on an orbital shaker in the dark for two days. Yeast cells were harvested by centrifugation at 11,000 rpm and washed three times with sterile distilled water. Genomic DNA was extracted from yeast cells using DNA Extraction Mini Kit (FAVORGEN, Taiwan) following the manufacturer's protocol. The ITS region and D1/D2 domains of LSU gene were amplified by polymerase chain reactions (PCR) using ITS1/ITS4 primers [117] and NL1/NL4 primers [118], respectively. The amplification of both D1/D2 domains and ITS region process consisted of an initial denaturation at 95 °C for 5 min, followed by 35 cycles of denaturation at 95 °C for 30 s, annealing at 52 °C for 45 s, an extension at 72 °C for 1 min and 72 °C for 10 min on a peqSTAR thermal cycler (PEQLAB Ltd., UK). PCR products were checked and purified by a PCR clean up Gel Extraction NucleoSpin® Gel and PCR Clean-up Kit (Macherey-Nagel, Germany). Final PCR products were sent to 1st Base Company Co., Ltd., (Kembangan, Malaysia) for sequencing. The obtained sequences were used to query GenBank via BLAST (http://blast.ddbj.nig.ac.jp/top-e.html, accessed on 25 August 2021).

Phylogenetic analysis was carried out based on the combined dataset of ITS and D1/D2 domains of LSU sequences. Sequences from this study along with those obtained from previous studies and the GenBank database were selected and provided in Table 2. Multiple sequence alignment was performed using MUSCLE [119]. A combination of D1/D2 domains of LSU and ITS alignment was deposited in TreeBASE under the study ID number 28785. A phylogenetic tree was constructed under maximum likelihood (ML) and Bayesian inference (BI) methods. The ML analysis was carried out using RAxML-HPC2 on XSEDE (8.2.10) in CIPRES Science Gateway V. 3.3 [120] using GTRCAT model with 25 categories and 1000 bootstrap (BS) replications. The optimum nucleotide substitution model was obtained using jModeltest v.2.3 [121] under the Akaike information criterion (AIC) method. The BI analysis was performed using MrBayes 3.2.6 software for Windows [122]. The selected optimal model of each gene is similar as GTR + I + G model. Six simultaneous Markov chains were run with one million generations and starting from random trees and keeping one tree every 100th generation until the average standard deviation of split frequencies was below 0.01. The value of burn-in was set to discard 25% of trees when calculating the posterior probabilities. Bayesian posterior probabilities (PP) were obtained from the 50% majority rule consensus of the trees kept. The tree topologies were visualized in FigTree v1.4.0 [123].

Table 2. DNA sequences used in the molecular phylogenetic analysis. Type strains indicated by "T".

Yeast Species	Strain	GenBank Acession Number		Reference
		ITS	D1/D2	
Wickerhamomyces alni	NRRL Y-11625[T]	-	EF550294	[1]
	CBS 6986	NR154966	KY110065	[124]
Wickerhamomyces anomalus	NRRL Y-366[T]	-	EF550341	[1]
	H1Wh	JQ857021	JQ856997	[17]
Wickerhamomyces arborarius	Bq 164[T]	NR_55000	FN908198	[36]
Wickerhamomyces bisporus	NRRL Y-1482[T]	-	EF550296	[1]
Wickerhamomyces bovis	NRRL YB-4184[T]	-	EF550298	[1]
	CBS 2616	NR154968	KY110109	[124]
Wickerhamomyces canadensis	NRRL Y-1888[T]	-	EF550300	[1]
	GoToruMP327	EF093299	EF016107	[125]

Table 2. *Cont.*

Yeast Species	Strain	GenBank Acession Number		Reference
		ITS	**D1/D2**	
Wickerhamomyces chambardii	NRRL Y-2378[T]	-	EF550344	[1]
	CBS 1900	NR154969	KY110114	[124]
Wickerhamomyces chaumierensis	CBS 8565[T]	HM156503	HM156533	[43]
Wickerhamomyces ciferrii	NRRL Y-1031[T]	-	EF550339	[1]
	UCDFST 83-22	MH153583	MH130275	[126]
Wickerhamomyces edaphicus	S-29[T]	AB436771	AB436763	[10]
	CBS 10408	KY105904	KY110120	[124]
Wickerhamomyces hampshirensis	NRRL YB-4128[T]	-	EF550334	[1]
Wickerhamomyces kurtzmanii	TF5-16-2[T]	MK573939	MK573939	[6]
Wickerhamomyces lannaensis	**SDBR-CMU-S3-15[T]**	**OK135750**	**MT639220**	**This study, [16]**
	SDBR-CMU-S2-02	**OK135752**	**MT623569**	**This study, [16]**
	SDBR-CMU-S2-06	**OK135753**	**MT613722**	**This study, [16]**
Wickerhamomyces lynferdii	NRRL Y-7723[T]	EF550342	EF550342	[1]
	BCRC 22676	NR111798	-	[127]
Wickerhamomyces menglaensis	NYNU 1673[T]	KY213818	KY213812	[51]
Wickerhamomyces mori	NYNU 1216[T]	JX204288	JX204287	[52]
	NYNU 1204	JX292100	JX292099	[52]
Wickerhamomyces mucosus	NRRL YB-1344[T]	-	EF550337	[1]
	CBS 6341	Z93877	KY110124	[124]
Wickerhamomyces myanmarensis	CBS 9786[T]	-	AB126678	[9]
	SU-263	MH236221	MH236219	[8]
Wickerhamomyces nanensis	**SDBR-CMU-S2-17[T]**	**OK143510**	**MT613875**	**This study, [16]**
	SDBR-CMU-S2-14	**OK143511**	**MT623571**	**This study, [16]**
Wickerhamomyces ochangensis	N7a-Y2[T]	NR154971	HM485464	[11]
	CBS 11843	KY105909	-	[124]
Wickerhamomyces onychis	NRRL Y-7123[T]	-	EF550279	[1]
	CBS 5587	KY105910	KY110125	[124]
Wickerhamomyces orientalis	KH-D1[T]	KF938677	KF938676	[60]
	12-101	KU253704	KU253703	[60]
Wickerhamomyces patagonicus	CRUB 1724[T]	FJ793131	FJ666399	[61]
	CBS 11398	NG057185	KY110126	[124]
Wickerhamomyces pijperi	NRRL YB-4309[T]	-	EF550335	[1]
	CBS 2887	HM156502	KY110127	[124]
Wickerhamomyces psychrolipolyticus	Y08-202-2[T]	-	LC333101	[65]
	Y08-202-2	-	LC333102	[65]
Wickerhamomyces queroliae	UFMG-T05-200[T]	EU580140	EU580140	[66]
Wickerhamomyces rabaulensis	NRRL Y-7945[T]	-	EF550303	[1]
	CBS 6797	KY105914	KY110128	[124]
Wickerhamomyces scolytoplatypi	NBRC 11029[T]	-	AB534166	[69]
	CBS 12186	KY105915	KY110130	[124]

Table 2. Cont.

Yeast Species	Strain	GenBank Acession Number		Reference
		ITS	**D1/D2**	
Wickerhamomyces siamensis	DMKU-RK359^T	NR111029	AB714248	[70]
	CBS 12570	KY105916	KY110131	[124]
Wickerhamomyces silvicola	NRRL Y-1678^T	-	EF550302	[1]
	GLMC 1708	MT156140	MT156324	[71]
Wickerhamomyces spegazzinii	JLU025^T	KJ832072	KJ832071	[73]
Wickerhamomyces strasburgensis	NRRL Y-2383^T	-	EF550333	[1]
Wickerhamomyces subpelliculosus	NRRL Y-1683^T	NR111336	EF550340	[1,127]
Wickerhamomyces sydowiorum	NRRL Y-7130^T	NR138219	EF550343	[1,36]
	NRRL Y-10996	FR690145	FR690073	[36]
Wickerhamomyces sylviae	PYCC6345^T	-	KF240728	[81]
	U92A1	-	KF240729	[81]
Wickerhamomyces tratensis	NBRC 107799^T	AB607029	AB607028	[82]
	CBS 12176	KY105935	KY110150	[124]
Wickerhamomyces xylosica	CBS 12320^T	NR160310	AB557867	[5]
	NT31	AB704715	NG064304	[5]
Wickerhamomyces xylosivorus	NBRC 111553^T	NR155013	LC202858	[4]
	14Y125	-	NG057186	[4]
Saccharomyces cerevisiae	NRRL Y-12632^T	AY046146	JQ689017	[128]
Spathaspora allomyrinae	CBS 13924^T	KP054268	KP054267	[129]

Note: sepecies obtained in this study are in bold.

3. Results

3.1. Phylogenetic Results

The sequences of five yeast strains were deposited in the GenBank database (Table 2). The alignment of a combination of ITS and D1/D2 domains of the LSU genes contained 1544 characters including gaps (ITS: 1−823 and D1/D2 domains of LSU: 824−1544). RAxML analysis of the combined dataset yielded a best scoring tree with a final ML optimization likelihood value of −12,120.4323. The matrix contained 776 distinct alignment patterns with 42.33% undetermined characters or gaps. Estimated base frequencies were recorded as follows: A = 0.2730, C = 0.1821, G = 0.2603, T = 0.2844; substitution rates AC = 1.0574, AG = 2.0209, AT = 1.4684, CG = 0.6712, CT = 4.4165, GT = 1.0000. The gamma distribution shape parameter alpha was equal to 0.2698 and the Tree-Length was equal to 4.4075. In addition, the final average standard deviation of the split frequencies at the end of the total MCMC generations was calculated as 0.00638 through BI analysis. Phylograms of the ML and BI analyses were similar in terms of topology (data not shown). Therefore, the phylogram obtained from the ML analysis was selected and presented for this study. The phylogram was comprised of 67 sequences of *Wickerhamomyces* strains (including 37 type strains obtained from either previous studies or the present study) and two sequences (*Saccharomyces cerevisiae* NRRL 12632 and *Spathaspora allomyrinae* CBS 13924) of the outgroup (Figure 3). Our phylogenetic analysis separated *Wickerhamomyces* by different species based on different topologies. Our analysis confirmed that *W. myanmarensis* (previously known as *P. myanmarensis*) belonged to the genus *Wickerhamomyces* according to the phylogenetic results of Arastehfar et al. [8] and Shimizu et al. [65]. Moreover, a phylogram clearly separated our yeast strains into two monophyletic clades with high support values (BS = 100% and PP = 1.0). The results indicated that our two yeast strains, SDBR-CMU-S2-17 and

SDBR-CMU-S2-14 (introduced as *W. nanensis*), were clearly distinguished from the previously known species of *Wickerhamomyces*. Moreover, three yeast strains in this study, SDBR-CMU-S2-02, SDBR-CMU-S2-15, and CMU-S3-06 (described here as *W. lannaensis*) formed a sister clade to *W. ochangensis* with high support (BS = 100% and PP = 1.0).

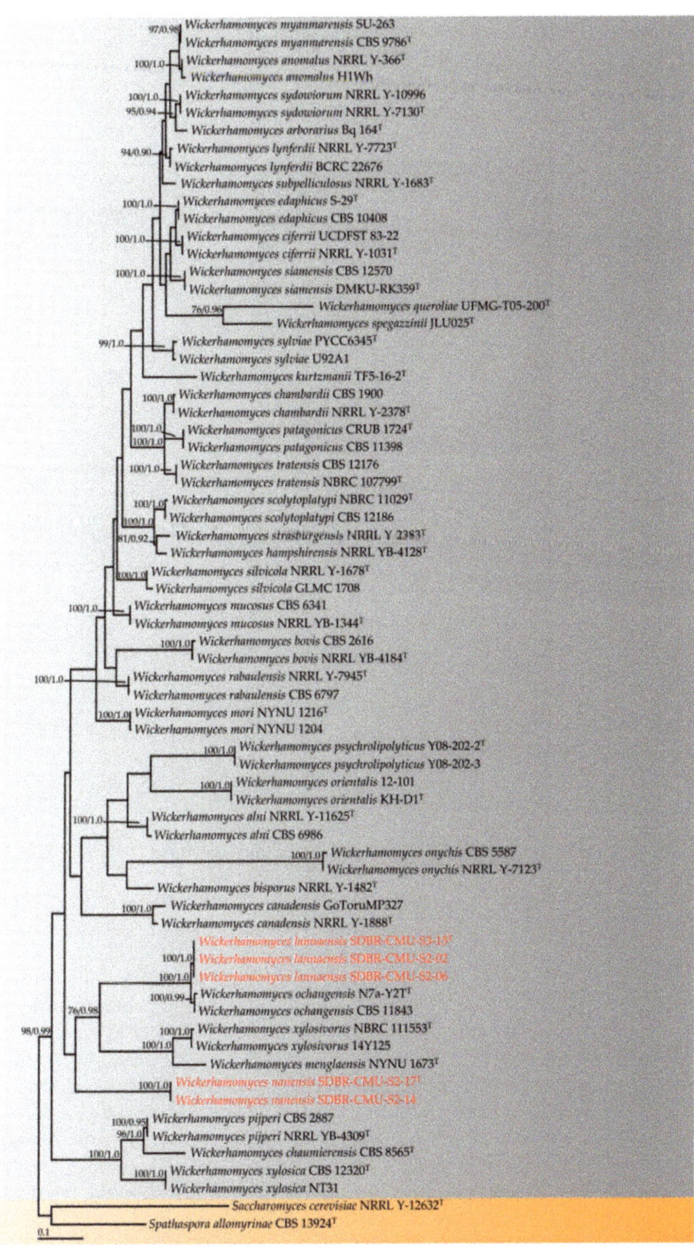

Figure 3. Phylogram derived from maximum likelihood analysis of 69 sequences of the combined ITS and D1/D2 sequences. *Saccharomyces cerevisiae* NRRL 12632 and *Spathaspora allomyrinae* CBS 13924 were used as the outgroup. The numbers above branches represent bootstrap percentages (**left**) and Bayesian posterior probabilities (**right**). Bootstrap values ≥ 75% and Bayesian posterior probabilities ≥ 0.90 are shown. Sequences obtained in this study are in red. Superscript "T" means the type strains.

3.2. Taxonomic Description of New Species

3.2.1. *Wickerhamomyces lannaensis* S. Nundaeng, J. Kumla, N. Suwannarach and S. Lumyong, sp. nov. (Figure 4)

Mycobank No.: 841356

Etymology: "*lannaensis*" refers to Lanna kingdom the historic name of northern Thailand, the collection locality of the type strain of the species.

Holotype: Thailand, Chiang Mai Province, Thep Sadej, Doi Saket District, in soil from Assam tea (*C. sinensis* var. *assamica*) plantation, May 2017, J. Kumla and N. Suwannarach, (holotype SDBR-CMU-S3-15^T, culture ex-type TBRC 15533)

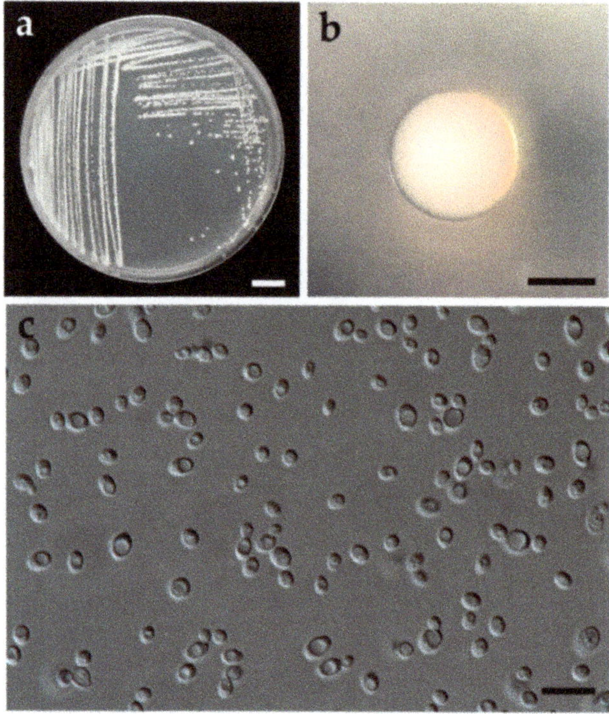

Figure 4. *Wickerhamomyces lannaensis* (holotype SDBR-CMU-S3-15). Culture (**a**), single colony (**b**) and budding cells (**c**) on YMA after two days at 30 °C. Scale bar a and b = 10 cm, c = 10 μm.

Description: The streak culture on YMA after two days at 30 °C is circular from (1–2 mm in diameter), white to cream color, smooth surface, dull-shining, entire margin, and raised elevation. After growth on YMA at 30 °C for two days, the cells are spheroidal to short ovoidal (3.6–3.8 × 2.4–2.6 μm), occur singly or in budding pairs. Pseudohyphae and true hyphae were absent. Ascospores were not obtained for individual strains and strain pairs on YMA, 5% MEA, PDA and V8 agar after incubation at 30 °C for one month. Urea hydrolysis and diazonium blue B reactions are negative. Fermentation tests, glucose is delayed positive, but galactose, maltose, sucrose, trehalose, melibiose, lactose, raffinose, and xylose are negative. D-glucose, D-xylose, rhamnose, cellobiose, salicin, inulin (weak), glycerol, D-glucitol, D-mannitol, D-glucono-1,5-lactone, D-gluconate, DL-lactate (weak), succinate, and ethanol are assimilated. No growth was observed in L-sorbose, N-acetyl glucosamine, D-ribose, L-arabinose, D-arabinose, sucrose, maltose, α,α-trehalose, α-methyl-D-glucoside, melibiose, lactose, raffinose, melezitose, soluble starch, erythritol, ribitol, galactitol, myo-inositol, 2-ketogluconic acid, 5-ketogluconic acid, D-glucuronate, D-galacturonic acid, citrate, methanol, and xylitol. For the assimilation of nitrogen compounds, growth on ammonium sulfate, potassium nitrate, sodium nitrite, ethylamine HCl, cadaverine, and creatine (weak) are positive and on L-lysine is latent positive.

Growth in the vitamin-free medium is weak positive. Growth was observed at 15 °C and 30 °C, but not at 35, 37, 40, 42 and 45 °C. Growth in the presence of 50% glucose is positive, but growth in the presence of 0.01% cycloheximide, 0.1% cycloheximide, 60% glucose, 10% NaCl with 5% glucose and 15% NaCl with 5% glucose are negative. Acid formation is negative.

Additional strains examined: Thailand, Nan Province, Muang District, Sri Na Pan, in soil from Assam tea (*C. sinensis* var. *assamica*) plantation, September 2016, J. Kumla and N. Suwannarach, SDBR-CMU-S2-02, SDBR-CMU-S2-06.

GenBank accession numbers: holotype SDBR-CMU-S3-15 (D1/D2: MT639220, ITS: OK135750); additional strains SDBR-CMU-S2-02 (D1/D2: MT623569, ITS: OK135752) and SDBR-CMU-S2-06 (D1/D2: MT613722, ITS: OK135753).

Note: Based on phylogenetic analyses, *W. lannaensis* formed a monophyletic clade in a well-supported clade and was found to be closely related to *W. ochangensis* (Figure 3). *Wickerhamomyces lannaensis* can be distinguished from *W. ochangensis* by its ability to assimilate inulin and creatine and its growth in 50% glucose medium [11]. Additionally, *W. ochangensis* was able to grow at a temperature of 37 °C, while *W. lannaensis* could not grow at 37 °C [11].

3.2.2. *Wickerhamomyces nanensis* J. Kumla, S. Nundaeng, N. Suwannarach and S. Lumyong, sp. nov. (Figure 5)

Mycobank No.: 841357

Etymology: "*nanensis*" refers to Nan Province of Thailand, the collection locality of the type strain of the species.

Holotype: Thailand, Nan Province, Muang District, Sri Na Pan, in soil from Assam tea (*C. sinensis* var. *assamica*) plantation, September 2016, J. Kumla, N. Suwannarach and S. Khuna, (holotype SDBR-CMU-S2-17^T, culture ex-type TBRC 15534)

Description: The streak culture on YMA after two days at 30 °C is circular from (1–2 mm in diameter), white to cream color, smooth surface, dull-shining, entire margin, and raised elevation. After growth on YMA at 30 °C for two days, the cells are spheroidal to short ovoidal (3.8–4.0 × 2.4–2.5 μm), occur singly or in budding pairs. Pseudohyphae (4.8–6.9 × 2.2–2.9 μm) were produced in Dalmau plate culture on 5% MEA and PDA after 7 days at 25 °C, but true hyphae are not obtained. Ascospores were not observed for individual strains and strain pairs on YMA, 5% MEA, PDA and V8 agar after incubation at 30 °C for one month. Urea hydrolysis and diazonium blue B reactions are negative. Fermentation test, glucose is delayed positive, but galactose, maltose, sucrose, trehalose, melibiose, lactose, raffinose, and xylose are not positive. D-glucose, D-galactose, cellobiose, salicin, glycerol, D-mannitol, D-glucono-1,5-lactone, DL-lactate (weak), succinate, citrate, and ethanol are assimilated. No growth was observed in L-sorbose, N-acetyl glucosamine, D-ribose, D-xylose, L-arabinose, D-arabinose, rhamnose, sucrose, maltose, α,α-trehalose, α-methyl-D-glucoside, melibiose, lactose, raffinose, melezitose, inulin, soluble starch, erythritol, ribitol, D-glucitol, galactitol, myo-inositol, 2-ketogluconic acid, 5-ketogluconic acid, D-gluconate, D-glucuronate, D-galacturonic acid, methanol, and xylitol. For the assimilation of nitrogen compounds, growth on ammonium sulfate, potassium nitrate (weak), sodium nitrite (weak), ethylamine HCl, l-lysine, and creatine (slow) are positive, but cadaverine is not. Growth in the vitamin-free medium is weak. Growth was observed at 15 °C and 30 °C, but not at 35, 37, 40, 42 and 45 °C. Growth in the presence of 50% glucose and acid formation are positive, but growth in the presence of 0.01% cycloheximide, 0.1% cycloheximide, 60% glucose, 10% NaCl with 5% glucose, and 15% NaCl with 5% glucose are negative.

Additional strain examined: Thailand, Nan Province, Muang District, Sri Na Pan, in soil from Assam tea (*C. sinensis* var. *assamica*) plantation, September 2016, J. Kumla and N. Suwannarach, SDBR-CMU-S2-14.

GenBank accession numbers: holotype SDBR-CMU-S2-17 (D1/D2: MT613875, ITS: OK143510); additional strain SDBR-CMU-S2-14 (D1/D2: MT623569, ITS: OK143511).

Note: Several morphological and biochemical characteristics of *W. nanensis* were similar to *W. chambardii*. However, *W. chambardii* differed from *W. nanensis* by its ascospore formation and could not assimilate D-mannitol [2]. Phylogenetic analyses clearly separated *W. nanensis* and *W. chambardii* as different species. Moreover, *W. nanensis* formed a monophyletic clade in a well-supported clade and was separated from other *Wickerhamomyces* species (Figure 3).

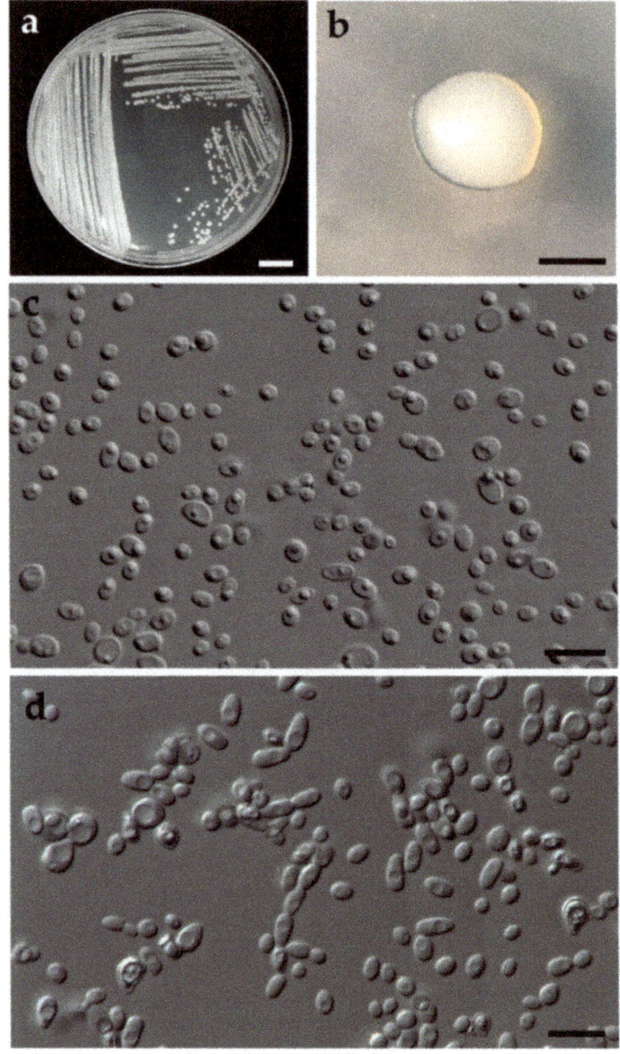

Figure 5. *Wickerhamomyces nanensis* (holotype SDBR-CMU-S3-15). Culture (**a**), single colony (**b**) and budding cells (**c**) on YMA after two days at 30 °C. Pseudohypahae (**d**) on 5% MAE agar after 7 days at 25 °C. Scale bar a and b = 10 cm, c and d = 10 μm.

3.3. New Combination

Wickerhamomyces myanmarensis (Nagats., H. Kawas. and T. Seki) J. Kumla, N. Suwannarach and S. Lumyong, comb. nov.

Mycobank No.: 841356

Basionym: *Pichia myanmarensis* Nagats., H. Kawas. and T. Seki, Int. J. Syst. Evol. Microbiol. 55: 1381, 2005.

Note: The combined ITS and D1/D2 phylogenetic analyses indicated that the type species, *P. myanmarensis*, belongs to the genus *Wickerhamomyces* and has a close phylogenetic relationship with *W. anomalus* (Figure 3). Accordingly, the phylogenetic results of Arastehfar et al. [8] and Shimizu et al. [65] found that *P. myanmarensis* was placed within the genus *Wickerhamomyces*.

3.4. Key to Species of Wickerhamomyces

A key to the identification of the *Wickerhamomyces* species introduced in the present study was derived from the key described by Kurtzman et al. [2]. Key characteristics are shown in Table 3.

1.   a. Melibiose is assimilated 2
       b. Melibiose is not assimilated 7
2.   (1) a. Raffinose is assimilated 3
       b. Raffinose is not assimilated .... *W. kurtzmanii*
3.   (2) a. Citrate is assimilated 4
       b. Citrate is not assimilated .... *W. orientalis*
4.   (3) a. Ribitol is assimilated 5
       b. Ribitol is not assimilated .... *W. spegazzinii*
5.   (4) a. Growth at 37 °C .... *W. edaphicus*
       b. Growth is absent at 37 °C 6
6.   (5) a. Ascospores observed on 5% MEA .... *W. sydowiorum*
       b. Ascospores not observed on 5% MEA .... *W. arborarius*
7.   (1) a. Raffinose is assimilated 8
       b. Raffinose is not assimilated 19
8.   (7) a. Nitrate is assimilated 9
       b. Nitrate is not assimilated 15
9.   (8) a. L-Rhamnose is assimilated 10
       b. L-Rhamnose is not assimilated 12
10.   (9) a. L-Arabinose is assimilated .... *W. ciferrii*
       b. L-Arabinose is not assimilated 11
11.   (10) a. Sucrose is assimilated .... *W. psychrolipolyticus*
       b. Sucrose is not assimilated .... *W. xylosivorus*
12.   (9) a. Growth in vitamin-free medium 13
       b. Growth is absent in vitamin-free medium .... *W. subpelliculosus*
13.   (12) a. Soluble starch is assimilated 14
       b. Soluble starch is not assimilated .... *W. lynferdii*
14.   (13) a. D-Arabinose is assimilated .... *W. myanmarensis*
       b. D-Arabinose is not assimilated .... *W. anomalus*
15.   (8) a. Ribitol is assimilated 16
       b. Ribitol is not assimilated 17
16.   (15) a. Galactose is assimilated .... *W. strasburgensis*
       b. Galactose is not assimilated .... *W. rabaulensis*
17.   (15) a. Growth in vitamin-free medium .... *W. patagonicus*
       b. Growth is absent in vitamin-free medium 18
18.   (17) a. Citrate is assimilated .... *W. onychis*
       b. Citrate is not assimilated .... *W. siamensis*
19.   (7) a. 2-Keto-D-gluconate is assimilated 20
       b. 2-Keto-D-gluconate is not assimilated 21

20. (19) a. D-Glucitol is assimilated . . . . *W. mucosus*
      b. D-Glucitol is not assimilated . . . . *.W. xylosicus*
21. (19) a. D-Arabinose is assimilated 22
      b. D-Arabinose is not assimilated 23
22. (21) a. Growth at 37 °C . . . . *W. sylviae*
      b. Growth is absent at 37 °C . . . . *W. mori*
23. (21) a. Galactose is assimilated 24
      b. Galactose is not assimilated 27
24. (23) a. L-Arabinose is assimilated . . . . *W. silvicola*
      b. L-Arabinose is not assimilated 25
25. (24) a. Sucrose is assimilated . . . . *W. scolytoplatypi*
      b. Sucrose is not assimilated 26
26. (24) a. D-Mannitol is assimilated . . . . *W. nanensis*
      b. D-Mannitol is not assimilated . . . .*W. chambardii*
27. (23) a. L-Sorbose is assimilated . . . . *W. pijperi*
      b. L- Sorbose is not assimilated 28
28. (27) a. D-Xylose is assimilated 29
      b. D-Xylose is not assimilated . . . . *W. tratensis*
29. (28) a. Sucrose is assimilated 30
      b. Sucrose is not assimilated 36
30. (29) a. Cellobiose is assimilated 31
      b. Cellobiose is not assimilated . . . . *W. queroliae*
31. (30) a. D-Glucitol is assimilated 32
      b. D-Glucitol is not assimilated . . . . *W. chaumierensis*
32. (31) a. Growth at 37 °C 33
      b. Growth is absent at 37 °C 34
33. (32) a. L-Arabinose is assimilated . . . . *W. bovis*
      b. L-Arabinose is not assimilated . . . . *W. canadensis*
34. (32) a. Nitrate is assimilated 35
      b. Nitrate is not assimilated . . . . *W. hampshirensis*
35. (34) a. True hyphae are formed . . . . *W. bisporus*
      b. True hyphae are not formed . . . . *W. alni*
36. (29) a. Citrate is assimilated . . . . *W. menglaensis*
      b. Citrate is not assimilated 37
37. (36) a. Growth at 37 °C . . . . *W. ochangensis*
      b. Growth is absent at 37 °C . . . . *W. lannaensis*

Table 3. Key characteristics of species assigned to the genus *Wickerhamomyces*.

Species	Ga	Sor	DXy	LAr	DAr	Rh	Su	Cel	Mlb	Raf	St	Rbl	DGlu	Man	Glt	2-ket	Cit	NO_3	-V	37 °C	Ascospores on 5% MEA	True Hyphae
W. alni	-	-	+	-	-	+	+	+	-	-	-	v	+	+	-	-	+	+	-	-	+	-
W. anomalus	v	-	v	v	-	-	+	+	-	+	+	v	+	+	-	-	+	+	+	v	+	-
W. arborarius	+	1/-	+	v	v	1/+	+	v	+	+	+	+	+	+	-	n	+	+	n	-	-	-
W. bisporus	-	-	+	w/-	-	+	+	+	-	-	-	v	w/+	v	-	-	+	+	-	-	n	+
W. bovis	-	-	+	+	-	v	+	+	-	-	v	-	+	+	-	-	+	-	-	+	+	-
W. canadensis	-	-	+	-	-	w/+	+	+	-	-	-	v	+	w/+	-	-	+	v	-	+	+	v
W. chambardii	+	-	-	-	-	-	-	+	-	-	-	-	-	-	-	-	w	-	-	-	+	-
W. chaumierensis	-	-	+	-	n	-	+	+	-	-	n	n	-	-	n	-	n	-	n	-	n	-
W. ciferrii	+	-	w/+	w/+	-	w/+	+	w/+	-	+	+	+	+	+	-	-	+	+	+	w/-	+	v
W. edaphicus	+	-	+	w/-	-	+	+	+	+	+	+	+	+	+	1/-	-	+	+	+	w	n	-
W. hampshirensis	-	-	+	-	-	s	+	+	-	-	-	w/+	+	v	-	-	s	-	-	-	+	-
W. kurtzmanii	-	-	-	-	-	-	+	-	w	-	-	-	-	w	-	n	-	+	-	-	-	-
W. lynferdii	+	-	-	-	-	-	+	+	-	+	-	+	+	+	-	-	+	+	+	-	+	-
W. menglaensis	-	-	w	w	-	w	-	+	-	-	w	-	+	+	-	-	+	+	+	n	-	-
W. mori	-	+	-	-	w	-	+	-	-	-	-	-	n	+	-	-	w	-	+	-	-	w
W. mucosus	-	+	+	-	v	-	+	+	-	-	+	-	+	w/+	-	+	-	-	-	-	+	-
W. myanmarensis	+	-	s	w	s	-	+	s	-	s	+	+	+	+	-	n	+	+	+	-	n	+
W. ochangensis	-	-	+	-	-	+	-	+	-	-	-	-	+	+	-	-	-	+	n	+	n	-
W. onychis	-	-	+	v	v	-	+	+	-	+	-	-	+	+	-	-	+	-	-	+	+	-
W. orientalis	+	n	w	w	w	w/-	+	+	w	w	-	w	n	w	n	n	-	-	-	+	-	+

Table 3. *Cont.*

Species	Ga	Sor	DXy	LAr	DAr	Rh	Su	Cel	Mlb	Raf	St	Rbl	DGlu	Man	Glt	2-ket	Cit	NO$_3$	-V	37 °C	Ascospores on 5% MEA	True Hyphae
																					Growth in/at *	
W. patago-nicus	w	-	+	-	-	+	n	w	-	w	w	-	w	-	-	n	-	-	+	-	n	-
W. pijperi	-	+	+	-	-	-	-	+	-	-	-	-	+	+	-	-	v	-	-	-	+	-
W. psy-chrolipolyti-cus	-	-	+	-	+	+	+	+	-	+	+	-	+	+	n	n	+	+	n	-	-	-
W. queroliae	-	-	+	+	-	+	+	-	-	-	-	+	+	+	-	-	w/s	+	-	+	-	-
W. rabaulensis	-	-	+	+	-	v	+	+	-	+	-	+	+	+	-	-	+	-	-	+	+	-
W. scolyto-platypi	+	-	s	-	-	s	+	+	-	-	+	s	+	+	-	-	-	+	-	-	+	-
W. siamensis	s	-	s	-	-	-	+	w	-	w	w	-	w	w	-	-	-	-	-	+	+	-
W. silvicola	+	v	+	+	-	+	v	+	-	-	-	+	+	v	-	-	v	+	-	v	+	v
W. spegazzinii	+	-	+	-	-	+	+	+	+	+	s	-	+	+	-	-	w	+	+	+	+	-
W. strasbur-gensis	+	-	+	+	-	+	+	+	-	+	-	+	+	+	-	-	+	-	-	v	+	-
W. subpel-liculosus	v	-	v	v	v	-	+	v	-	+	v	v	+	+	-	-	+	+	-	v	+	v
W. sydowio-rum	+	-	v	+	-	+	+	+	+	+	v	+	+	+	-	-	+	+	+	-	+	-
W. sylviae	v	-	s/-	+	+	+,-	w/-	s/-	-	-	w/+	s/-	-	-	-	-	-	v	+	+	-	-
W. tratensis	-	-	-	-	-	-	-	v	-	-	-	-	v	v	-	-	-	n	n	+	n	-
W. xylosicus	-	+	+	-	-	-	+	+	-	-	-	-	-	+	-	w	-	-	n	-	+	-
W. xylosivorus	w/-	-	+	-	-	+	-	+	-	w	-	-	+	-	n	-	-	+	+	n	-	-
W. lannaensis	-	-	+	-	-	+	-	+	-	-	-	-	+	+	-	-	-	+	w	-	-	-
W. nanensis	+	-	-	-	-	-	-	+	-	-	-	-	-	+	-	-	+	w	w	-	-	-

* Ga = Galactose, Sor = L-Sorbose, DXy = D-Xylose, LAr = L-Arabinose, DAr = D-Arabinose, Rh = L-Rhamnose, Su = Sucrose, Cel = Cellobiose, Mlb = Melibiose, Raf = Raffinose, St = Soluble starch, Rbl = Ribitol, DGlu = D-glucitol, Man = D-Mannitol, Glt = Galactitol, 2-ket = 2-ketogluconic acid, Cit = Citrate, NO$_3$ = Potassium nitrate, -V = vitamin-free medium, 37 °C = Growth at 37 °C. "+" = strong growth or produce, "-" = absence of growth or not produce, "w" = weak growth, "v" = strain variable response, "l" = latent positive, "s" = slow positive, and "n" = no data.

4. Discussion

Traditional methods of identification and characterization for the *Wickerhamomyces* species are based primarily on phenotypical characteristics. These are further recognized as relevant morphological, biochemical, and physiological characteristics [41,128]. However, identification can be difficult because some species have similar appearances, and some biochemical characteristics are consistent across a number of species. In accordance with this evidence, previous species of *Wickerhamomyces* were originally classified into various yeast genera [1,2,4,5,50,70,75]. In 2008, the genus *Wickerhamomyces* was proposed by Kurtzman et al. [1], wherein this genus was clearly separated from other yeast genera based on phylogenetic evidence. Subsequently, some previously identified species were then transferred from the genera *Candida*, *Hansenula*, *Pichia*, and *Williopsis* [1,4,5,50,54,70]. Therefore, molecular phylogenetic analysis is necessary to concretely identify the *Wickerhamomyces* species. Species of the genus *Wickerhamomyces* are known to be widely distributed throughout the world and have been isolated from various habitats as shown in Table 1. Prior to conducting our study, *Wickerhamomyces* consisted of 35 accepted and published species according to molecular phylogenetic analysis. Our phylogenetic results were similar to those of Arastehfar et al. [8] and Shimizu et al. [65] who indicated that *P. myanmarensis* should be placed in the genus *Wickerhamomyces*. Consequently, we have proposed that this yeast species be named *W. myanmarensis*.

Yeast diversity has been investigated in various habitats throughout different regions of Thailand [5,10,15,16,18,19,46,70,78]. *Wickerhamomyces anomalus* was first species reported in Thailand in 2002 [20]. In 2009, the first new species, *W. edaphicus* has been discovered in Thailand [10]. Until now, a total of eight *Wickerhamomyces* species have been found [5,10,20,46,68,70,78,82]. However, *W. siamensis*, *W. tratensis*, and *W. xylosicus* were only known to be from Thailand [5,70,82]. In this study, two new *Wickerhamomyces* species, namely *W. lannaensis* and *W. nanensis*, that were isolated from soil collected from Assam tea plantations in northern Thailand were proposed based on identification through molecular phylogenetic and phenotypic (morphological, biochemical, and physiological characteristics) analyses. Therefore, effective identification of the *Wickerhamomyces* species has increased the number of species found in Thailand to 10 species and has led to 38 global species. This present discovery has increased the number of species of yeast known to be from Thailand and is considered important in terms of stimulating deeper investigations of yeast varieties in Thailand. Ultimately, these findings will help researchers gain a better understanding of the distribution and ecology of *Wickerhamomyces*.

Many species of the genus *Wickerhamomyces* have been investigated, and some strains have been used in a variety of biotechnology, food, and beverage industries, as well as in medical and agricultural fields [83–102]. Despite the fact that many *Wickerhamomyces* species can survive in a variety of environments, climate change has had an impact on both the terrestrial biome and the aquatic environment. These environments are known to serve as habitats for a number of microorganisms [130–134] and may have an impact on the global diversity and distribution of *Wickerhamomyces*. Therefore, in addition to studying the diversity and distribution of newly identified species, future research should focus on the effects of climate change on *Wickerhamomyces*.

Author Contributions: Conceptualization, S.N., J.K., N.S. and S.L. (Saisamorn Lumyong); methodology, S.N., J.K., N.S. and S.L. (Savitree Limtong); software, N.S. and J.K.; validation, N.S., J.K., S.L. (Savitree Limtong) and S.L. (Saisamorn Lumyong); formal analysis, S.N., J.K., N.S.; investigation, S.N., J.K. and N.S.; resources, J.K., N.S. and S.K.; data curation, N.S., J.K. and N.S.; writing—original draft, S.N., J.K. and N.S.; writing—review and editing, J.K., S.N., N.S., S.L. (Savitree Limtong) and S.L. (Saisamorn Lumyong); supervision, S.L. (Saisamorn Lumyong). All authors have read and agreed to the published version of the manuscript.

Funding: The authors gratefully acknowledge the financial support provided from Chiang Mai University, Thailand.

Institutional Review Board Statement: Not applicable.

Informed Consent Statement: Not applicable.

Data Availability Statement: The DNA sequence data obtained from this study have been deposited in GenBank under accession numbers; D1/D2 domains (MT639220, MT623569, MT613722, MT613875 and MT623569) and ITS (OK135750, OK135752, OK135753, OK143510 and OK143511).

Acknowledgments: The authors are grateful to Russell Kirk Hollis for kind help in the English correction.

Conflicts of Interest: The authors declare no conflict of interest.

References

1. Kurtzman, C.P.; Robnett, C.J.; Basehoar-Powers, E. Phylogenetic relationships among species of *Pichia*, *Issatchenkia* and *Williopsis* determined from multigene sequence analysis, and the proposal of *Barnettozyma* gen. nov., *Lindnera* gen. nov. and *Wickerhamomyces* gen. nov. *FEMS. Yeast. Res.* **2008**, *8*, 939–954. [CrossRef] [PubMed]

2. Kurtzman, C.P. Wickerhamomyces Kurtzman, Robnett & Basehoar-Powers. In *The Yeasts*; Kurtzman, C.P., Fell, J.W., Boekhout, T., Eds.; Elsevier: Amsterdam, The Netherlands, 2011; pp. 899–917.

3. Yarrow, D. Methods for the isolation, maintenance and identification of yeasts. In *The Yeasts*; Kurtzman, C.P., Fell, J.W., Eds.; Elsevier: Amsterdam, The Netherlands, 1998; pp. 77–100.

4. Kobayashi, R.; Kanti, A.; Kawasaki, H. Three novel species of d-xylose-assimilating yeasts, *Barnettozyma xylosiphila* sp. nov., *Barnettozyma xylosica* sp. nov. and *Wickerhamomyces xylosivorus* f.a., sp. nov. *Int. J. Syst. Evol. Microbiol.* **2017**, *67*, 3971–3976. [CrossRef]

5. Limtong, S.; Nitiyon, S.; Kaewwichian, R.; Jindamorakot, S.; Am-In, S.; Yongmanitchai, W. *Wickerhamomyces xylosica* sp. nov. and *Candida phayaonensis* sp. nov., two xylose-assimilating yeast species from soil. *Int. J. Syst. Evol. Microbiol.* **2012**, *62*, 2786–2792. [CrossRef]

6. Zhou, Y.; Jia, B.S.; Han, P.J.; Wang, Q.M.; Li, A.H.; Zhou, Y.G. *Wickerhamomyces kurtzmanii* sp. nov. an ascomycetous yeast isolated from crater lake water, Da Hinggan Ling Mountain, China. *Curr. Microbiol.* **2019**, *76*, 1537–1544. [CrossRef] [PubMed]

7. Index Fungorum. Available online: http://www.indexfungorum.org (accessed on 25 July 2021).

8. Arastehfar, A.; Bakhtiari, M.; Daneshnia, F.; Fang, W.; Sadati, S.K.; Al-Hatmi, A.M.; Groenewald, M.; Sharifi-Mehr, H.; Liao, W.; Pan, W.; et al. First fungemia case due to environmental yeast *Wickerhamomyces myanmarensis*: Detection by multiplex qPCR and antifungal susceptibility. *Future Microbiol.* **2019**, *14*, 267–274. [CrossRef] [PubMed]

9. Nagatsuka, Y.; Kawasaki, H.; Seki, T. *Pichia myanmarensis* sp. nov., a novel cation-tolerant yeast isolated from palm sugar in Myanmar. *Int. J. Syst. Evol. Microbiol.* **2005**, *55*, 1379–1382. [CrossRef] [PubMed]

10. Limtong, S.; Yongmanitchai, W.; Kawasaki, H.; Fujiyama, K. *Wickerhamomyces edaphicus* sp. nov. and *Pichia jaroonii* sp. nov., two ascomycetous yeast species isolated from forest soil in Thailand. *FEMS Yeast Res.* **2009**, *9*, 504–510. [CrossRef] [PubMed]

11. Shin, K.S.; Bae, K.S.; Lee, K.H.; Park, D.S.; Kwon, G.S.; Lee, J.B. *Wickerhamomyces ochangensis* sp. nov., an ascomycetous yeast isolated from the soil of a potato field. *Int. J. Syst. Evol. Microbiol.* **2011**, *61*, 2543–2546. [CrossRef]

12. Phaff, H.J.; Miller, M.W.; Miranda, M. *Hansenula alni*, a new heterothallic species of yeast from exudates of alder trees. *Int. J. Syst. Evol.* **1979**, *29*, 60–63. [CrossRef]

13. Hansen, E.C. Sur la germination des spores chez les *Saccharomyces*. *Ann. Microgr.* **1891**, *3*, 449–474.

14. Sláviková, E.; Vadkertiová, R.; Vránová, D. Yeasts colonizing the leaf surfaces. *J. Basic Microbiol.* **2007**, *47*, 344–350. [CrossRef]

15. Into, P.; Pontes, A.; Sampaio, J.P.; Limtong, S. Yeast diversity associated with the phylloplane of corn plants cultivated in Thailand. *Microorganisms* **2020**, *8*, 80. [CrossRef] [PubMed]

16. Kumla, J.; Nundaeng, S.; Suwannarach, N.; Lumyong, S. Evaluation of multifarious plant growth promoting trials of yeast isolated from the soil of Assam tea (*Camellia sinensis* var. assamica) plantations in Northern Thailand. *Microorganisms* **2020**, *8*, 1168. [CrossRef]

17. Carrasco, M.; Rozas, J.M.; Barahona, S.; Alcaíno, J.; Cifuentes, V.; Baeza, M. Diversity and extracellular enzymatic activities of yeasts isolated from King George Island, the sub-Antarctic region. *BMC Microbiol.* **2012**, *12*, 251. [CrossRef]

18. Kaewkrajay, C.; Chanmethakul, T.; Limtong, S. Assessment of diversity of culturable marine yeasts associated with corals and zoanthids in the gulf of Thailand, South China Sea. *Microorganisms* **2020**, *8*, 474. [CrossRef] [PubMed]

19. Satianpakiranakorn, P.; Khunnamwong, P.; Limtong, S. Yeast communities of secondary peat swamp forests in Thailand and their antagonistic activities against fungal pathogens cause of plant and postharvest fruit diseases. *PLoS ONE* **2020**, *15*, e0230269. [CrossRef] [PubMed]

20. Limtong, S.; Sintara, S.; Suwannarit, P.; Lotong, N. Yeast diversity in Thai traditional alcoholic starter. *Kasetsart J. (Nat. Sci.)* **2002**, *36*, 149–158.

21. Hoondee, P.; Wattanagonniyom, T.; Weeraphan, T.; Tanasupawat, S.; Savarajara, A. Occurrence of oleaginous yeast from mangrove forest in Thailand. *World J. Microbiol. Biotechnol.* **2019**, *35*, 108. [CrossRef] [PubMed]

22. Tepeeva, A.N.; Glushakova, A.M.; Kachalkin, A.V. Yeast communities of the Moscow city soils. *Microbiology* **2018**, *87*, 407–415. [CrossRef]

23. Koricha, A.D.; Han, D.Y.; Bacha, K.; Bai, F.Y. Diversity and distribution of yeasts in indigenous fermented foods and beverages of Ethiopia. *J. Sci. Food Agric.* **2020**, *100*, 3630–3638. [CrossRef] [PubMed]

24. Vadkertiová, R.; Molnárová, J.; Vránová, D.; Sláviková, E. Yeasts and yeast-like organisms associated with fruits and blossoms of different fruit trees. *Can. J. Microbiol.* **2012**, *58*, 1344–1352. [CrossRef] [PubMed]

25. PrasannaKumar, C.; Velmurugan, S.; Subramanian, K.; Pugazhvendan, S.R.; Nagaraj, D.S.; Khan, K.F.; Sadiappan, B.; Manokaran, S.; Hemalatha, K.R. DNA barcoding analysis of more than 1000 marine yeast isolates reveals previously unrecorded species. *BioRxiv* **2020**, 1–39.

26. Abu-Mejdad, N.M.J.A.; Al-Badran, A.I.; Al-Saadoon, A.H.; Minati, M.H. A new report on gene expression of three killer toxin genes with antimicrobial activity of two killer toxins in Iraq. *Bull. Natl. Res. Cent.* **2020**, *44*, 162. [CrossRef]

27. Horváth, E.; Sipiczki, M.; Csoma, H.; Miklós, I. Assaying the effect of yeasts on growth of fungi associated with disease. *BMC Microbiol.* **2020**, *20*, 320. [CrossRef]

28. Hamoudi-Belarbi, L.; Nouri, L.H.; Belkacemi, K. Effectiveness of convective drying to conserve indigenous yeasts with high volatile profile isolated from algerian fermented raw bovine milk (Rayeb). *J. Food Sci. Technol.* **2016**, *36*, 476–484. [CrossRef]

29. You, L.; Wang, S.; Zhou, R.; Hu, X.; Chu, Y.; Wang, T. Characteristics of yeast flora in Chinese strong-flavoured liquor fermentation in the Yibin region of China. *J. Inst. Brew.* **2016**, *122*, 517–523. [CrossRef]

30. Borling Welin, J.; Lyberg, K.; Passoth, V.; Olstorpe, M. Combined moist airtight storage and feed fermentation of barley by the yeast *Wickerhamomyces anomalus* and a lactic acid bacteria consortium. *Front. Plant. Sci.* **2015**, *6*, 270. [CrossRef] [PubMed]

31. Pires, J.F.; de Souza Cardoso, L.; Schwan, R.F.; Silva, C.F. Diversity of microbiota found in coffee processing wastewater treatment plant. *World J. Microbiol. Biotechnol.* **2017**, *33*, 211. [CrossRef] [PubMed]

32. Delgado-Ospina, J.; Triboletti, S.; Alessandria, V.; Serio, A.; Sergi, M.; Paparella, A.; Rantsiou, K.; Chaves-López, C. Functional biodiversity of yeasts isolated from Colombian fermented and dry cocoa beans. *Microorganisms* **2020**, *8*, 1086. [CrossRef] [PubMed]

33. da Conceição, L.E.F.R.; Saraiva, M.A.F.; Diniz, R.H.S.; Oliveira, J.; Barbosa, G.D.; Alvarez, F.; Correa, L.F.; Mezadri, H.; Coutrim, M.X.; Afonso, R.J.; et al. Biotechnological potential of yeast isolates from cachaça: The Brazilian spirit. *J. Ind. Microbiol. Biotechnol.* **2015**, *42*, 237–246. [CrossRef] [PubMed]

34. Da Cunha, A.C.; Gomes, L.S.; Godoy-Santos, F.; Faria-Oliveira, F.; Teixeira, J.A.; Sampaio, G.M.S.; Trópia, M.J.M.; Castro, I.M.; Lucas, C.; Brandão, R.L. High-affinity transport, cyanide-resistant respiration, and ethanol production under aerobiosis underlying efficient high glycerol consumption by *Wickerhamomyces anomalus*. *J. Ind. Microbiol. Biotechnol.* **2019**, *46*, 709–723. [CrossRef] [PubMed]

35. Cunha, A.C.; Santos, R.A.; Riaño-Pachon, D.M.; Squina, F.M.; Oliveira, J.V.; Goldman, G.H.; Souza, A.T.; Gomes, L.S.; Godoy-Santos, F.; Teixeira, J.A. Draft genome sequence of *Wickerhamomyces anomalus* LBCM1105, isolated from cachaça fermentation. *Genet. Mol. Biol.* **2020**, *43*, 1–5. [CrossRef] [PubMed]

36. James, S.A.; Barriga, E.J.C.; Barahona, P.P.; Harrington, T.C.; Lee, C.F.; Bond, C.J.; Roberts, I.N. *Wickerhamomyces arborarius* fa, sp. nov., an ascomycetous yeast species found in arboreal habitats on three different continents. *Int. J. Syst. Evol.* **2014**, *64*, 1057–1061. [CrossRef]

37. Beck, O. Eine neue Endomyces-Art, Endomyces bisporus. *Ann. Mycol.* **1922**, *20*, 219–227.

38. Linnakoski, R.; Lasarov, I.; Veteli, P.; Tikkanen, O.P.; Viiri, H.; Jyske, T.; Kasanen, R.; Duong, T.A.; Wingfield, M.J. Filamentous fungi and yeasts associated with mites phoretic on Ips typographus in eastern Finland. *Forests* **2021**, *12*, 743. [CrossRef]

39. Dohet, L.; Gregoire, J.C.; Berasategui, A.; Kaltenpoth, M.; Biedermann, P.H. Bacterial and fungal symbionts of parasitic *Dendroctonus* bark beetles. *FEMS Microbiol. Ecol.* **2016**, *92*, fiw129. [CrossRef] [PubMed]

40. Uden, V.N.; Carmo Souza, L.D. Yeast from the bovine caecum. *J. Gen. Microbiol.* **1957**, *16*, 385–395. [CrossRef] [PubMed]

41. Wickerham, L.J. *Taxonomy of Yeasts*; U.S. Department of Agriculture: Washington, DC, USA, 1951; pp. 1–56.

42. Ramirez, C. Note sur deux nouvelles espèces de levures isolées de divers milieux. *Rev. Mycol.* **1954**, *19*, 98–102.

43. Groenewald, M.; Robert, V.; Smith, M.T. Five novel *Wickerhamomyces*-and *Metschnikowia*-related yeast species, *Wickerhamomyces chaumierensis* sp. nov., *Candida pseudoflosculorum* sp. nov., *Candida danieliae* sp. nov., *Candida robnettiae* sp. nov. and *Candida eppingiae* sp. nov., isolated from plants. *Int. J. Syst. Evol.* **2011**, *61*, 2015–2022. [CrossRef] [PubMed]

44. Lodder, J. Über einige durch das "Centraalbureau voor Schimmelcultures" neuerworbene sporogene Hefearten. *Zentralbl. Bakteriol. Parasitenkd. Abt. II* **1932**, *86*, 227–253.

45. Sitepu, I.R.; Enriquez, L.L.; Nguyen, V.; Doyle, C.; Simmons, B.A.; Singer, S.W.; Fry, R.; Simmons, C.W.; Boundy-Mills, K. Ethanol production in switchgrass hydrolysate by ionic liquid-tolerant yeasts. *Bioresour. Technol. Rep.* **2019**, *7*, 100275. [CrossRef]

46. Limtong, S.; Kaewwichian, R. The diversity of culturable yeasts in the phylloplane of rice in Thailand. *Ann. Microbiol.* **2015**, *65*, 667–675. [CrossRef]

47. Patel, K.; Patel, F.R. Optimization of culture conditions for biosurfactant production by *Wickerhamomyces edaphicus* isolated from mangrove region of Mundra, Kutch, Gujarat. *Indian J. Sci. Technol.* **2020**, *13*, 1935–1943. [CrossRef]

48. Kurtzman, C.P. Two new species of *Pichia* from arboreal habitats. *Mycologia* **1987**, *79*, 410–417. [CrossRef]

49. Suh, S.O.; Zhou, J. Yeasts associated with the curculionid beetle *Xyloterinus politus*: *Candida xyloterini* sp. nov., *Candida palmyrensis* sp. nov. and three common ambrosia yeasts. *Int. J. Syst. Evol. Microbiol.* **2010**, *60*, 1702–1708. [CrossRef]

50. van der Walt, J.P.; Johannsen, E. *Hansenula lynferdii* sp. nov. *Antonie Van Leeuwenhoek* **1975**, *41*, 13–16. [CrossRef]

51. Chai, C.Y.; Huang, L.N.; Cheng, H.; Liu, W.J.; Hui, F.L. *Wickerhamomyces menglaensis* fa, sp. nov., a yeast species isolated from rotten wood. *Int. J. Syst. Evol. Microbiol.* **2019**, *69*, 1509–1514. [CrossRef] [PubMed]

52. Hui, F.L.; Chen, L.; Chu, X.Y.; Niu, Q.H.; Ke, T. *Wickerhamomyces mori* sp. nov., an anamorphic yeast species found in the guts of wood-boring insect larvae. *Int. J. Syst. Evol. Microbiol.* **2013**, *63*, 1174–1178. [CrossRef] [PubMed]

53. Wlckerham, L.J.; Kurtzman, C.P. Two new saturn-spored species of *Pichia*. *Mycologia* **1971**, *63*, 1013–1018. [CrossRef]
54. Yarrow, D. *Pichia onychis* sp. n. Antonie Van Leeuwenhoek. *J. Microbiol. Serol.* **1965**, *31*, 465–467.
55. Cioch-Skoneczny, M.; Satora, P.; Skoneczny, S.; Skotniczny, M. Biodiversity of yeasts isolated during spontaneous fermentation of cool climate grape musts. *Arch. Microbiol.* **2021**, *203*, 153–162. [CrossRef]
56. Abu-Mejdad, N.M.J.A.; Al-Badran, A.I.; Al-Saadoon, A.H. New record of ascomycetous yeasts strains from soil in Basrah, Iraq. *Drug Invent. Today.* **2019**, *11*, 3073–3080.
57. Bah, A.; Ferjani, R.; Fhoula, I.; Gharbi, Y.; Najjari, A.; Boudabous, A.; Ouzari, H.I. Microbial community dynamic in tomato fruit during spontaneous fermentation and biotechnological characterization of indigenous lactic acid bacteria. *Ann. Microbiol.* **2019**, *69*, 41–49. [CrossRef]
58. Ooi, T.S.; Sepiah, M.; Khairul Bariah, S. Diversity of yeast species identified during spontaneous shallow box fermentation of cocoa beans in Malaysia. *Int. J. Eng. Innov. Technol.* **2016**, *3*, 379–385.
59. Maciel, N.O.; Johann, S.; Brandão, L.R.; Kucharíková, S.; Morais, C.G.; Oliveira, A.P.; Freitas, G.J.; Borelli, B.M.; Pellizzari, F.M.; Santos, D.A.; et al. Occurrence, antifungal susceptibility, and virulence factors of opportunistic yeasts isolated from Brazilian beaches. *Mem. Inst. Oswaldo Cruz* **2019**, *114*, 114. [CrossRef] [PubMed]
60. Nasr, S.; Nguyen, H.D.; Soudi, M.R.; Fazeli, S.A.S.; Sipiczki, M. *Wickerhamomyces orientalis* fa, sp. nov.: An ascomycetous yeast species belonging to the *Wickerhamomyces* clade. *Int. J. Syst. Evol. Microbiol.* **2016**, *66*, 2534–2539. [CrossRef]
61. de García, V.; Brizzio, S.; Libkind, D.; Rosa, C.A.; van Broock, M. *Wickerhamomyces patagonicus* sp. nov., an ascomycetous yeast species from Patagonia, Argentina. *Int. J. Syst. Evol. Microbiol.* **2010**, *60*, 1693–1696. [CrossRef] [PubMed]
62. van Walt, J.P.; Tscheuschner, I.T. Three new yeasts. *Antonie Van Leeuwenhoek* **1957**, *23*, 184–190. [CrossRef]
63. Nielsen, D.S.; Teniola, O.D.; Ban-Koffi, L.; Owusu, M.; Andersson, T.S.; Holzapfel, W.H. The microbiology of Ghanaian cocoa fermentations analysed using culture-dependent and culture-independent methods. *Int. J. Food Microbiol.* **2007**, *114*, 168–186. [CrossRef] [PubMed]
64. Abdel-Sater, M.A.; Moubasher, A.H.; Zeinab, S.M. Diversity of yeasts and filamentous fungi in five fresh fruit juices in Egypt. *Curr. Res. Environ.* **2017**, *7*, 356–386. [CrossRef]
65. Shimizu, Y.; Konno, Y.; Tomita, Y. *Wickerhamomyces psychrolipolyticus* fa, sp. nov., a novel yeast species producing two kinds of lipases with activity at different temperatures. *Int. J. Syst. Evol.* **2020**, *70*, 1158–1165. [CrossRef] [PubMed]
66. Rosa, C.A.; Morais, P.B.; Lachance, M.-A.; Santos, R.O.; Melo, W.G.; Viana, R.H.; Bragança, M.A.; Pimenta, R.S. *Wickerhamomyces queroliae* sp. nov. and *Candida jalapaonensis* sp. nov., two yeast species isolated from Cerrado ecosystem in North Brazil. *Int. J. Syst. Evol. Microbiol.* **2009**, *59*, 1232–1236. [CrossRef] [PubMed]
67. Soneda, M.; Uchida, S. A survey on the yeasts. *Bull. Nat. Science Museum* **1971**, *14*, 451–453.
68. Junyapate, K.; Jindamorakot, S.; Limtong, S. *Yamadazyma ubonensis* fa, sp. nov., a novel xylitol-producing yeast species isolated in Thailand. *Antonie Van Leeuwenhoek* **2014**, *105*, 471–480. [CrossRef]
69. Ninomiya, S.; Mikata, K.; Kajimura, H.; Kawasaki, H. Two novel ascomycetous yeast species, *Wickerhamomyces scolytoplatypi* sp. nov. and *Cyberlindnera xylebori* sp. nov., isolated from ambrosia beetle galleries. *Int. J. Syst. Evol. Microbiol.* **2013**, *63*, 2706–2711. [CrossRef] [PubMed]
70. Kaewwichian, R.; Kawasaki, H.; Limtong, S. *Wickerhamomyces siamensis* sp. nov., a novel yeast species isolated from the phylloplane in Thailand. *Int. J. Syst. Evol. Microbiol.* **2013**, *63*, 1568–1573. [CrossRef] [PubMed]
71. Bien, S.; Damm, U. Prunus trees in Germany—A hideout of unknown fungi? *Mycol. Prog.* **2020**, *19*, 667–690. [CrossRef]
72. Hyun, S.H.; Min, J.H.; Lee, H.B.; Kim, H.K.; Lee, J.S. Characteristics of two unrecorded yeasts from wildflowers in Ulleungdo, Korea. *Kor. J. Mycol.* **2014**, *42*, 170–173. [CrossRef]
73. Masiulionis, V.E.; Pagnocca, F.C. *Wickerhamomyces spegazzinii* sp. nov., an ascomycetous yeast isolated from the fungus garden of *Acromyrmex lundii* nest (Hymenoptera: Formicidae). *Int. J. Syst. Evol. Microbiol.* **2016**, *66*, 2141–2145. [CrossRef] [PubMed]
74. Ramírez, C.; Boidin, J. *Saccharomyces chambardi*, nouvelle espèce de levure isolée de liqueur tannante. *De La Société Linnéenne De Lyon* **1954**, *23*, 151–152. [CrossRef]
75. Kurtzman, C.P. Synonomy of the yeast genera *Hansenula* and *Pichia* demonstrated through comparisons of deoxyribonucleic acid relatedness. *Antonie Van Leeuwenhoek* **1984**, *50*, 209–217. [CrossRef]
76. Moubasher, A.H.; Abdel-Sater, M.A.; Soliman, Z. Yeasts and filamentous fungi inhabiting guts of three insect species in Assiut, Egypt. *Mycosphere* **2017**, *8*, 1297–1316. [CrossRef]
77. Scott, D.B.; Van der Walt, J.P. *Hansenula sydowiorum* sp. n. *Antonie Van Leeuwenhoek* **1970**, *36*, 45–48. [CrossRef] [PubMed]
78. Boonmak, C.; Jindamorakot, S.; Kawasaki, H.; Yongmanitchai, W.; Suwanarit, P.; Nakase, T.; Limtong, S. *Candida siamensis* sp. nov., an anamorphic yeast species in the *Saturnispora* clade isolated in Thailand. *FEMS Yeast Res.* **2009**, *9*, 668–672. [CrossRef] [PubMed]
79. Samagaci, L.; Ouattara, H.; Niamké, S.; Lemaire, M. *Pichia* kudrazevii and *Candida nitrativorans* are the most well-adapted and relevant yeast species fermenting cocoa in Agneby-Tiassa, a local Ivorian cocoa producing region. *Int. Food Res. J.* **2016**, *89*, 773–780. [CrossRef] [PubMed]
80. Echeverrigaray, S.; Scariot, F.J.; Foresti, L.; Schwarz, L.V.; Rocha, R.K.M.; da Silva, G.P.; Moreira, J.P.; Delamare, A.P.L. Yeast biodiversity in honey produced by stingless bees raised in the highlands of southern Brazil. *Int. J. Food Microbiol.* **2021**, *347*, 109200. [CrossRef] [PubMed]

81. Francesca, N.; Carvalho, C.; Almeida, P.M.; Sannino, C.; Settanni, L.; Sampaio, J.P.; Moschetti, G. *Wickerhamomyces sylviae* f.a., sp. nov., an ascomycetous yeast species isolated from migratory birds. *Int. J. Syst. Evol. Microbiol.* **2013**, *63*, 4824–4830. [CrossRef] [PubMed]

82. Nakase, T.; Jindamorakot, S.; Am-In, S.; Ninomiya, S.; Kawasaki, H. *Wickerhamomyces tratensis* sp. nov. and *Candida namnaoensis* sp. nov., two novel ascomycetous yeast species in the *Wickerhamomyces* clade found in Thailand. *J. Gen. Appl. Microbiol.* **2012**, *58*, 145–152. [CrossRef]

83. Ravasio, D.; Carlin, S.; Boekhout, T.; Groenewald, M.; Vrhovsek, U.; Walther, A.; Wendland, J. Adding flavor to beverages with non-conventional yeasts. *Fermentation* **2018**, *4*, 15. [CrossRef]

84. Yan, S.; Xiangsong, C.; Xiang, X. Improvement of the aroma of lily rice wine by using aroma-producing yeast strain *Wickerhamomyces anomalus* HN006. *AMB Express* **2019**, *9*, 89. [CrossRef]

85. Li, W.; Shi, C.; Guang, J.; Ge, F.; Yan, S. Development of Chinese chestnut whiskey: Yeast strains isolation, fermentation system optimization, and scale-up fermentation. *AMB Express* **2021**, *11*, 17. [CrossRef] [PubMed]

86. Czarnecka, M.; Żarowska, B.; Połomska, X.; Restuccia, C.; Cirvilleri, G. Role of biocontrol yeasts *Debaryomyces hansenii* and *Wickerhamomyces anomalus* in plants' defence mechanisms against *Monilinia fructicola* in apple fruits. *Food Microbiol.* **2019**, *83*, 1–8. [CrossRef]

87. Limtong, S.; Into, P.; Attarat, P. Biocontrol of rice seedling rot disease caused by *Curvularia lunata* and *Helminthosporium oryzae* by epiphytic yeasts from plant leaves. *Microorganisms* **2020**, *8*, 647. [CrossRef] [PubMed]

88. Lima, J.R.; Gondim, D.M.F.; Oliveira, J.T.A.; Oliveira, F.S.A.; Gonçalves, L.R.B.; Viana, F.M.P. Use of killer yeast in the management of postharvest papaya anthracnose. *Postharvest Biol. Technol.* **2013**, *83*, 58–64. [CrossRef]

89. Fernandez-San Millan, A.; Farran, I.; Larraya, L.; Ancin, M.; Arregui, L.M.; Veramendi, J. Plant growth-promoting traits of yeasts isolated from Spanish vineyards: Benefits for seedling development. *Microbiol. Res.* **2020**, *237*, 126480. [CrossRef]

90. Walker, G.M. *Pichia anomala*: Cell physiology and biotechnology relative to other yeasts. *Antonie Van Leeuwenhoek* **2011**, *99*, 25–34. [CrossRef] [PubMed]

91. Passoth, V.; Olstorpe, M.; Schnürer, J. Past, present and future research directions with *Pichia anomala*. *Antonie Van Leeuwenhoek* **2011**, *99*, 121–125. [CrossRef] [PubMed]

92. Golubev, W.I. Antifungal activity of *Wickerhamomyces silvicola*. *Microbiology* **2015**, *84*, 610–615. [CrossRef]

93. Golubev, W.I. Taxonomic specificity of the sensitivity to the *Wickerhamomyces bovis* fungistatic mycocin. *Microbiology* **2016**, *85*, 444–448. [CrossRef]

94. Silva, C.F.; Schwan, R.F.; Dias, Ë.S.; Wheals, A.E. Microbial diversity during maturation and natural processing of coffee cherries of *Coffea arabica* in Brazil. *Int. J. Food Microbiol.* **2000**, *60*, 251–260. [CrossRef]

95. Zhou, N.; Schifferdecker, A.J.; Gamero, A.; Compagno, C.; Boekhout, T.; Piškur, J.; Knecht, W. *Kazachstania gamospora* and *Wickerhamomyces subpelliculosus*: Two alternative baker's yeasts in the modern bakery. *Int. J. Food Microbiol.* **2017**, *250*, 45–58. [CrossRef] [PubMed]

96. Adelabu, B.; Kareem, S.; Oluwafemi, F.; Adeogun, A. Consolidated bioprocessing of ethanol from corn straw by *Saccharomyces diastaticus* and *Wikerhamomyces chambardii*. *Food Appl. Biosci. J.* **2018**, *6*, 1–17.

97. Adelabu, B.; Kareem, S.O.; Adeogun, A.I.; Wakil, S.M. Optimization of cellulase enzyme from sorghum straw by yeasts isolated from plant feeding-termite *Zonocerus variegatus*. *Food Appl. Biosci. J.* **2019**, *7*, 81–101.

98. Dejwatthanakomol, C.; Anuntagool, J.; Morikawa, M.; Thaniyavarn, J. Production of biosurfactant by *Wickerhamomyces anomalus* PY189 and its application in lemongrass oil encapsulation. *Sci. Asia.* **2016**, *42*, 252–258. [CrossRef]

99. Fernandes, N.D.A.T.; de Souza, A.C.; Simoes, L.A.; Dos Reis, G.M.F.; Souza, K.T.; Schwan, R.F.; Dias, D.R. Eco-friendly biosurfactant from *Wickerhamomyces anomalus* CCMA 0358 as larvicidal and antimicrobial. *Microbiol. Res.* **2020**, *241*, 126571. [CrossRef]

100. Samadlouie, H.R.; Nurmohamadi, S.; Moradpoor, F.; Gharanjik, S. Effect of low-cost substrate on the fatty acid profiles of *Mortierella alpina* CBS 754.68 and *Wickerhamomyces siamensis* SAKSG. *Biotechnol. Biotechnol. Equip.* **2018**, *32*, 1228–1235. [CrossRef]

101. Palladino, F.; Rodrigues, R.C.; Cadete, R.M.; Barros, K.O.; Rosa, C.A. Novel potential yeast strains for the biotechnological production of xylitol from sugarcane bagasse. *Biofuels Bioprod. Biorefining* **2021**, *15*, 690–702. [CrossRef]

102. Seymour, F.R.; Slodki, M.E.; Plattner, R.D.; Stodola, R.M. Methylation and acetolysis of extracellular D-mannans from yeast. *Carbohydr. Res.* **1976**, *48*, 225–237. [CrossRef]

103. Dufour, J.P.; Verstrepen, K.; Derdelinckx, G. Brewing yeasts. In *Yeasts in Food*; Boekhout, T., Robert, V., Eds.; Elsevier: Amsterdam, The Netherlands, 2003; pp. 347–388.

104. Timke, M.; Wang-Lieu, N.Q.; Altendorf, K.; Lipski, A. Fatty acid analysis and spoilage potential of biofilms from two breweries. *J. Appl. Microbiol.* **2005**, *99*, 1108–1122. [CrossRef] [PubMed]

105. Bonjean, B.; Guillaume, L.D. Yeasts in bread and baking products. In *Yeasts in Food*; Boekhout, T., Robert, V., Eds.; Elsevier: Amsterdam, The Netherlands, 2003; pp. 289–307.

106. Lanciotti, R.; Sinigaglia, M.; Gardini, F.; Guerzoni, M.E. *Hansenula anomala* as spoilage agent of cream-filled cakes. *Microbiol. Res.* **1998**, *153*, 145–148. [CrossRef]

107. Paula, C.R.; Krebs, V.L.; Auler, M.E.; Ruiz, L.S.; Matsumoto, F.E.; Silva, E.H.; Diniz, E.M.; Vaz, F.A. Nosocomial infection in newborns by *Pichia anomala* in a Brazilian intensive care unit. *Med. Mycol.* **2006**, *44*, 479–484. [CrossRef]

108. Linton, C.J.; Borman, A.M.; Cheung, G.; Holmes, A.D.; Szekely, A.; Palmer, M.D.; Bridge, P.D.; Campbell, C.K.; Johnson, E.M. Molecular identification of unusual pathogenic yeast isolates by large ribosomal subunit gene sequencing: 2 years of experience at the United Kingdom Mycology Reference Laboratory. *J. Clin. Microbiol.* **2007**, *45*, 1152–1158. [CrossRef] [PubMed]

109. Oliveira, V.K.P.; Ruiz, L.d.S.; Oliveira, N.A.J.; Moreira, D.; Hahn, R.C.; Melo, A.S.d.A.; Nishikaku, A.S.; Paula, C.R. Fungemia caused by *Candida* species in a children's public hospital in the city of São Paulo, Brazil: Study in the period 2007–2010. *Rev. Do Inst. De Med. Trop. De São Paulo* **2014**, *56*, 301–305. [CrossRef] [PubMed]

110. Kamoshita, M.; Matsumoto, Y.; Nishimura, K.; Katono, Y.; Murata, M.; Ozawa, Y.; Shimmura, S.; Tsubota, K. *Wickerhamomyces anomalus* fungal keratitis responds to topical treatment with antifungal micafungin. *J. Infect. Chemother.* **2015**, *21*, 141–143. [CrossRef] [PubMed]

111. Dutra, V.R.; Silva, L.F.; Oliveira, A.N.M.; Beirigo, E.F.; Arthur, V.M.; da Silva, R.B.; Ferreira, T.B.; Andrade-Silva, L.; Silva, M.V.; Fonseca, F.M.; et al. Fatal case of fungemia by *Wickerhamomyces anomalus* in a pediatric patient diagnosed in a teaching hospital from Brazil. *J. Fungi* **2020**, *6*, 147. [CrossRef] [PubMed]

112. De Hoog, G.S. Risk assessment of fungi reported from humans and animals. *Mycoses* **1996**, *39*, 407–417. [CrossRef] [PubMed]

113. Hyde, K.D.; Norphanphoun, C.; Chen, J.; Dissanayake, A.J.; Doilom, M.; Hongsanan, S.; Jayawardena, R.S.; Jeewon, R.; Perera, R.H.; Thongbai, B.; et al. Thailand's amazing diversity: Up to 96% of fungi in northern Thailand may be novel. *Fungal Divers.* **2018**, *93*, 215–239. [CrossRef]

114. Kumla, J.; Suwannarach, N.; Wannathes, W. *Hymenagaricus saisamornae* sp. nov. (Agaricales, Basidiomycota) from northern Thailand. *Chiang Mai J. Sci.* **2021**, *48*, 827–836.

115. Nakase, T.; Suzuki, M. *Bullera intermedia* sp. nov. and *Sporobolomyces oryzicola* sp. nov. isolated from dead leaves of *Oryza sativa*. *J. Gen. Appl. Microbiol.* **1986**, *32*, 149–155. [CrossRef]

116. Hagler, A.N.; Ahearn, D.G. Rapid diazonium blue B test to detect basidiomycetous yeasts. *Int. J. Syst. Evol.* **1981**, *31*, 204–208. [CrossRef]

117. White, T.J.; Bruns, T.; Lee, S.; Taylor, J. Amplification and direct sequencing of fungal ribosomal RNA genes for phylogenetics. In *PCR Protocols: A Guide to Methods and Applications*; Innis, M.A., Gelfand, D.H., Sninsky, J.J., White, T.J., Eds.; Academic Press: San Diego, CA, USA, 1990; pp. 315–322.

118. Kurtzman, C.P. DNA relatedness among saturn-spored yeasts assigned to the genera *Williopsis* and *Pichia*. *Antonie Van Leeuwenhoek* **1991**, *60*, 13–19. [CrossRef] [PubMed]

119. Edgar, R.C. MUSCLE: A multiple sequence alignment method with reduced time and space complexity. *BMC Bioinform* **2004**, *5*, 113. [CrossRef]

120. Miller, M.A.; Pfeiffer, W.; Schwartz, T. Creating the cipres science gateway for inference of large phylogenetic trees. In Proceedings of the 2010 Gateway Computing EnvironmentsWorkshop (GCE), New Orleans, LA, USA, 14 November 2010; IEEE: Manhattan, NY, USA; pp. 1–8.

121. Darriba, D.; Taboada, G.L.; Doallo, R.; Posada, D. jModelTest 2: More models, new heuristics and parallel computing. *Nat. Methods.* **2012**, *9*, 772. [CrossRef] [PubMed]

122. Ronquist, F.; Huelsenbeck, J.P. MrBayes 3: Bayesian phylogenetic inference under mixed models. *Bioinformatics* **2003**, *19*, 1572–1574. [CrossRef]

123. Rambaut, A. FigTree Tree Figure Drawing Tool Version 131, Institute of Evolutionary 623 Biology, University of Edinburgh. Available online: http://treebioedacuk/software/figtree/ (accessed on 20 August 2021).

124. Vu, D.; Groenewald, M.; Szöke, S.; Cardinali, G.; Eberhardt, U.; Stielow, B.; de Vries, M.; Verkleij, G.J.M.; Crous, P.W.; Boekhout, T.; et al. DNA barcoding analysis of more than 9 000 yeast isolates contributes to quantitative thresholds for yeast species and genera delimitation. *Stud. Mycol.* **2016**, *85*, 91–105. [CrossRef]

125. Rivera, F.N.; Gómez, Z.; González, E.; López, N.; Rodríguez, C.H.H.; Zúñiga, G. Yeasts associated with bark beetles of the genus *Dendroctonus erichson* (Coleoptera: Curculionidae: Scolytinae): Molecular identification and biochemical characterization. In Proceedings of the Third Workshop on Genetics of Bark Beetles and Associated Microorganisms, Asheville, NC, USA, 20 May 2006; pp. 45–48.

126. Sitepu, I.; Enriquez, L.; Nguyen, V.; Fry, R.; Simmons, B.; Singer, S.; Simmons, C.; Boundy-Mills, K.L. Ionic liquid tolerance of yeasts in family Dipodascaceae and genus *Wickerhamomyces*. *Appl. Biochem. Biotechnol.* **2020**, *191*, 1580–1593. [CrossRef] [PubMed]

127. Schoch, C.L.; Robbertse, B.; Robert, V.; Vu, D.; Cardinali, G.; Irinyi, L.; Meyer, W.; Nilsson, R.H.; Hughes, K.; Miller, A.N.; et al. Finding needles in haystacks: Linking scientific names, reference specimens and molecular data for fungi. *Database* **2014**, *2014*, bau061. [CrossRef] [PubMed]

128. Kurtzman, C.P.; Robnett, C.J. Phylogenetic relationships among yeasts of the "*Saccharomyces* complex" determined from multigene sequence analyses. *FEMS Yeast Res.* **2003**, *3*, 417–432. [CrossRef]

129. Wang, Y.; Ren, Y.C.; Zhang, Z.T.; Ke, T.; Hui, F.L. *Spathaspora allomyrinae* sp. nov., a D-xylose-fermenting yeast species isolated from a scarabeid beetle *Allomyrina dichotoma*. *Int. J. Syst. Evol.* **2016**, *66*, 2008–2012. [CrossRef]

130. Barnett, J.A. A history of research on yeasts 2: Louis Pasteur and his contemporaries, 1850–1880. *Yeast* **2000**, *16*, 755–771. [CrossRef]

131. Cavicchioli, R.; Ripple, W.J.; Timmis, K.N.; Azam, F.; Bakken, L.R.; Baylis, M.; Behrenfeld, M.J.; Boetius, A.; Boyd, P.W.; Classen, A.T.; et al. Scientists' warning to humanity: Microorganisms and climate change. *Nat. Rev. Microbiol.* **2019**, *17*, 569–586. [CrossRef] [PubMed]

132. Jain, P.K.; Purkayastha, S.D.; De Mandal, S.; Passari, A.K.; Govindarajan, R.K. Effect of climate change on microbial diversity and its functional attributes. In *Recent Advancements in Microbial Diversity*; De Mandal, S., Bhatt, P., Eds.; Academic Press: San Diego, CA, USA, 2020; pp. 315–331.
133. van der Putten, W.H. Climate change, aboveground-belowground interactions, and species' range shifts. *Annu. Rev. Ecol. Evol. Syst.* **2012**, *43*, 365–383. [CrossRef]
134. Wookey, P.A.; Aerts, R.; Bardgett, R.D.; Baptist, F.; BraThen, K.A.; Cornelissen, J.H.C.; Shaver, G.R. Ecosystem feedbacks and cascade processes: Understanding their role in the responses of Arctic and alpine ecosystems to environmental change. *Glob. Change Biol.* **2009**, *15*, 1153–1172. [CrossRef]

Journal of
Fungi

Article

Species Diversity and Distribution Characteristics of *Calonectria* in Five Soil Layers in a *Eucalyptus* Plantation

LingLing Liu [1,2], WenXia Wu [1] and ShuaiFei Chen [1,*]

1   China Eucalypt Research Centre (CERC), Chinese Academy of Forestry (CAF), Zhanjiang 524022, Guangdong Province, China; liulinglingfp@126.com (L.L.); wuwenxia_hainan@126.com (W.W.)
2   Nanjing Forestry University (NJFU), Nanjing 210037, Jiangsu Province, China
*   Correspondence: shuaifei.chen@gmail.com; Tel.: +86-759-338-1022

Abstract: The genus *Calonectria* includes pathogens of various agricultural, horticultural, and forestry crops. Species of *Calonectria* are commonly collected from soils, fruits, leaves, stems, and roots. Some species of *Calonectria* isolated from soils are considered as important plant pathogens. Understanding the species diversity and distribution characteristics of *Calonectria* species in different soil layers will help us to clarify their long-term potential harm to plants and their patterns of dissemination. To our knowledge, no systematic research has been conducted concerning the species diversity and distribution characteristics of *Calonectria* in different soil layers. In this study, 1000 soil samples were collected from five soil layers (0–20, 20–40, 40–60, 60–80, and 80–100 cm) at 100 sampling points in one 15-year-old *Eucalyptus urophylla* hybrid plantation in southern China. A total of 1037 isolates of *Calonectria* present in all five soil layers were obtained from 93 of 100 sampling points. The 1037 isolates were identified based on DNA sequence comparisons of the translation elongation factor 1-alpha (*tef1*), β-tubulin (*tub2*), calmodulin (*cmdA*), and histone H3 (*his3*) gene regions, as well as the combination of morphological characteristics. These isolates were identified as *C. hongkongensis* (665 isolates; 64.1%), *C. aconidialis* (250 isolates; 24.1%), *C. kyotensis* (58 isolates; 5.6%), *C. ilicicola* (47 isolates; 4.5%), *C. chinensis* (2 isolates; 0.2%), and *C. orientalis* (15 isolates; 1.5%). With the exception of *C. orientalis*, which resides in the *C. brassicae* species complex, the other five species belonged to the *C. kyotensis* species complex. The results showed that the number of sampling points that yielded *Calonectria* and the number (and percentage) of *Calonectria* isolates obtained decreased with increasing depth of the soil. More than 84% of the isolates were obtained from the 0–20 and 20–40 cm soil layers. The deeper soil layers had comparatively lower numbers but still harbored a considerable number of *Calonectria*. The diversity of five species in the *C. kyotensis* species complex decreased with increasing soil depth. The genotypes of isolates in each *Calonectria* species were determined by *tef1* and *tub2* gene sequences. For each species in the *C. kyotensis* species complex, in most cases, the number of genotypes decreased with increasing soil depth. The 0–20 cm soil layer contained all of the genotypes of each species. To our knowledge, this study presents the first report of *C. orientalis* isolated in China. This species was isolated from the 40–60 and 60–80 cm soil layers at only one sampling point, and only one genotype was present. This study has enhanced our understanding of the species diversity and distribution characteristics of *Calonectria* in different soil layers.

Keywords: fungal ecology; multi-gene phylogeny; plant pathogen; soil-borne fungi; tree disease

Citation: Liu, L.; Wu, W.; Chen, S. Species Diversity and Distribution Characteristics of *Calonectria* in Five Soil Layers in a *Eucalyptus* Plantation. *J. Fungi* **2021**, 7, 857. https://doi.org/10.3390/jof7100857

Academic Editors: Anush Kosakyan, Rodica Catana and Alona Biketova

Received: 31 August 2021
Accepted: 7 October 2021
Published: 13 October 2021

Publisher's Note: MDPI stays neutral with regard to jurisdictional claims in published maps and institutional affiliations.

1. Introduction

Species in the genus *Calonectria* (*Hypocreales*, *Nectriaceae*) are phytopathogenic fungi that cause serious losses to plant crops in tropical and subtropical regions of the world [1–6]. Many species of *Calonectria* are important pathogens of agricultural, horticultural, and forestry crops and these species occur in approximately 335 plant species in nearly 100 plant families [1]. Species of *Calonectria* have been isolated from soils, fruits, leaves, stems, and roots [1,4,7–14]. The fungi are best known as foliar, shoot, and root pathogens [1,2,4,5], and they are commonly associated with disease symptoms, including seedling damping-off, seedling rot, cutting rot, leaf spots, leaf blight, shoot blight, crown cankers, stem lesions, collar and root rots, and tuber rot [1,14–23].

Some species of *Calonectria* isolated from soils are important plant pathogens. *Calonectria ilicicola* is a soil-borne fungal pathogen of worldwide importance that causes black rot disease in peanut and red crown rot in soybean [21,24–28]. Recently, we isolated five *Calonectria* species, namely *C. aconidialis*, *C. auriculiformis*, *C. hongkongensis*, *C. pseudoreteaudii*, and *C. reteaudii*, from soils in a plantation of *Eucalyptus* trees [14]. Inoculation results showed that all five species caused leaf spot, leaf blight, and seedling rot to the tested *Eucalyptus* genotypes within three days [14].

Previous research results indicated a high level of species diversity of *Calonectria* in southern China, especially in soils [9,11,13,14,23]. Currently, a total of 125 *Calonectria* species have been described using DNA sequence-based phylogenetic analyses and morphological comparisons [5,13,29–35]. A total of 25 species of *Calonectria* have been identified and described in China based on DNA sequence data [5,9,11,13,14,36]. Of these species, 17 have been isolated from soils, with 11 from soils under plantation *Eucalyptus* trees [5,9,11,14].

Some *Calonectria* species can survive in soil for long periods, and microsclerotia are the primary survival structures [37]. Microsclerotia of some *Calonectria* species can survive in the absence of hosts for 15 years or more [38,39]. *Calonectria* microsclerotia have been recorded at depths of up to 66 cm below the soil surface [40]. Long-term survival and deep soil presence of microsclerotia are serious threats to the management of diseases caused by *Calonectria* species.

Understanding the diversity and distribution characteristics of *Calonectria* species in different soil layers will help us to clarify their potential long-term harm to plants and potential dissemination patterns. Very little research has been conducted concerning the distribution characteristics of microsclerotia in soils, and the few published studies have focused only on the surface soil [38,41]. In the past several years, studies have been conducted to understand *Calonectria* species diversity in forest soils [9–11,13,14,36], but all of the soil samples obtained for *Calonectria* isolation were collected from the 0–20 cm soil layer. In this study, a relatively large number of soil samples were collected from five different soil layers up to 100 cm depth in one 15-year-old *Eucalyptus urophylla* hybrid plantation. Isolates of *Calonectria* from this plantation were obtained and identified. The aims of this study were as follows: (1) to understand the species diversity of *Calonectria* in different soil layers; and (2) to understand the distribution characteristics of each *Calonectria* species in different soil layers.

2. Materials and Methods

2.1. Study Site, Soil Sampling, and Calonectria Isolation

This study was performed in a *Eucalyptus urophylla* hybrid plantation (21°15′31.74″ N, 110°06′35″ E; altitude 90 m) located in the South China Experimental Nursery, China Eucalypt Research Centre (CERC), ZhanJiang, GuangDong Province, China. The *Eucalyptus* plantation is located on the northern edge of the tropics, with a maritime monsoon climate [42]. The average annual precipitation is 1777 mm, and the period from May to October accounts for 84.1% of the annual precipitation. The annual average temperature is 23.4 °C (http://en.weather.com.cn; accessed date: 10 August 2021). The soil type is Rhodi-Udic Ferralosols, according to the Chinese Soil Taxonomy Classification [42,43]. The area of the *Eucalyptus* plantation is about 6 ha (400 × 150 m), and the planting density of *Eucalyptus* trees is 3 × 2 m. The *Eucalyptus* trees were 15 years old.

One hundred points in the *Eucalyptus* plantation were selected for soil sampling. The 100 points were randomly distributed in the plantation, and the distance between adjacent sampling points was 10 m. Soil samples were collected from five layers at each sampling point: 0–20, 20–40, 40–60, 60–80, and 80–100 cm. Two soil samples were collected in each soil layer for each sampling point. In total, 1000 soil samples were collected from the 100 sampling points. Each of the soil samples was placed in a resealable plastic bag and transferred to the laboratory for *Calonectria* isolation. The soil samples were collected from July to August 2020.

For *Calonectria* isolation, the collected soil was transferred into a plastic cylinder sampling cup (diameter = 4.5 cm, height = 5 cm, and volume = 80 mL) (Chengdu Rich Science Industry Co., Ltd., Chengdu, China); the soil sample occupied two-thirds of the volume of the whole sampling cup volume. The soil sample was moistened by spraying with sterile water and stirred evenly with a sterilized bamboo stick. *Medicago sativa* (alfalfa) seeds were scattered onto the soil surface after it was surface-disinfested (30 s in 75% ethanol and washed several times with sterile water) in the sampling cup. The sampling cup with soil and alfalfa seeds was incubated at 25 °C under 12 h of daylight and 12 h of darkness. After one week, sporulating conidiophores with typical morphological characteristics of *Calonectria* species [1] were produced on infected alfalfa tissue. Using a dissection microscope (AxioCam Stemi 2000C, Carl Zeiss, Germany), the single conidial mass was scattered onto 2% malt extract agar (MEA) (20 g malt extract powder and 20 g agar powder per liter of water: malt extract powder was obtained from Beijing Shuangxuan microbial culture medium products factory, Beijing, China; the agar powder was obtained from Beijing Solarbio Science & Technology Co., Ltd., Beijing, China) using a sterile needle. After incubation at 25 °C for three to four hours, the germinated conidia were individually transferred onto fresh MEA under the dissection microscope and incubated at 25 °C for one week to obtain single-conidium cultures. For each soil sample, the soil was transferred into two plastic sampling cups for *Calonectria* isolation.

2.2. DNA Extraction, PCR Amplification, and Sequencing

All isolates obtained in this study were used for DNA extraction and sequence comparisons. DNA was extracted from 10-day-old cultures. Mycelia were collected using a sterilized scalpel and transferred to 2-mL Eppendorf tubes. The total genomic DNA was extracted using the CTAB protocol described by van Burik and co-authors [44]. The extracted DNA was dissolved in 30 μL TE buffer (1 M Tris-HCl and 0.5 M EDTA, pH 8.0), and 2.5 μL RNase (10 mg/mL) was added at 37 °C for one hour to degrade RNA. Finally, the DNA concentration was measured using a NanoDrop 2000 spectrometer (Thermo Fisher Scientific, Waltham, MA, USA).

According to previous research results, sequences of partial gene regions of translation elongation factor 1-alpha (*tef1*) and β-tubulin (*tub2*), as well as calmodulin (*cmdA*) and histone H3 (*his3*), were used to successfully identify *Calonectria* species [5,14]. These four partial gene regions were amplified using the primer pairs EF1-728F/EF2, T1/CYLTUB1R, CAL-228F/CAL-2Rd, and CYLH3F/CYLH3R, respectively. The PCR procedure was conducted as described by Liu and Chen [36] and Wang and Chen [23].

To obtain accurate sequences for each of the sequenced isolates, all of the PCR products were sequenced in both forward and reverse directions using the same primers used for PCR amplification by the Beijing Genomics Institute, Guangzhou, China. All of the sequences obtained in this study were edited using MEGA v. 7.0 software [45] and were deposited in GenBank (https://www.ncbi.nlm.nih.gov; accessed date: 18 September 2021). The *tef1* and *tub2* gene regions were sequenced for all *Calonectria* isolates. The isolates were genotyped by the *tef1* and *tub2* sequences. Based on the genotypes generated by *tef1* and *tub2* sequences, up to eight isolates for each *tef1-tub2* genotype were selected for sequencing the *cmdA* and *his3* gene regions.

2.3. Multi-Gene Phylogenetic Analyses, Morphology, and Species Identification

A standard nucleotide BLAST search was conducted using the *tef1*, *tub2*, *cmdA*, and *his3* sequences to preliminarily identify the species from which the isolates were obtained in this study. Sequences of *tef1*, *tub2*, *cmdA*, and *his3* gene regions obtained in this study were compared with sequences of type specimen strains of published *Calonectria* species. Sequences of all of the published species in the relevant species complexes were used for sequence comparisons and phylogenetic analyses. The datasets of Liu and co-authors [5] were used as templates for analyses, while sequences of other recently described *Calonectria* species [13,32–35] were also used for sequence comparisons.

Sequences of each of the *tef1*, *tub2*, *cmdA*, and *his3* gene regions, as well as the combination of these four gene regions, were aligned using the online version of MAFFT v. 7 (http://mafft.cbrc.jp/alignment/server; accessed date: 7 August 2021) with the alignment strategy FFT-NS-i (Slow; interactive refinement method). Sequence alignments were manually edited using MEGA v. 7.0 software [45] after initial alignments.

For *Calonectria* species, maximum parsimony (MP) and maximum likelihood (ML) are frequently used for phylogenetic analyses [5,9,12,14]. Both MP and ML were used for phylogenetic analyses of sequence datasets of each of the four genes and the combination of the four gene regions in order to test whether the analysis results between the two methods were consistent. The MP and ML analyses were conducted by the methods described by Liu and Chen [36]. Phylogenetic trees were viewed by MEGA v. 7.0 [45]. Sequence data of two isolates of *Curvicladiella cignea* (CBS 109167 and CBS 109168) were used as outgroups [5].

The isolates selected for sequencing *tef1*, *tub2*, *cmdA*, and *his3* gene regions were used for morphological studies. Size of macroconidia and width of vesicles are the most typical asexual characteristics used for morphological comparisons for species of *Calonectria* [5,9,11,13,14,29,36]. In order to induce asexual structures, isolates were cultured on 2% MEA in Petri dishes at 25 °C for 10 days. Sterile water was then added to the Petri dishes, and a sterilized, soft-bristled paintbrush was used to dislodge the mycelium from the agar surface. The water was then removed, and the dishes were placed upside down and incubated at 25 °C for 2–3 days. This resulted in asexual structures being produced on the surface of the cultures for some *Calonectria* isolates, a pattern that has been noted for *Calonectria pteridis* by Graça and co-authors [46] and for *Calonectria pentaseptata* (synonymized as a synonym of *C. pseudoreteaudii* in Liu and co-authors [5]) by Wang and Chen [23]. Fifty measurements of macroconidia and vesicles were measured for the selected isolates that produced abundant macroconidia and vesicles.

2.4. Calonectria Species Diversity in Different Soil Layers

After all of the *Calonectria* isolates were identified, the number of isolates present in each identified species was counted. The species diversity associated with soil layers was computed. The distribution characteristics of each *Calonectria* species in each soil layer were recorded, including the number of sampling points from which each *Calonectria* species was obtained and the number of isolates of each *Calonectria* species in each of the five soil layers.

2.5. Genotyping of Isolates within Each Calonectria Species

After all of the *Calonectria* isolates were identified, we examined the genotype diversity of each identified *Calonectria* species in the five different soil layers. The genotypes of isolates within each species were determined based on *tef1* and *tub2* sequences, and the number of isolates belonging to each genotype was recorded.

2.6. Genotype Diversity of Calonectria Species in Different Soil Layers

Based on the results of genotype analysis of each isolate determined by the sequences of *tef1* and *tub2* gene regions, the numbers of genotypes of each *Calonectria* species in different soil layers were counted. To investigate possible evolutionary relationships among the observed *tef1*–*tub2* genotypes for the *Calonectria* species identified in this study with the most dominant species, minimum spanning networks (MSN) were constructed using Bruvo's distance with the R packages poppr and ape [47,48].

3. Results

3.1. Soil Sampling and Calonectria Isolation

One thousand soil samples from 100 sample points were collected from the *E. urophylla* hybrid plantation, with 200 soil samples from each of the five soil layers. For each soil sample, two plastic sampling cups with soil and alfalfa seeds were used for the incubation of *Calonectria*. After the conidia were transferred onto fresh MEA and incubated at 25 °C, more than 90% of the conidia germinated within four hours. For each sampling cup, one to four single conidia were transferred onto fresh MEA to obtain one to four single-conidium cultures. In total, *Calonectria* fungi were isolated from 93 sampling points in the plantation; the totals were 92, 40, 20, 7, and 5 from the 0–20, 20–40, 40–60, 60–80, and 80–100 cm soil layers, respectively (Supplementary Table S1, Supplementary Figure S1). One thousand and thirty-seven isolates of *Calonectria* were obtained, with 564 (54.4%), 310 (29.9%), 107 (10.3%), 28 (2.7%), and 28 isolates (2.7%) from the 0–20, 20–40, 40–60, 60–80, and 80–100 cm soil layers, respectively, and 84.3% of the isolates were distributed in the 0–20 and 20–40 cm soil layers (Table 1, Supplementary Table S2, Figure 1). From the results, it was clear that the number of sampling points that yielded *Calonectria* and the number (and percentage) of *Calonectria* isolates obtained decreased with increasing soil depth (Supplementary Figure S1, Figure 1).

Table 1. Number of isolates obtained for each *Calonectria* species from each soil layer.

Soil Layer	*C. hongkongensis*	*C. aconidialis*	*C. kyotensis*	*C. ilicicola*	*C. chinensis*	*C. orientalis*	All six *Calonectria* species	Percentage
0–20 cm	373	140	33	16	2	0	564	54.4%
20–40 cm	203	74	14	19	0	0	310	29.9%
40–60 cm	61	20	7	8	0	11	107	10.3%
60–80 cm	8	8	4	4	0	4	28	2.7%
80–100 cm	20	8	0	0	0	0	28	2.7%
All five soil layers	665	250	58	47	2	15	1037	
Percentage	64.1%	24.1%	5.6%	4.5%	0.2%	1.5%		

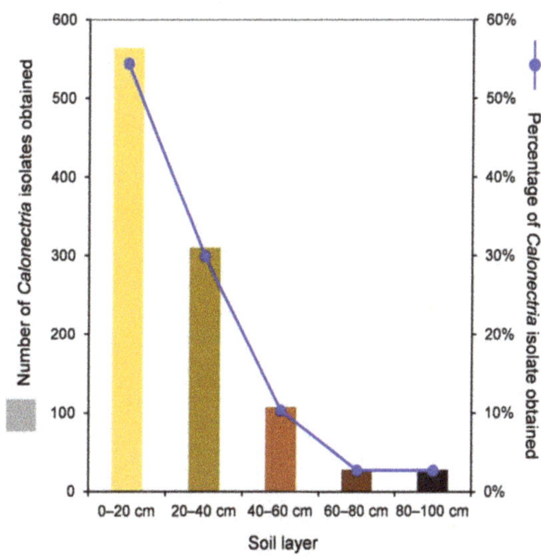

Figure 1. Numbers and percentages of *Calonectria* isolates obtained in each of the five soil layers.

3.2. Sequencing

The *tef1* and *tub2* genes were amplified for all the 1037 isolates obtained in this study (Supplementary Table S2). Twenty-two genotypes were generated based on *tef1* and *tub2* gene sequences (Table 2). Depending on the isolate number of each *tef1-tub2* genotype, one to eight isolates of each genotype were selected; finally, 85 isolates in total were selected to sequence the *cmdA* and *his3* gene regions (Table 3). The sequence fragments were approximately 500, 565, 685, and 440 bp for the *tef1*, *tub2*, *cmdA*, and *his3* gene regions, respectively.

Table 2. Isolate numbers of each genotype from each *Calonectria* species.

Calonectria Species	Number of Genotypes Determined by *tef1* and *tub2* Gene Sequences	Genotype Determined by *tef1* and *tub2* Gene Sequences	Number of Isolates of Each Genotype	Number of isolates of Each *Calonectria* Species
C. hongkongensis	11	AA	561	665
		AB	1	
		AC	4	
		AD	7	
		AE	2	
		AF	20	
		AG	15	
		AH	4	
		BA	15	
		CA	5	
		DA	31	
C. aconidialis	3	AA	156	250
		AB	9	
		AC	85	
C. kyotensis	3	AA	33	58
		AB	19	
		BA	6	
C. ilicicola	3	AA	26	47
		AB	9	
		BB	12	
C. chinensis	1	AA	2	2
C. orientalis	1	AA	15	15
All six *Calonectria* species	22		1037	1037

3.3. Multi-Gene Phylogenetic Analyses, Morphology, and Species Identification

The standard nucleotide BLAST search results conducted using the *tef1*, *tub2*, *cmdA*, and *his3* sequences showed that the isolates obtained in the current study belonged to two species complexes of *Calonectria*, namely, the C. *kyotensis* species complex and the C. *brassicae* species complex. The 85 *Calonectria* isolates with four gene regions sequenced were used for phylogenetic analyses (Table 3). Based on the recently published results in Liu and co-authors [5] and Crous and co-authors [34], sequences of *tef1*, *tub2*, *cmdA*, and *his3* of published species in the C. *kyotensis* species complex and C. *brassicae* species complex, respectively, were used for sequence comparisons and phylogenetic analyses (Table 4).

The partition homogeneity test (PHT) comparing the *tef1*, *tub2*, *cmdA*, and *his3* gene combination datasets generated a *p*-value of 0.001, indicating that the accuracy of the combined datasets did not suffer relative to the individual partitions [60]. Thus, sequences of the four loci were combined for analyses. Between the MP and ML trees, the overall topologies were similar for the phylogenetic trees based on *tef1*, *tub2*, *cmdA*, and *his3* individually and the combination datasets, but the relative positions of some *Calonectria* species slightly differed. The five ML trees are presented in Figure 2 and Supplementary Figures S2–S5. The numbers of taxa and parsimony-informative characters, statistical values of the MP analyses, and parameters of the best-fit substitution models of ML analyses are provided in Table 5.

Table 3. Isolates sequenced and used for phylogenetic analyses and morphological studies in this study.

Identity	Genotype [1]	Isolate No. [2]	Sampling Point No. [3]	Soil Layer	Sample and Isolate Information [4]	Collectors	GenBank Accession No. [5]			
							tef1	tub2	cmdA	his3
C. aconidialis	AAAA	CSF20325	6	0–20 cm	20200711-1-(3)_0–20 cm_A_R2_SC2	S.F. Chen, L.L. Liu, J.L. Han, Y Liu, and X.Y. Liang	OK167700	OK168737	OK169148	OK169232
C. aconidialis	AAAA	CSF21348	98	0–20 cm	20200816-1-(6)_0–20 cm_A_R2_SC1	L.L. Liu, J.L. Han, and L.S. Sun	OK167855	OK168892	OK169151	OK169235
C. aconidialis	AACA	CSF20378	9	0–20 cm	20200711-1-(6)_0–20 cm_A_R2_SC1	S.F. Chen, L.L. Liu, J.L. Han, Y Liu, and X.Y. Liang	OK167701	OK168738	OK169149	OK169233
C. aconidialis	AACA	CSF20447	11	0–20 cm	20200715-1-(1)_0–20 cm_B_R2_SC2	S.F. Chen, L.L. Liu, J.L. Han, L.S. Sun, and W.W. Li	OK167704	OK168741	OK169150	OK169234
C. aconidialis	ABBA	CSF20985 [6]	68	20–40 cm	20200811-1-(4)_0–40 cm_B_R1_SC3	L.L. Liu, J.L. Han, and L.S. Sun	OK167856	OK168893	OK169152	OK169236
C. aconidialis	ABBA	CSF21262	93	20–40 cm	20200816-1-(1)_0–40 cm_B_R1_SC1	L.L. Liu, J.L. Han, and L.S. Sun	OK167857	OK168894	OK169153	OK169237
C. aconidialis	ABBA	CSF21266	93	20–40 cm	20200816-1-(1)_0–40 cm_B_R2_SC2	L.L. Liu, J.L. Han, and L.S. Sun	OK167861	OK168898	OK169154	OK169238
C. aconidialis	ABBA	CSF21349	98	0–20 cm	20200816-1-(6)_0–20 cm_A_R2_SC2	L.L. Liu, J.L. Han, and L.S. Sun	OK167864	OK168901	OK169155	OK169239
C. aconidialis	ACAA	CSF20257	1	0–20 cm	20200709-1-(1)_0–20 cm_A_R1_SC1	S.F. Chen, L.L. Liu, J.L. Han, Y Liu, and X.Y. Liang	OK167865	OK168902	OK169156	OK169240
C. aconidialis	ACAA	CSF20323 [6]	6	0–20 cm	20200711-1-(3)_0–20 cm_A_R1_SC1	S.F. Chen, L.L. Liu, J.L. Han, Y Liu, and X.Y. Liang	OK167866	OK168903	OK169157	OK169241
C. aconidialis	ACAA	CSF20376 [6]	9	0–20 cm	20200711-1-(6)_0–20 cm_A_R1_SC1	S.F. Chen, L.L. Liu, J.L. Han, Y Liu, and X.Y. Liang	OK167868	OK168905	OK169158	OK169242
C. aconidialis	ACAA	CSF21346	98	0–20 cm	20200816-1-(6)_0–20 cm_A_R1_SC1	L.L. Liu, J.L. Han, and L.S. Sun	OK167946	OK168983	OK169159	OK169243
C. chinensis	AAAA	CSF20756 [6]	52	0–20 cm	20200809-1-(2)_0–20 cm_A_R2_SC1	L.L. Liu, J.L. Han, and L.S. Sun	OK168055	OK169092	OK169184	OK169268
C. chinensis	AAAA	CSF20759 [6]	52	0–20 cm	20200809-1-(2)_0–20 cm_A_R2_SC4	L.L. Liu, J.L. Han, and L.S. Sun	OK168056	OK169093	OK169185	OK169269
C. hongkongensis	AAAA	CSF20258	1	0–20 cm	20200709-1-(1)_0–20 cm_A_R1_SC2	S.F. Chen, L.L. Liu, J.L. Han, Y Liu, and X.Y. Liang	OK167035	OK168072	OK169109	OK169194
C. hongkongensis	AAAA	CSF20271	2	0–20 cm	20200709-1-(2)_0–20 cm_A_R1_SC1	S.F. Chen, L.L. Liu, J.L. Han, Y Liu, and X.Y. Liang	OK167044	OK168081	OK169110	OK169195
C. hongkongensis	AAAA	CSF20291	3	0–20 cm	20200709-1-(3)_0–20 cm_A_R2_SC1	S.F. Chen, L.L. Liu, J.L. Han, Y Liu, and X.Y. Liang	OK167056	OK168093	OK169111	OK169196
C. hongkongensis	AAAA	CSF21370	100	0–20 cm	20200816-1-(8)_0–20 cm_A_R2_SC2	L.L. Liu, J.L. Han, and L.S. Sun	OK167588	OK168625	OK169112	OK169197
C. hongkongensis	ABA-	CSF20758	52	0–20 cm	20200809-1-(2)_0–20 cm_A_R2_SC3	L.L. Liu, J.L. Han, and L.S. Sun	OK167596	OK168633	OK169113	–[7]
C. hongkongensis	ACAA	CSF20524	17	0–20 cm	20200715-1-(7)_0–20 cm_B_R1_SC1	S.F. Chen, L.L. Liu, J.L. Han, L.S. Sun, and W.W. Li	OK167597	OK168634	OK169114	OK169198
C. hongkongensis	ACAA	CSF20525	17	0–20 cm	20200715-1-(7)_0–20 cm_B_R2_SC2	S.F. Chen, L.L. Liu, J.L. Han, L.S. Sun, and W.W. Li	OK167598	OK168635	OK169115	OK169199
C. hongkongensis	ACAB	CSF21368	100	0–20 cm	20200816-1-(8)_0–20 cm_A_R1_SC2	L.L. Liu, J.L. Han, and L.S. Sun	OK167599	OK168636	OK169116	OK169200
C. hongkongensis	ACAB	CSF21372	100	0–20 cm	20200816-1-(8)_0–20 cm_B_R1_SC2	L.L. Liu, J.L. Han, and L.S. Sun	OK167600	OK168637	OK169117	OK169201
C. hongkongensis	ADAA	CSF20412	10	0–20 cm	20200711-1-(7)_0–20 cm_B_R1_SC1	S.F. Chen, L.L. Liu, J.L. Han, Y Liu, and X.Y. Liang	OK167601	OK168638	OK169118	OK169202
C. hongkongensis	ADAA	CSF20454	11	20–40 cm	20200715-1-(1)_0–40 cm_A_R2_SC3	S.F. Chen, L.L. Liu, J.L. Han, L.S. Sun, and W.W. Li	OK167602	OK168639	OK169119	OK169203
C. hongkongensis	ADAA	CSF20834	60	0–20 cm	20200810-1-(4)_0–20 cm_B_R1_SC1	L.L. Liu, J.L. Han, and L.S. Sun	OK167604	OK168641	OK169120	OK169204
C. hongkongensis	ADAA	CSF21304	96	0–20 cm	20200816-1-(4)_0–20 cm_A_R2_SC1	L.L. Liu, J.L. Han, and L.S. Sun	OK167607	OK168644	OK169121	OK169205
C. hongkongensis	AEAA	CSF20923	65	0–20 cm	20200811-1-(1)_0–20 cm_A_R1_SC1	L.L. Liu, J.L. Han, and L.S. Sun	OK167608	OK168645	OK169122	OK169206
C. hongkongensis	AEAA	CSF20924 [6]	65	0–20 cm	20200811-1-(1)_0–20 cm_A_R1_SC2	L.L. Liu, J.L. Han, and L.S. Sun	OK167609	OK168646	OK169123	OK169207
C. hongkongensis	AFAA	CSF20259	1	0–20 cm	20200709-1-(1)_0–20 cm_A_R2_SC1	S.F. Chen, L.L. Liu, J.L. Han, Y Liu, and X.Y. Liang	OK167610	OK168647	OK169124	OK169208
C. hongkongensis	AFAA	CSF20309	4	0–20 cm	20200711-1-(1)_0–20 cm_A_R1_SC1	S.F. Chen, L.L. Liu, J.L. Han, Y Liu, and X.Y. Liang	OK167611	OK168648	OK169125	OK169209
C. hongkongensis	AFAA	CSF20470	12	0–20 cm	20200715-1-(2)_0–20 cm_A_R2_SC1	S.F. Chen, L.L. Liu, J.L. Han, L.S. Sun, and W.W. Li	OK167615	OK168652	OK169126	OK169210
C. hongkongensis	AFAA	CSF21233	90	0–20 cm	20200815-1-(3)_0–20 cm_B_R2_SC2	L.L. Liu, J.L. Han, and L.S. Sun	OK167629	OK168666	OK169127	OK169211
C. hongkongensis	AGAA	CSF20380	9	0–20 cm	20200711-1-(6)_0–20 cm_B_R1_SC1	S.F. Chen, L.L. Liu, J.L. Han, Y Liu, and X.Y. Liang	OK167630	OK168667	OK169128	OK169212
C. hongkongensis	AGAA	CSF20441	11	0–20 cm	20200715-1-(1)_0–20 cm_A_R1_SC2	S.F. Chen, L.L. Liu, J.L. Han, L.S. Sun, and W.W. Li	OK167631	OK168668	OK169129	OK169213
C. hongkongensis	AGAA	CSF20528	17	40–60 cm	20200715-1-(7)_0–60 cm_A_R1_SC1	S.F. Chen, L.L. Liu, J.L. Han, L.S. Sun, and W.W. Li	OK167632	OK168669	OK169130	OK169214
C. hongkongensis	AGAA	CSF21018	71	0–20 cm	20200811-1-(7)_0–20 cm_B_R1_SC1	L.L. Liu, J.L. Han, and L.S. Sun	OK167644	OK168681	OK169131	OK169215
C. hongkongensis	AHAA	CSF20760	52	0–20 cm	20200809-1-(2)_0–20 cm_B_R1_SC1	L.L. Liu, J.L. Han, and L.S. Sun	OK167645	OK168682	OK169132	OK169216
C. hongkongensis	AHAA	CSF20761 [6]	52	0–20 cm	20200809-1-(2)_0–20 cm_B_R1_SC2	L.L. Liu, J.L. Han, and L.S. Sun	OK167646	OK168683	OK169133	OK169217
C. hongkongensis	AHAA	CSF21155	82	0–20 cm	20200813-1-(2)_0–20 cm_B_R2_SC1	L.L. Liu, J.L. Han, and L.S. Sun	OK167647	OK168684	OK169134	OK169218
C. hongkongensis	AHAA	CSF21156	82	0–20 cm	20200813-1-(2)_0–20 cm_B_R2_SC2	L.L. Liu, J.L. Han, and L.S. Sun	OK167648	OK168685	OK169135	OK169219
C. hongkongensis	BAAA	CSF20472	12	0–20 cm	20200715-1-(2)_0–20 cm_B_R1_SC1	S.F. Chen, L.L. Liu, J.L. Han, L.S. Sun, and W.W. Li	OK167649	OK168686	OK169136	OK169220
C. hongkongensis	BAAA	CSF20734	51	0–20 cm	20200809-1-(1)_0–20 cm_A_R1_SC1	L.L. Liu, J.L. Han, and L.S. Sun	OK167652	OK168689	OK169137	OK169221
C. hongkongensis	BAAA	CSF21183	86	0–20 cm	20200814-1-(2)_0–20 cm_B_R1_SC1	L.L. Liu, J.L. Han, and L.S. Sun	OK167657	OK168694	OK169138	OK169222
C. hongkongensis	BAAA	CSF21359	99	0–20 cm	20200816-1-(7)_0–20 cm_A_R2_SC1	L.L. Liu, J.L. Han, and L.S. Sun	OK167660	OK168697	OK169139	OK169223
C. hongkongensis	CAAA	CSF20353 [6]	7	0–20 cm	20200711-1-(4)_0–20 cm_A_R2_SC2	S.F. Chen, L.L. Liu, J.L. Han, Y Liu, and X.Y. Liang	OK167664	OK168701	OK169140	OK169224
C. hongkongensis	CAAA	CSF20358	7	20–40 cm	20200711-1-(4)_0–40 cm_B_R1_SC1	S.F. Chen, L.L. Liu, J.L. Han, Y Liu, and X.Y. Liang	OK167665	OK168702	OK169141	OK169225
C. hongkongensis	CAAA	CSF20359	7	20–40 cm	20200711-1-(4)_0–40 cm_B_R1_SC2	S.F. Chen, L.L. Liu, J.L. Han, Y Liu, and X.Y. Liang	OK167666	OK168703	OK169142	OK169226
C. hongkongensis	CAAA	CSF20360 [6]	7	20–40 cm	20200711-1-(4)_0–40 cm_B_R1_SC3	S.F. Chen, L.L. Liu, J.L. Han, Y Liu, and X.Y. Liang	OK167667	OK168704	OK169143	OK169227
C. hongkongensis	DAAA	CSF20334	6	20–40 cm	20200711-1-(3)_0–40 cm_B_R1_SC1	S.F. Chen, L.L. Liu, J.L. Han, Y Liu, and X.Y. Liang	OK167669	OK168706	OK169144	OK169228
C. hongkongensis	DAAA	CSF20383 [6]	9	0–20 cm	20200711-1-(6)_0–20 cm_B_R2_SC2	S.F. Chen, L.L. Liu, J.L. Han, Y Liu, and X.Y. Liang	OK167673	OK168710	OK169145	OK169229
C. hongkongensis	DAAA	CSF20444	11	0–20 cm	20200715-1-(1)_0–20 cm_B_R1_SC1	S.F. Chen, L.L. Liu, J.L. Han, L.S. Sun, and W.W. Li	OK167678	OK168715	OK169146	OK169230
C. hongkongensis	DAAA	CSF21367	100	0–20 cm	20200816-1-(8)_0–20 cm_A_R1_SC1	L.L. Liu, J.L. Han, and L.S. Sun	OK167699	OK168736	OK169147	OK169231

Table 3. *Cont.*

Identity	Genotype [1]	Isolate No. [2]	Sampling Point No. [3]	Soil Layer	Sample and Isolate Information [4]	Collectors	GenBank Accession No. [5]			
							tef1	*tub2*	*cmdA*	*his3*
C. ilicicola	AAAB	CSF20594	29	0–20 cm	20200727-1-(5)_0–20 cm_A_R2_SC1	L.L. Liu, J.L. Han, and L.S. Sun	OK168008	OK169045	OK169172	OK169256
C. ilicicola	AAAB	CSF21126	80	20–40 cm	20200812-1-(8)_0–40 cm_A_R2_SC1	L.L. Liu, J.L. Han, and L.S. Sun	OK168010	OK169047	OK169173	OK169257
C. ilicicola	AAAB	CSF21219	89	0–20 cm	20200815-1-(2)_0–20 cm_A_R2_SC1	L.L. Liu, J.L. Han, and L.S. Sun	OK168014	OK169051	OK169174	OK169258
C. ilicicola	AAAB	CSF21310 [6]	96	20–40 cm	20200816-1-(4)_0–40 cm_A_R1_SC1	L.L. Liu, J.L. Han, and L.S. Sun	OK168016	OK169053	OK169175	OK169259
C. ilicicola	ABAA	CSF20618 [6]	32	0–20 cm	20200729-1-(2)_0–20 cm_A_R1_SC1	L.L. Liu, J.L. Han, L.S. Sun, Y Liu, and X.Y. Liang	OK168034	OK169071	OK169176	OK169260
C. ilicicola	ABAA	CSF20620	32	0–20 cm	20200729-1-(2)_0–20 cm_A_R2_SC1	L.L. Liu, J.L. Han, L.S. Sun, Y Liu, and X.Y. Liang	OK168036	OK169073	OK169177	OK169261
C. ilicicola	ABAA	CSF20624	32	20–40 cm	20200729-1-(2)_0–40 cm_A_R1_SC1	L.L. Liu, J.L. Han, L.S. Sun, Y Liu, and X.Y. Liang	OK168038	OK169075	OK169178	OK169262
C. ilicicola	ABAA	CSF20703	45	0–20 cm	20200731-1-(2)_0–20 cm_B_R1_SC1	L.L. Liu, J.L. Han, and L.S. Sun	OK168042	OK169079	OK169179	OK169263
C. ilicicola	BBAA	CSF20853	61	20–40 cm	20200810-1-(5)_0–40 cm_A_R1_SC8	L.L. Liu, J.L. Han, and L.S. Sun	OK168043	OK169080	OK169180	OK169264
C. ilicicola	BBBA	CSF21052 [6]	74	0–20 cm	20200812-1-(2)_0–20 cm_A_R1_SC1	L.L. Liu, J.L. Han, and L.S. Sun	OK168044	OK169081	OK169181	OK169265
C. ilicicola	BBBA	CSF21198	87	0–20 cm	20200814-1-(3)_0–20 cm_A_R2_SC2	L.L. Liu, J.L. Han, and L.S. Sun	OK168047	OK169084	OK169182	OK169266
C. ilicicola	BBBA	CSF21292	95	0–20 cm	20200816-1-(3)_0–20 cm_A_R1_SC1	L.L. Liu, J.L. Han, and L.S. Sun	OK168053	OK169090	OK169183	OK169267
C. kyotensis	AAAA	CSF20372	8	0–20 cm	20200711-1-(5)_0–20 cm_B_R2_SC1	S.F. Chen, L.L. Liu, J.L. Han, Y Liu, and X.Y. Liang	OK167950	OK168987	OK169160	OK169244
C. kyotensis	AAAA	CSF20443	11	0–20 cm	20200715-1-(1)_0–20 cm_A_R2_SC2	S.F. Chen, L.L. Liu, J.L. Han, L.S. Sun, and W.W. Li	OK167952	OK168989	OK169161	OK169245
C. kyotensis	AAAA	CSF21350	98	0–20 cm	20200816-1-(6)_0–20 cm_B_R1_SC1	L.L. Liu, J.L. Han, and L.S. Sun	OK167981	OK169018	OK169163	OK169247
C. kyotensis	AAAB	CSF20518	16	0–20 cm	20200715-1-(6)_0–20 cm_B_R2_SC1	S.F. Chen, L.L. Liu, J.L. Han, L.S. Sun, and W.W. Li	OK167953	OK168990	OK169162	OK169246
C. kyotensis	ABAA	CSF21191 [6]	86	40–60 cm	20200814-1-(2)_0–60 cm_B_R2_SC1	L.L. Liu, J.L. Han, and L.S. Sun	OK167998	OK169035	OK169167	OK169251
C. kyotensis	ABAB	CSF20260	1	0–20 cm	20200709-1-(1)_0–20 cm_A_R2_SC2	S.F. Chen, L.L. Liu, J.L. Han, Y Liu, and X.Y. Liang	OK167983	OK169020	OK169164	OK169248
C. kyotensis	ABAB	CSF20432	10	40–60 cm	20200711-1-(7)_0–60 cm_B_R2_SC1	S.F. Chen, L.L. Liu, J.L. Han, Y Liu, and X.Y. Liang	OK167988	OK169025	OK169166	OK169250
C. kyotensis	ABBA	CSF20338	6	20–40 cm	20200711-1-(3)_0–40 cm_B_R2_SC1	S.F. Chen, L.L. Liu, J.L. Han, Y Liu, and X.Y. Liang	OK167986	OK169023	OK169165	OK169249
C. kyotensis	BAAA	CSF20275	2	20–40 cm	20200709-1-(2)_0–40 cm_A_R1_SC1	S.F. Chen, L.L. Liu, J.L. Han, Y Liu, and X.Y. Liang	OK168002	OK169039	OK169168	OK169252
C. kyotensis	BAAA	CSF20276 [6]	2	20–40 cm	20200709-1-(2)_0–40 cm_A_R1_SC2	S.F. Chen, L.L. Liu, J.L. Han, Y Liu, and X.Y. Liang	OK168003	OK169040	OK169169	OK169253
C. kyotensis	BAAA	CSF21111	78	0–20 cm	20200812-1-(6)_0–20 cm_B_R2_SC1	L.L. Liu, J.L. Han, and L.S. Sun	OK168006	OK169043	OK169170	OK169254
C. kyotensis	BAAA	CSF21335 [6]	97	0–20 cm	20200816-1-(5)_0–20 cm_A_R1_SC2	L.L. Liu, J.L. Han, and L.S. Sun	OK168007	OK169044	OK169171	OK169255
C. orientalis	AAAA	CSF20602	31	40–60 cm	20200729-1-(1)_0–60 cm_A_R1_SC1	L.L. Liu, J.L. Han, L.S. Sun, Y Liu, and X.Y. Liang	OK168057	OK169094	OK169186	OK169270
C. orientalis	AAAA	CSF20603	31	40–60 cm	20200729-1-(1)_0–60 cm_A_R1_SC2	L.L. Liu, J.L. Han, L.S. Sun, Y Liu, and X.Y. Liang	OK168058	OK169095	OK169187	OK169271
C. orientalis	AAAA	CSF20606	31	40–60 cm	20200729-1-(1)_0–60 cm_B_R1_SC1	L.L. Liu, J.L. Han, L.S. Sun, Y Liu, and X.Y. Liang	OK168061	OK169098	OK169188	OK169272
C. orientalis	AAAA	CSF20607	31	40–60 cm	20200729-1-(1)_0–60 cm_B_R1_SC2	L.L. Liu, J.L. Han, L.S. Sun, Y Liu, and X.Y. Liang	OK168062	OK169099	OK169189	OK169273
C. orientalis	AAAA	CSF20610	31	40–60 cm	20200729-1-(1)_0–60 cm_B_R2_SC1	L.L. Liu, J.L. Han, L.S. Sun, Y Liu, and X.Y. Liang	OK168064	OK169101	OK169190	OK169274
C. orientalis	AAAA	CSF20611	31	40–60 cm	20200729-1-(1)_0–60 cm_B_R2_SC2	L.L. Liu, J.L. Han, L.S. Sun, Y Liu, and X.Y. Liang	OK168065	OK169102	OK169191	OK169275
C. orientalis	AAAA	CSF20614 [6]	31	60–80 cm	20200729-1-(1)_0–80 cm_B_R1_SC1	L.L. Liu, J.L. Han, L.S. Sun, Y Liu, and X.Y. Liang	OK168068	OK169105	OK169192	OK169276
C. orientalis	AAAA	CSF20615	31	60–80 cm	20200729-1-(1)_0–80 cm_B_R1_SC2	L.L. Liu, J.L. Han, L.S. Sun, Y Liu, and X.Y. Liang	OK168069	OK169106	OK169193	OK169277

[1] Genotype within each *Calonectria* species, determined by sequences of the *tef1*, *tub2*, *cmdA*, and *his3* regions; "-" means not available. [2] CSF: Culture collection located at China Eucalypt Research Centre (CERC), Chinese Academy of Forestry, ZhanJiang, GuangDong Province, China. [3] Number of 100 sampling points in this study. [4] Information associated with sample point and isolate, for example, "20200711-1-(3)_0–20 cm_A_R2_SC2" indicated sample number "20200711-1-(3), soil layer (0–20 cm), sample plastic bag (A), plastic sampling cup (R2), single conidium 2 (SC2). [5] *tef1* = translation elongation factor 1-alpha; *tub2* = β-tubulin; *cmdA* = calmodulin; *his3* = histone H3. [6] Isolates used for measuring macroconidia and vesicles in the current study. [7] "–" represents the relative locus was not successfully amplified in the current study.

Table 4. Isolates from other studies used in the phylogenetic analyses in this study.

Species Code [1]	Species	Isolates No. [2,3]	Other Collection Number [3]	Hosts	Area of Occurrence	Collector	GenBank Accession No. [4]				References
							tef1	tub2	cmdA	his3	
Species in *Calonectria kyotensis* species complex											
B4	*C. aconidialis*	CMW 35174[T]	CBS 136086; CERC 1850	Soil in *Eucalyptus* plantation	HaiNan, China	X. Mou and S.F. Chen	MT412695	**OK357463**	MT335165	MT335404	[5,9]
		CMW 35384	CBS 136091; CERC 1886	Soil in *Eucalyptus* plantation	HaiNan, China	X. Mou and S.F. Chen	MT412696	**OK357464**	MT335166	MT335405	[5,9]
B5	*C. aeknauliensis*	CMW 48253[T]	CBS 143559	Soil in *Eucalyptus* plantation	Aek Nauli, North Sumatra, Indonesia	M.J. Wingfield	MT412710	**OK357465**	MT335180	MT335419	[5,12]
		CMW 48254	CBS 143560	Soil in *Eucalyptus* plantation	Aek Nauli, North Sumatra, Indonesia	M.J. Wingfield	MT412711	**OK357466**	MT335181	MT335420	[5,12]
B8	*C. asiatica*	CBS 114073[T]	CMW 23782; CPC 3900	Debris leaf litter	Prathet Thai, Thailand	N.L. Hywel-Jones	AY725705	AY725616	AY725741	AY725658	[29,49]
B17	*C. brassicicola*	CBS 112841[T]	CMW 51206; CPC 4552	Soil at *Brassica* sp.	Indonesia	M.J. Wingfield	KX784689	KX784619	KX784561	N/A [5]	[30]
B19	*C. bumicola*	CMW 48257[T]	CBS 143575	Soil in *Eucalyptus* plantation	Aek Nauli, North Sumatra, Indonesia	M.J. Wingfield	MT412736	**OK357467**	MT335205	MT335445	[5,12]
B20	*C. canadiana*	CMW 23673[T]	CBS 110817; STE-U 499	*Picea* sp.	Canada	S. Greifenhagen	MT412737	MT412958	MT335206	MT335446	[1,5,17,50]
		CERC 8952	–	Soil	HeNan, China	S.F. Chen	MT412821	MT413035	MT335290	MT335530	[5,36]
B23	*C. chinensis*	CMW 23674[T]	CBS 114827; CPC 4101	Soil	Hong Kong, China	E.C.Y. Liew	MT412751	MT412972	MT335220	MT335460	[5,29,49]
		CMW 30986	CBS 112744; CPC 4104	Soil	Hong Kong, China	E.C.Y. Liew	MT412752	MT412973	MT335221	MT335461	[5,29,49]
B26	*C. cochinchinensis*	CMW 49915[T]	CBS 143567	Soil in *Hevea brasiliensis* plantation	Duong Minh Chau, Tay Ninh, Vietnam	N.Q. Pham, Q.N. Dang, and T.Q. Pham	MT412756	MT412977	MT335225	MT335465	[5,12]
		CMW 47186	CBS 143568	Soil in *Acacia auriculiformis* plantation	Song May, Dong Nai, Vietnam	N.Q. Pham and T.Q. Pham	MT412757	MT412978	MT335226	MT335466	[5,12]
B29	*C. colombiensis*	CMW 23676[T]	CBS 112220; CPC 723	Soil in *E. grandis* trees	La Selva, Colombia	M.J. Wingfield	MT412759	MT412980	MT335228	MT335468	[5,49]
		CMW 30985	CBS 112221; CPC 724	Soil in *E. grandis* trees	La Selva, Colombia	M.J. Wingfield	MT412760	MT412981	MT335229	MT335469	[5,49]
B31	*C. curvispora*	CMW 23693[T]	CBS 116159; CPC 765	Soil	Tamatave, Madagascar	P.W. Crous	MT412763	**OK357468**	MT335232	MT335472	[1,5,9,29,51]
		CMW 48245	CBS 143565	Soil in *Eucalyptus* plantation	Aek Nauli, North Sumatra, Indonesia	M.J. Wingfield	MT412764	N/A	MT335233	MT335473	[5,12]
B46	*C. heveicola*	CMW 49913[T]	CBS 143570	Soil in *H. brasiliensis* plantation	Bau Bang, Binh Duong, Vietnam	N.Q. Pham, Q.N. Dang, and T.Q. Pham	MT412786	MT413004	MT335255	MT335495	[5,12]
		CMW 49928	CBS 143571	Soil	Bu Gia Map National Park, Binh Phuoc, Vietnam	N.Q. Pham, Q.N. Dang, and T.Q. Pham	MT412811	MT413025	MT335280	MT335520	[5,12]
B48	*C. hongkongensis*	CBS 114828[T]	CMW 51217; CPC 4670	Soil	Hong Kong, China	M.J. Wingfield	MT412789	MT413007	MT335258	MT335498	[5,49]
		CERC 3570	CMW 47271	Soil in *Eucalyptus* plantation	BeiHai, Guangxi, China	S.F. Chen, J.Q. Li, and G.Q. Li	MT412791	MT413009	MT335260	MT335500	[5,11]
B51	*C. ilicicola*	CMW 30998[T]	CBS 190.50; IMI 299389; STE-U 2482	*Solanum tuberosum*	Bogor, Java, Indonesia	K.B. Boedijn and J. Reitsma	MT412797	**OK357469**	MT335266	MT335506	[1,5,29,52]
B52	*C. indonesiae*	CMW 23683[T]	CBS 112823; CPC 4508	*Syzygium aromaticum*	Warambunga, Indonesia	M.J. Wingfield	MT412798	MT413015	MT335267	MT335507	[5,49]
		CBS 112840	CMW 51205; CPC 4554	*S. aromaticum*	Warambunga, Indonesia	M.J. Wingfield	MT412799	MT413016	MT335268	MT335508	[5,49]
B55	*C. kyotensis*	CBS 114525[T]	ATCC 18834; CMW 51824; CPC 2367	*Robinia pseudoacacia*	Japan	T. Terashita	MT412802	MT413019	MT335271	MT335511	[1,5,30,53]
		CBS 114550	CMW 51825; CPC 2351	Soil	China	M.J. Wingfield	MT412777	MT412995	MT335246	MT335486	[5,30]
B57	*C. lantauensis*	CERC 3302[T]	CBS 142888; CMW 47252	Soil	LiDao, Hong Kong, China	M.J. Wingfield and S.F. Chen	MT412803	**OK357470**	MT335272	MT335512	[5,11]
		CERC 3301	CBS 142887; CMW 47251	Soil	LiDao, Hong Kong, China	M.J. Wingfield and S.F. Chen	MT412804	**OK357471**	MT335273	MT335513	[5,11]

Table 4. *Cont.*

Species Code [1]	Species	Isolates No. [2,3]	Other Collection Number [3]	Hosts	Area of Occurrence	Collector	tef1	tub2	cmdA	his3	References
								GenBank Accession No. [4]			
B58	*C. lateralis*	CMW 31412[T]	CBS 136629	Soil in *Eucalyptus* plantation	GuangXi, China	X. Zhou, G. Zhao, and F. Han	MT412805	MT413020	MT335274	MT335514	[5,9]
B66	*C. malesiana*	CMW 23687[T]	CBS 112752; CPC 4223	Soil	Northern Sumatra, Indonesia	M.J. Wingfield	MT412817	MT413031	MT335286	MT335526	[5,49]
		CBS 112710	CMW 51199; CPC 3899	Leaf litter	Prathet, Thailand	N.L. Hywel-Jones	MT412818	MT413032	MT335287	MT335527	[5,49]
B80	*C. pacifica*	CMW 16726[T]	A1568; CBS 109063; IMI 354528; STE-U 2534	*Araucaria heterophylla*	Hawaii, USA	M. Aragaki	MT412842	**OK357472**	MT335311	MT335551	[1,5,49,50]
		CMW 30988	CBS 114038	*Ipomoea aquatica*	Auckland, New Zealand	C.F. Hill	MT412843	**OK357473**	MT335312	MT335552	[1,5,29,49]
B86	*C. penicilloides*	CMW 23696[T]	CBS 174.55; STE-U 2388	*Prunus* sp.	Hatizyo Island, Japan	M. Ookubu	MT412869	MT413081	MT335338	MT335578	[1,5,54]
B112	*C. sumatrensis*	CMW 23698[T]	CBS 112829; CPC 4518	Soil	Northern Sumatra, Indonesia	M.J. Wingfield	MT412913	**OK357474**	MT335382	MT335622	[5,49]
		CMW 30987	CBS 112934; CPC 4516	Soil	Northern Sumatra, Indonesia	M.J. Wingfield	MT412914	**OK357475**	MT335383	MT335623	[5,49]
B113	*C. syzygiicola*	CBS 112831[T]	CMW 51204; CPC 4511	*Syzygium aromaticum*	Sumatra, Indonesia	M.J. Wingfield	KX784736	KX784663	N/A	N/A	[30]
B116	*C. uniseptata*	CBS 413.67[T]	CMW 23678; CPC 2391; IMI 299577	*Paphiopedilum callosum*	Celle, Germany	W. Gerlach	GQ267307	GQ267208	GQ267379	GQ267248	[30]
B120	*C. yunnanensis*	CERC 5339[T]	CBS 142897; CMW 47644	Soil in *Eucalyptus* plantation	YunNan, China	S.F. Chen and J.Q. Li	MT412927	MT413134	MT335396	MT335636	[5,11]
		CERC 5337	CBS 142895; CMW 47642	Soil in *Eucalyptus* plantation	YunNan, China	S.F. Chen and J.Q. Li	MT412928	MT413135	MT335397	MT335637	[5,11]
B124	*C. singaporensis*	CBS 146715[T]	MUCL 048320	leaf litter submerged in a small stream	Mac Ritchie Reservoir, Singapore	C. Decock	MW890086	MW890124	MW890042	MW890055	[34]
		CBS 146713	MUCL 048171	leaf litter submerged in a small stream	Mac Ritchie Reservoir, Singapore	C. Decock	MW890084	MW890123	MW890040	MW890053	[34]
Species in *Calonectria brassicae* species complex											
B12	*C. brachiatica*	CMW 25298[T]	CBS 123700	*Pinus maximinoi*	Buga, Colombia	M.J. Wingfield	MT412726	MT412948	MT335195	MT335435	[5,7]
		CMW 25302	–	*Pinus ecunumanii*	Buga, Colombia	M.J. Wingfield	MT412727	MT412949	MT335196	MT335436	[5,7]
B16	*C. brassicae*	CBS 111869[T]	CPC 2409	*Argyreia splendens*	Indonesia	F. Bugnicourt	MT412733	MT412955	MT335202	MT335442	[1,5,29,30]
B25	*C. clavata*	CMW 23690[T]	ATCC 66389; CBS 114557; CPC 2536; P078-1543	*Callistemon viminalis*	Lake Placid, Florida, USA	C.P. Seymour and E.L. Barnard	MT412754	MT412975	MT335223	MT335463	[1,5,29,55]
		CMW 30994	CBS 114666; CPC 2537; P078-1261	Root debris in peat	Lee County, Florida, USA	D. Ferrin	MT412755	MT412976	MT335224	MT335464	[1,5,29,55]
B34	*C. duoramosa*	CBS 134656[T]	–	Soil in tropical rainforest	Monte Dourado, Pará, Brazil	R.F. Alfenas	KM395853	KM395940	KM396027	KM396110	[10]
		LPF453	–	Soil in *Eucalyptus* plantation	Monte Dourado, Pará, Brazil	R.F. Alfenas	KM395854	KM395941	KM396028	KM396111	[10]
B35	*C. ecuadorae*	CMW 23677[T]	CBS 111406; CPC 1635	Soil	Ecuador	M.J. Wingfield	MT412773	MT412991	MT335242	MT335482	[5,29,56]
		CBS 111706	CMW 51821; CPC 1636	Soil	Ecuador	M.J. Wingfield	MT412771	MT412989	MT335240	MT335480	[5,31]
B43	*C. gracilis*	CBS 111807[T]	AR2677; CMW 51189; STE-U 2634	*Manilkara zapota*	Pará, Brazil	F. Carneiro de Albuquerque	GQ267323	AF232858	GQ267407	DQ190646	[1,30,31,56,57]
		CBS 111284	CMW 51175; CPC 1483	Soil	Imbrapa, Brazil	P.W. Crous	GQ267324	DQ190567	GQ267408	DQ190647	[1,30,31,56,57]
B77	*C. octoramosa*	CBS 111423[T]	CMW 51819; CPC 1650	Soil	Ecuador	M.J. Wingfield	MT412834	MT413048	MT335303	MT335543	[5,31]
B78	*C. orientalis*	CMW 20291[T]	CBS 125260	Soil	Langam, Indonesia	M.J. Wingfield	MT412835	MT413049	MT335304	MT335544	[5,29]
		CMW 20273	CBS 125259	Soil	Teso East, Indonesia	M.J. Wingfield	MT412836	MT413050	MT335305	MT335545	[5,29]

Table 4. *Cont.*

Species Code [1]	Species	Isolates No. [2,3]	Other Collection Number [3]	Hosts	Area of Occurrence	Collector	GenBank Accession No. [4]				References
							tef1	*tub2*	*cmdA*	*his3*	
B82	*C. paraensis*	CBS 134669[T]	LPF430	Soil in *Eucalyptus* plantation	Monte Dourado, Pará, Brazil	R.F. Alfenas	KM395837	KM395924	KM396011	KM396094	[10]
		LPF429	–	Soil in tropical rainforest	Monte Dourado, Pará, Brazil	R.F. Alfenas	KM395841	KM395928	KM396015	KM396098	[10]
B83	*C. parvispora*	CBS 111465[T]	CPC 1902	Soil	Brazil	A.C. Alfenas	MT412845	MT413057	MT335314	MT335554	[5,31]
		CMW 30981	CBS 111478; CPC 1921	Soil	Brazil	A.C. Alfenas	MT412844	MT413056	MT335313	MT335553	[5,29,30]
B84	*C. pauciphialidica*	CMW 30980[T]	CBS 111394; CPC 1628	Soil	Ecuador	M.J. Wingfield	MT412846	MT413058	MT335315	MT335555	[5,29,56]
B88	*C. pini*	CMW 31209[T]	CBS 123698	*Pinus patula*	Buga, Valle del Cauca, Colombia	C.A. Rodas	MT412870	MT413082	MT335339	MT335579	[5,29]
		CBS 125523	CMW 31210	*Pinus patula*	Buga, Valle del Cauca, Colombia	C.A. Rodas	GQ267345	GQ267225	GQ267437	GQ267274	[29]
B91	*C. pseudobrassicae*	CBS 134662[T]	LPF280	Soil in *Eucalyptus* plantation	Santana, Pará, Brazil	A.C. Alfenas	KM395849	KM395936	KM396023	KM396106	[10]
		CBS 134661	LPF260	Soil in *Eucalyptus* plantation	Santana, Pará, Brazil	A.C. Alfenas	KM395848	KM395935	KM396022	KM396105	[10]
B92	*C. pseudoecuadoriae*	CBS 111402[T]	CMW 51179; CPC 1639	Soil	Ecuador	M.J. Wingfield	KX784723	KX784652	KX784589	N/A	[30,31]
B105	*C. quinqueramosa*	CBS 134654[T]	LPF065	Soil in *Eucalyptus* plantation	Monte Dourado, Pará, Brazil	R.F. Alfenas	KM395855	KM395942	KM396029	KM396112	[10]
		CBS 134655	LPF281	Soil in *Eucalyptus* plantation	Santana, Pará, Brazil	A.C. Alfenas	KM395856	KM395943	KM396030	KM396113	[10]
B107	*C. robigophila*	CBS 134652[T]	LPF192	*Eucalyptus* sp. leaf	Açailandia, Maranhao, Brazil	R.F. Alfenas	KM395850	KM395937	KM396024	KM396107	[10]
		CBS 134653	LPF193	*Eucalyptus* sp. leaf	Açailandia, Maranhao, Brazil	R.F. Alfenas	KM395851	KM395938	KM396025	KM396108	[10]
Outgroups											
	Curvicladiella cignea	CBS 109167[T]	CPC 1595; MUCL 40269	Decaying leaf	French Guiana	C. Decock	KM231867	KM232002	KM231287	KM231461	[56,58,59]
		CBS 109168	CPC 1594; MUCL 40268	Decaying seed	French Guiana	C. Decock	KM231868	KM232003	KM231286	KM231460	[56,58,59]

[1] Codes B1 to B120 of the 120 accepted *Calonectria* species resulting from Liu and co-authors [5], "B124" indicated *C. singaporensis* described in Crous and co-authors [34]. [2] T: ex-type isolates of the species. [3] AR: Amy Y. Rossman working collection; ATCC: American Type Culture Collection, Virginia, USA; CBS: Westerdijk Fungal Biodiversity Institute, Utrecht, The Netherlands; CERC: China Eucalypt Research Centre, ZhanJiang, GuangDong Province, China; CMW: Culture collection of the Forestry and Agricultural Biotechnology Institute FABI, University of Pretoria, Pretoria, South Africa; CPC: Pedro Crous working collection housed at Westerdijk Fungal Biodiversity Institute; IMI: International Mycological Institute, MUCL: Mycotheque, Laboratoire de Mycologie Systematique st Appliqee, l'Universite, Louvian-la-Neuve, Belgium; STE-U: Department of Plant Pathology, University of Stellenbosch, South Africa; "–" represents no other collection number. [4] *tef1*: translation elongation factor 1-alpha; *tub2*: β-tubulin; *cmdA*: calmodulin; *his3*: histone H3; for GenBank Accession No. in bold, the sequences were submitted in this study. [5] N/A represents data that is not available.

Table 5. Statistical values of datasets for maximum parsimony and maximum likelihood analyses in this study.

Dataset	No. of Taxa	No. of bp [1]	Maximum Parsimony						
			PIC [2]	No. of Trees	Tree Length	CI [3]	RI [4]	RC [5]	HI [6]
tef1	157	522	241	110	588	0.697	0.973	0.678	0.303
tub2	156	597	256	1000	694	0.635	0.967	0.614	0.365
cmdA	156	697	277	1000	617	0.676	0.969	0.655	0.324
his3	153	478	166	973	602	0.570	0.960	0.547	0.430
tef1/tub2/cmdA/his3	157	2303	944	150	2671	0.609	0.962	0.586	0.391

Dataset	Maximum likelihood							
	Subst. mode [7]	NST [8]	Rate matrix					Rates
tef1	TIM2+G	6	1.8670	3.4436	1.8670	1.0000	5.0336	Gamma
tub2	TPM3uf+I+G	6	1.4137	4.7965	1.0000	1.4137	4.7965	Gamma
cmdA	TrN+G	6	1.0000	3.5934	1.0000	1.0000	7.2024	Gamma
his3	GTR+I+G	6	2.5191	8.8466	5.6820	2.1055	15.5239	Gamma
tef1/tub2/cmdA/his3	GTR+I+G	6	1.5966	4.2868	1.3927	0.9904	5.5003	Gamma

[1] bp = base pairs. [2] PIC = number of parsimony informative characters. [3] CI = consistency index. [4] RI = retention index. [5] RC = rescaled consistency index. [6] HI = homoplasy index. [7] Subst. model = best fit substitution model. [8] NST = number of substitution rate categories.

The phylogenetic analyses showed that the 85 *Calonectria* isolates were clustered in six groups (Group A, Group B, Group C, Group D, Group E, and Group F) based on *tef1*, *tub2*, *cmdA*, *his3*, and combined *tef1/tub2/cmdA/his3* analyses (Figure 2, Supplementary Figures S2–S5). The analyses showed that isolates in Groups A, B, C, D, and E belonged to the *C. kyotensis* species complex. Isolates in Groups A, B, C, and E were clustered with or were closest to *C. hongkongensis*, *C. kyotensis*, *C. chinensis*, and *C. ilicicola*, respectively, based on the *tef1*, *tub2*, *cmdA*, *his3*, and combined *tef1/tub2/cmdA/his3* trees (Figure 2, Supplementary Figures S2–S5). Therefore, isolates in Groups A, B, C, and E were identified as *C. hongkongensis*, *C. kyotensis*, *C. chinensis*, and *C. ilicicola*, respectively. Isolates in Group D were clustered in two sub-groups, sub-group D1 and sub-group D2, in the *tub2* tree. Isolates in sub-group D1 were clustered with or were closest to *C. aconidialis*; isolates in sub-group D2 were clustered with *C. asiatica* (Supplementary Figure S3); isolates in Group D were clustered with or were closest to *C. aconidialis* based on the *tef1*, *cmdA*, *his3*, and combined *tef1/tub2/cmdA/his3* trees (Figure 2, Supplementary Figures S2, S4, and S5). Isolates in Group D were identified as *C. aconidialis*. Isolates in Group F belonged to the *C. brassicae* species complex. These isolates were consistently only clustered with *C. orientalis* based on the *tef1*, *tub2*, *his3*, and combined *tef1/tub2/cmdA/his3* trees and were clustered with both *C. orientalis* and *C. brassicae* in the *cmdA* tree (Figure 2, Supplementary Figures S2–S5). Isolates in Group F were identified as *C. orientalis*.

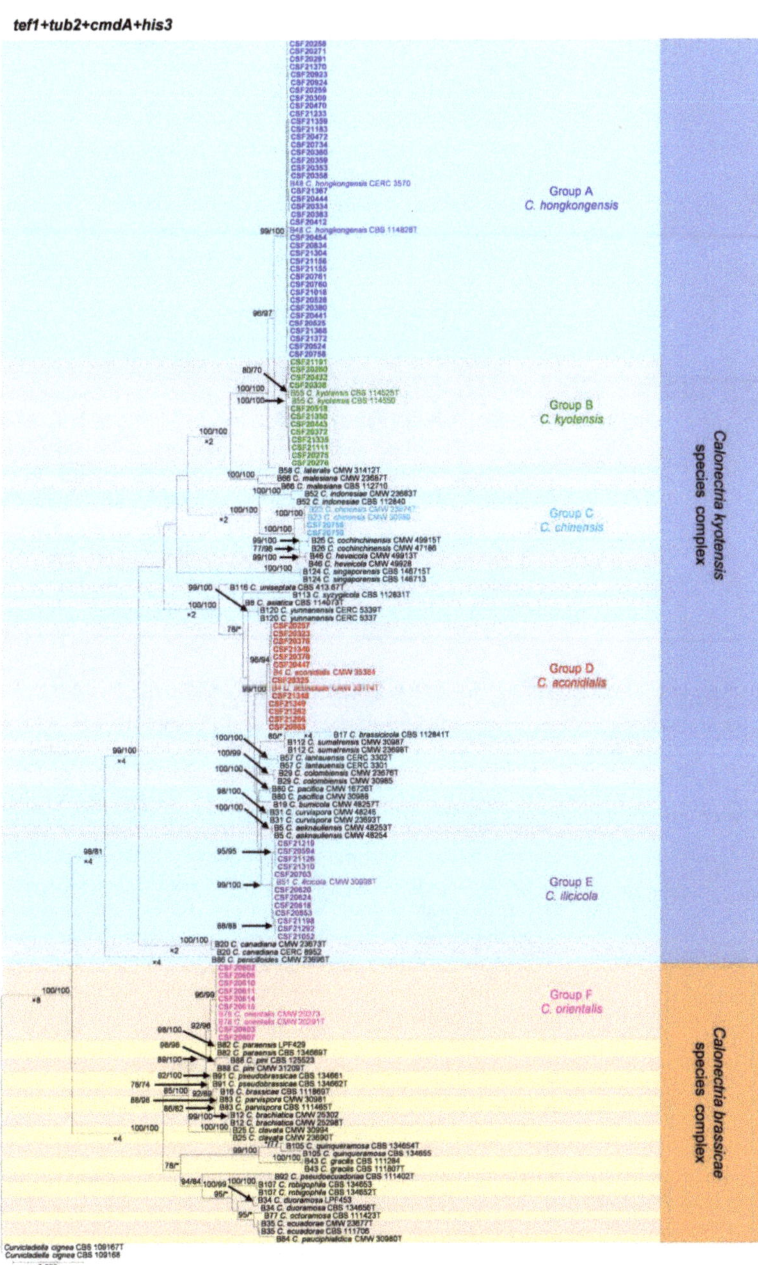

Figure 2. Phylogenetic tree of *Calonectria* species based on maximum likelihood (ML) analyses of the dataset of combined *tef1*, *tub2*, *cmdA*, and *his3* gene sequences in this study. Bootstrap support values ≥ 70% are presented above the branches as follows: ML/MP. Bootstrap values < 70% or absent are marked with "*". Isolates highlighted in six different colors, and bold were obtained in this study. Ex-type isolates are marked with "T". The "B" species codes are consistent with the recently published results in Liu and co-authors [5]. *Curvicladiella cignea* (CBS 109167 and CBS 109168) was used as the outgroup taxon.

Based on the results of phylogenetic analyses and induction of asexual structures, 17 isolates representing six *Calonectria* species were selected for macroconidia and vesicle morphological comparisons (Tables 3 and 6). These representative isolates could be classified into two groups based on the shape of the vesicles. Isolates of *C. aconidialis*, *C. chinensis*, *C. hongkongensis*, *C. ilicicola*, and *C. kyotensis* produce sphaeropedunculate vesicles, while the vesicles of *C. orientalis* are typically clavate. With the exception of *C. ilicicola* isolates, which produce 1(–3) septate macroconidia, isolates of the other five species all produced one septate macroconidium (Table 6). The shape of the vesicle and the number of macroconidia septations for each of the six *Calonectria* species found in this study were consistent with the described strains of relevant species in previous studies [1,9,29,49] (Table 6).

The morphological comparisons showed that significant variation existed in the size of macroconidia or width of vesicles among some isolates of each species of *C. aconidialis*, *C. hongkongensis*, and *C. kyotensis* identified in this study. For example, the macroconidia of *C. aconidialis* isolate CSF20985 were much longer than those of the other two tested *C. aconidialis* isolates CSF20323 and CSF20376. In *C. hongkongensis*, the macroconidia of isolate CSF20383 were longer than those of the other four isolates; the vesicles of isolate CSF20924 were wider than those of other isolates. In *C. kyotensis*, the macroconidia of isolate CSF20276 were much longer than those of isolate CSF21191 (Table 6).

The measurement results further showed that macroconidia size and vesicle width of isolates of some species obtained in this study were not always similar to the originally described strains of the same *Calonectria* species. For example, the macroconidia lengths of isolates of *C. chinensis* and *C. orientalis* obtained in this study were much shorter than the originally described strains of the relevant species [29,49] (Table 6).

3.4. Calonectria Species Diversity in Different Soil Layers

Based on the sequence comparisons of *tef1*, *tub2*, *cmdA*, and *his3* sequences, the 1037 *Calonectria* isolates were identified as *C. hongkongensis* (665 isolates; 64.1%), *C. aconidialis* (250 isolates; 24.1%), *C. kyotensis* (58 isolates; 5.6%), *C. ilicicola* (47 isolates; 4.5%), *C. chinensis* (2 isolates; 0.2%), and *C. orientalis* (15 isolates; 1.5%) (Table 1). *Calonectria hongkongensis* was dominant, followed by *C. aconidialis*. Each of the two dominant species was isolated from more than or close to 50% of all of the sampling points, and the two species accounted for 88.2% of all of the *Calonectria* isolates obtained in this study (Table 1, Supplementary Table S1, Figure 3). Both *C. chinensis* and *C. orientalis* were only isolated from one sampling point; *C. chinensis* was only isolated from the 0–20 cm soil layer, and only two isolates were obtained; *C. orientalis* was isolated from the soil layers of 40–60 and 60–80 cm, and 11 and 4 isolates in the two soil layers were obtained, respectively (Table 1, Supplementary Table S1, Figure 3).

With the exception of *C. orientalis* in the *C. brassicae* species complex, the diversity of species in the *C. kyotensis* species complex decreased with increasing soil depth. Five, four, four, four, and two species were identified in the soil layers of 0–20, 20–40, 40–60, 60–80, and 80–100 cm, respectively (Table 1, Supplementary Table S1).

For each of the five species in the *C. kyotensis* species complex, the number of sampling points containing *Calonectria* decreased with increasing depth of the soil, with the exception of *C. hongkongensis* in soil layers of 60–80 cm (2 sampling points) and 80–100 cm (4 sampling points) (Supplementary Table S1, Figure 4A); the number of isolates obtained decreased with increasing soil depth, with the exception of *C. hongkongensis* in the 60–80 cm soil layer (8 isolates) and 80–100 cm (20 isolates) as well as *C. ilicicola* in the 0–20 cm (16 isolates) and 20–40 cm (19 isolates) soil layers (Table 1, Figure 4B). Most isolates were obtained from the soil layers 0–20 and 20–40 cm, accounting for 86.6%, 85.6%, 81%, 74.5%, and 100% of all of the obtained isolates within each species of *C. hongkongensis*, *C. aconidialis*, *C. kyotensis*, *C. ilicicola*, and *C. chinensis*, respectively (Figure 5).

Table 6. Morphological comparisons of *Calonectria* isolates and species obtained in the current study.

Species	Isolate/Species	Macroconidia (L × W) [1,2,3]	Macroconidia Average (L × W) [1,2]	Macroconidia Septation	Vesicle Width [1,2,3]	Vesicle Width Average [1]
C. aconidialis	Isolate CSF20323 (this study)	(35–)39.5–45.5(–48) × (4–)4–4.5(–5)	42.5 × 4.5	1	(3.5–)4.5–6(–6.5)	5
	Isolate CSF20376 (this study)	(34.5–)38.5–45(–47.5) × (4–)4.5–5(–5.5)	41.5 × 4.5	1	(4–)4.5–11(–13)	8
	Isolate CSF20985 (this study)	(41–)46.5–51.5(–54) × (4–)4.5–5(–5.5)	49 × 5	1	(4.5–)5–6.5(–9.5)	6
	Species (this study)	(34.5–)40–48.5(–54) × (4–)4.5–5(–5.5)	44.5 × 4.5	1	(3.5–)4–8.5(–13)	6
	Species [9]	N/A [4]	N/A	N/A	N/A	N/A
C. chinensis	Isolate CSF20756 (this study)	(35.5–)40–45(–49) × (3.5–)4–4.5(–4.5)	42.5 × 4	1	(3.5–)3.5–9(–11.5)	6.5
	Isolate CSF20759 (this study)	(34.5–)37.5–43(–46) × (3.5–)4–4.5(–5)	40.5 × 4	1	(3–)5–10.5(–12)	8
	Species (this study)	(34.5–)38.5–44(–49) × (3.5–)4–4.5(–5)	41.5 × 4	1	(3–)4–10(–12)	7
	Species [49]	(38–)41–48(–56) × (3.5–)4(–4.5)	45 × 4	1	6–9	N/A
C. hongkongensis	Isolate CSF20353 (this study)	(33.5–)36–42(–48) × (3.5–)4–4.5(–4.5)	39 × 4	1	(4–)5–8.5(–10.5)	6.5
	Isolate CSF20360 (this study)	(34–)35.5–40(–43.5) × (3.5–)4–4.5(–5)	37.5 × 4	1	(4.5–)5.5–9(–11)	7.5
	Isolate CSF20383 (this study)	(37.5–)42.5–48(–50.5) × (4–)4–4.5(–5)	45.5 × 4.5	1	(4–)6–10.5(–11)	8.5
	Isolate CSF20761 (this study)	(32–)34.5–39.5(–43) × (3.5–)3.5–4(–4.5)	37 × 4	1	(4–)5.5–8(–9.5)	6.5
	Isolate CSF20924 (this study)	(35–)37.5–44(–45.5) × (3.5–)4–4.5(–5)	40.5 × 4	1	(6–)9–13(–14.5)	11
	Species (this study)	(32–)36–44(–50.5) × (3.5–)4–4.5(–5)	40 × 4	1	(4–)5.5–10.5(–14.5)	8
	Species [49]	(38–)45–48(–53) × 4(–4.5)	46.5 × 4	1	8–14	N/A
C. ilicicola	Isolate CSF20618 (this study)	(52.5–)56.5–66(–71.5) × (6–)6.5–7.5(–8)	61.5 × 7	1(–3)	(8–)9–11(–11.5)	10
	Isolate CSF21052 (this study)	(31–)50.5–69(–78) × (3–)5–7(–7.5)	59.5 × 6	1(–3)	(3.5–)5–8(–11)	6.5
	Isolate CSF21310 (this study)	(50–)55–62.5(–67) × (5.5–)6–7(–7.5)	58.5 × 6.5	(1–)3	(4–)6.5–10(–11.5)	8.5
	Species (this study)	(31–)53.5–66(–78) × (3–)6–7(–8)	60 × 6.5	1(–3)	(3.5–)6–10(–11.5)	8
	Species [1]	(45–)70–82(–90) × (4–)5–6.5(–7)	62 × 6	(1–)3	(6–)7–10(–12)	N/A
C. kyotensis	Isolate CSF20276 (this study)	(33.5–)36.5–44(–51) × (3.5–)4–4.5(–4.5)	40.5 × 4	1	(6.5–)8.5–11.5(–12.5)	10
	Isolate CSF21191(this study)	(29.5–)32.5–38.5(–42.5) × (3.5–)4–4.5(–5)	35.5 × 4	1	(5–)7.5–10.5(–11.5)	9
	Isolate CSF21335 (this study)	(32–)35.5–40(–43) × (3.5–)4–4.5(–5)	38 × 4	1	(5–)8–10(–11)	9
	Species (this study)	(29.5–)34.5–41.5(–51) × (3.5–)4–4.5(–5)	38 × 4	1	(5–)7.5–10.5(–12.5)	9
	Species [1]	(35–)45–50(–55) × 3–4(–5)	40 × 3.5	1	6–12	N/A
C. orientalis	Isolate CSF20614 (this study)	(30.5–)35–40(–43.5) × (4.5–)5–5.5(–5.5)	37.5 × 5	1	(3–)4–6.5(–7.5)	5
	Species [29]	(43–)46–50(–53) × 4(–5)	48 × 4	1	5–10	N/A

[1] All of the measurements are in μm. Fifty macroconidia and vesicles were measured for each isolate, with the exception of the vesicle of isolate CSF20618, for which 25 vesicles were measured because of the limited number of vesicles produced. [2] L × W = length × width. [3] Measurements are presented in the format ((minimum–) (average—standard deviation)—(average + standard deviation) (–maximum)). [4] N/A represents data that are not available.

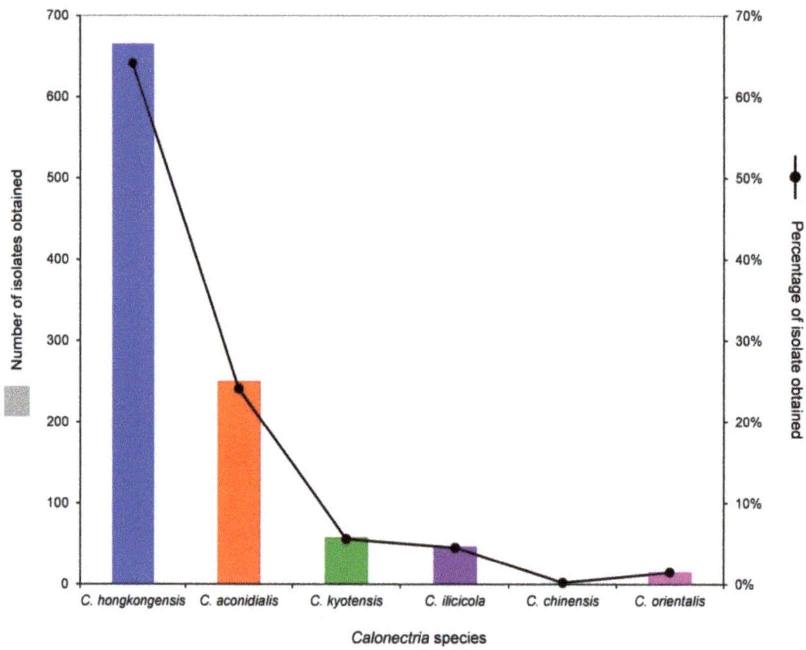

Figure 3. Numbers and percentages of isolates obtained for each *Calonectria* species from all soil samples collected.

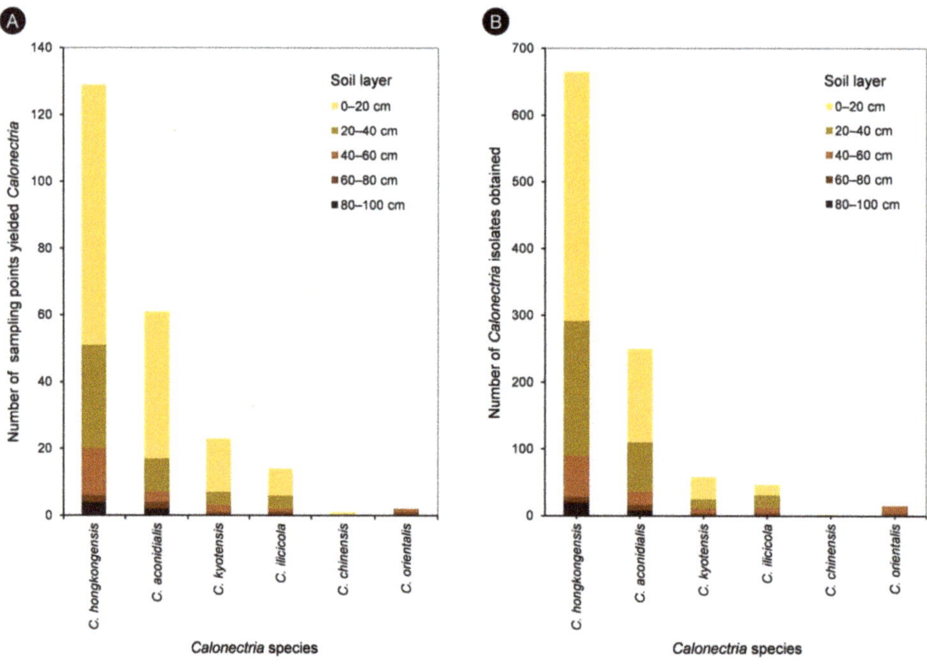

Figure 4. Number of sampling points yielded different *Calonectria* species in each of the five soil layers (**A**), and numbers of isolates obtained for different *Calonectria* species in each of the five soil layers (**B**).

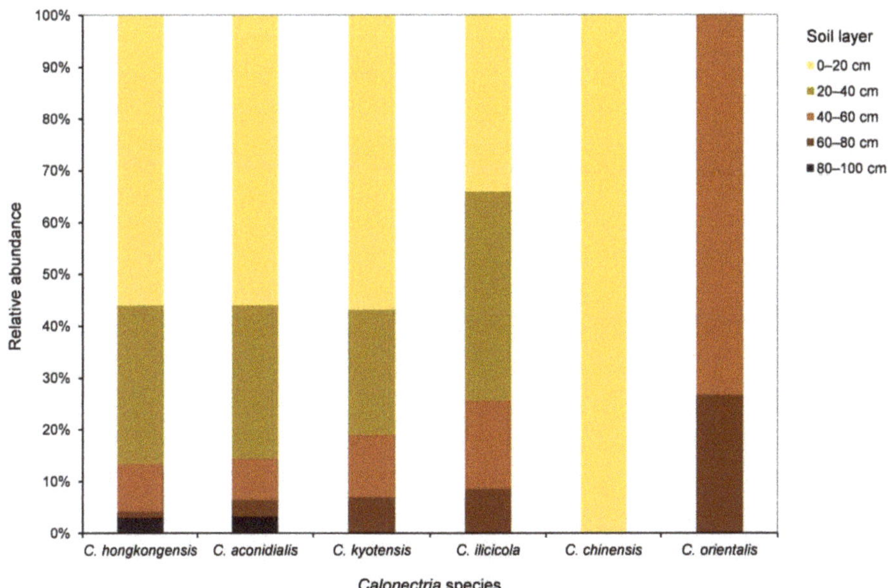

Figure 5. Relative abundances of each *Calonectria* species in each of the five soil layers. Relative abundance was based on the proportional frequencies of isolates of each *Calonectria* species in each soil layer.

3.5. Genotyping of Isolates within Each Calonectria Species

For the 1037 *Calonectria* isolates obtained and identified in this study, the genotype results based on *tef1* and *tub2* sequences indicated that 11, 3, 3, 3, 1, and 1 genotype(s) existed in *C. hongkongensis*, *C. aconidialis*, *C. kyotensis*, *C. ilicicola*, *C. chinensis*, and *C. orientalis*, respectively (Table 2). The isolates presenting the dominant genotype (genotype AA) accounted for 84.4%, 62.4%, 56.9%, 55.3%, 100%, and 100% of all of the isolates obtained from *C. hongkongensis*, *C. aconidialis*, *C. kyotensis*, *C. ilicicola*, *C. chinensis*, and *C. orientalis*, respectively (Table 2).

3.6. Genotype Diversity of Calonectria Species in Different Soil Layers

The *tef1-tub2* genotypes of each *Calonectria* species in each soil layer are listed in Table 7 and are shown in Figure 6. For each species in the *C. kyotensis* species complex, the results showed that the number of genotypes decreased with increasing soil depth, with the exception of *C. hongkongensis* and *C. aconidialis* in the 60–80 cm (one genotype) and 80–100 cm (two genotypes) soil layers (Table 7, Figure 6A,B); the 0–20 cm soil layer contained all of the genotypes of each species in the *C. kyotensis* complex (Table 7, Figure 6A–E). For the genotype with the most isolates of each species in the *C. kyotensis* complex, the majority of isolates were obtained from 0–20 cm soil layer, with the exception of *C. ilicicolla* (Table 7, Figure 6A–E). Only one genotype of *C. orientalis* was present in the 40–60 and 60–80 cm soil layers (Table 7, Figure 6F).

Table 7. Isolate numbers of each genotype in each soil layer for each *Calonectria* species.

Calonectria Species	Soil Layer	Genotype Determined by *tef1* Gene Sequences	Number of Isolates Based on *tef1* Genotype	Genotype Determined by *tub2* Gene Sequence	Number of Isolates Based on *tub2* Genotype	Genotype Determined by *tef1* and *tub2* Gene Sequences	Number of Isolates Based on *tef1* and *tub2* Genotype	Number of Isolates in Each Soil Layer for Each Species
C. hongkongensis	0–20 cm	A	346	A	337	AA	310	373
		B	15	B	1	AB	1	
		C	1	C	4	AC	4	
		D	11	D	5	AD	5	
				E	2	AE	2	
				F	9	AF	9	
				G	11	AG	11	
				H	4	AH	4	
						BA	15	
						CA	1	
						DA	11	
	20–40 cm	A	186	A	197	AA	180	203
		C	4	D	2	AD	2	
		D	13	F	4	AF	4	
						CA	4	
						DA	13	
	40–60 cm	A	58	A	50	AA	47	61
		D	3	F	7	AF	7	
				G	4	AG	4	
						DA	3	
	60–80 cm	A	8	A	8	AA	8	8
	80–100 cm	A	16	A	20	AA	16	20
		D	4			DA	4	
C. aconidialis	0–20 cm	A	140	A	98	AA	98	140
				B	1	AB	1	
				C	41	AC	41	
	20–40 cm	A	74	A	40	AA	40	74
				B	8	AB	8	
				C	26	AC	26	
	40–60 cm	A	20	A	6	AA	6	20
				C	14	AC	14	
	60–80 cm	A	8	A	8	AA	8	8
	80–100 cm	A	8	A	4	AA	4	8
				C	4	AC	4	
C. kyotensis	0–20 cm	A	31	A	27	AA	25	33
		B	2	B	6	AB	6	
						BA	2	
	20–40 cm	A	10	A	12	AA	8	14
		B	4	B	2	AB	2	
						BA	4	
	40–60 cm	A	7	B	7	AB	7	7
	60–80 cm	A	4	B	4	AB	4	4
	80–100 cm	–	–	–	–	–	–	0
C. ilicicola	0–20 cm	A	9	A	4	AA	4	16
		B	7	B	12	AB	5	
						BB	7	
	20–40 cm	A	18	A	14	AA	14	19
		B	1	B	5	AB	4	
						BB	1	
	40–60 cm	A	8	A	8	AA	8	8
	60–80 cm	B	4	B	4	BB	4	4
	80–100 cm	–	–	–	–	–	–	0
C. chinensis	0–20 cm	A	2	A	2	AA	2	2
	20–40 cm	–	–	–	–	–	–	0
	40–60 cm	–	–	–	–	–	–	0
	60–80 cm	–	–	–	–	–	–	0
	80–100 cm	–	–	–	–	–	–	0
C. orientalis	0–20 cm	–	–	–	–	–	–	0
	20–40 cm	–	–	–	–	–	–	0
	40–60 cm	A	11	A	11	AA	11	11
	60–80 cm	A	4	A	4	AA	4	4
	80–100 cm	–	–	–	–	–	–	0

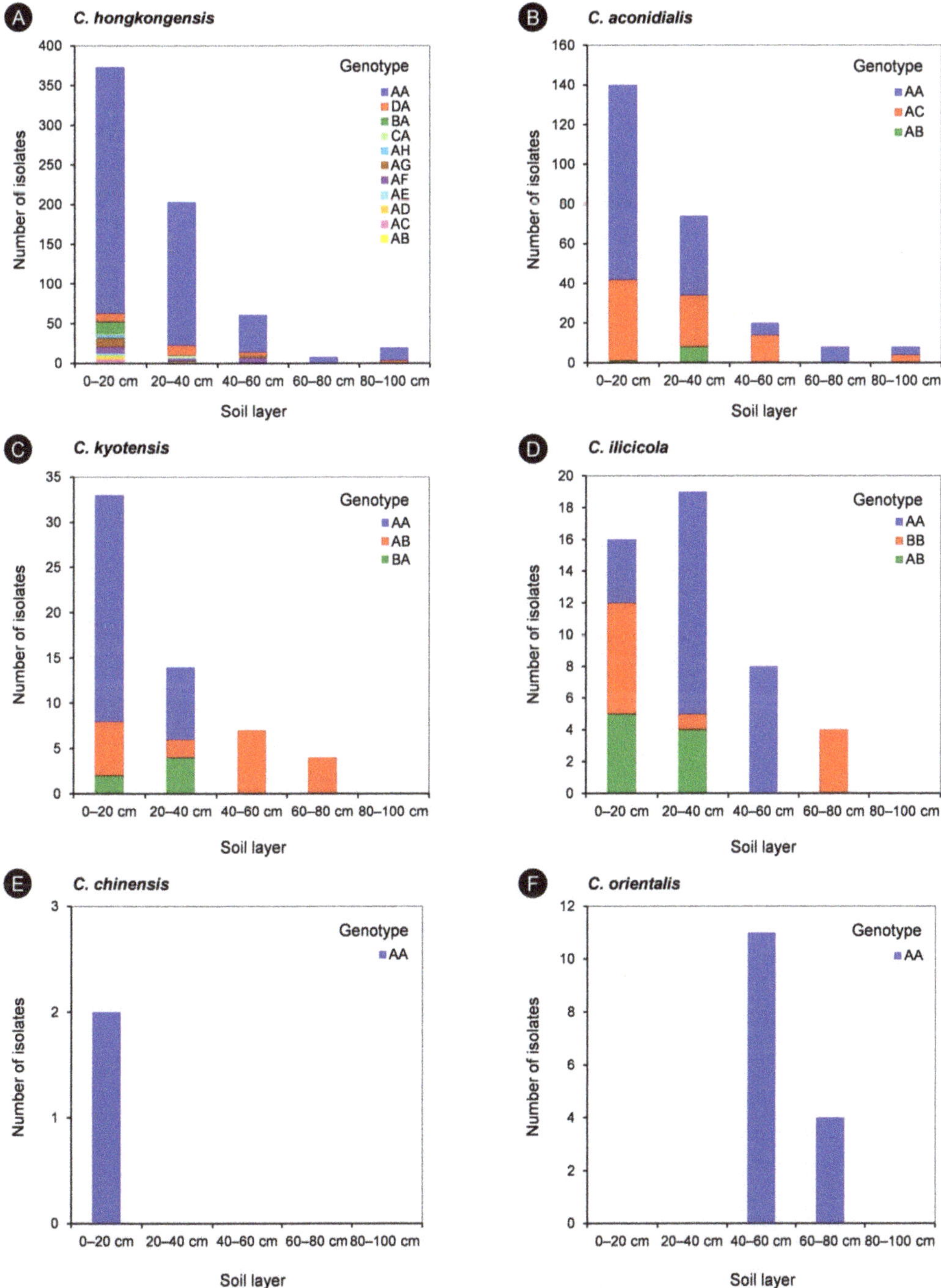

Figure 6. The isolate numbers of each genotype of each *Calonectria* species in five soils layers. The genotypes were determined by DNA sequences of *tef1* and *tub2* gene regions. (**A**): *C. hongkongensis*; (**B**): *C. aconidialis*; (**C**): *C. kyotensis*; (**D**) *C. ilicicola*; (**E**): *C. chinensis*; (**F**): *C. orientalis*.

The minimum spanning network (MSN) analysis was conducted for *C. hongkongensis*, which was considered as the dominant species identified in this study. The analysis revealed that most isolates of *C. hongkongensis* were genotype AA (561 isolates), followed by genotypes DA (31 isolates) and AF (20 isolates); genotype AA was present in the isolates from all five soil layers; genotypes AB, AC, AE, AH, and BA were present only in the isolates from the 0–20 cm soil layer, and the other genotypes were present in isolates from two to four soil layers. Isolates from the 0–20 cm soil layer contained all of the genotypes (Figure 7).

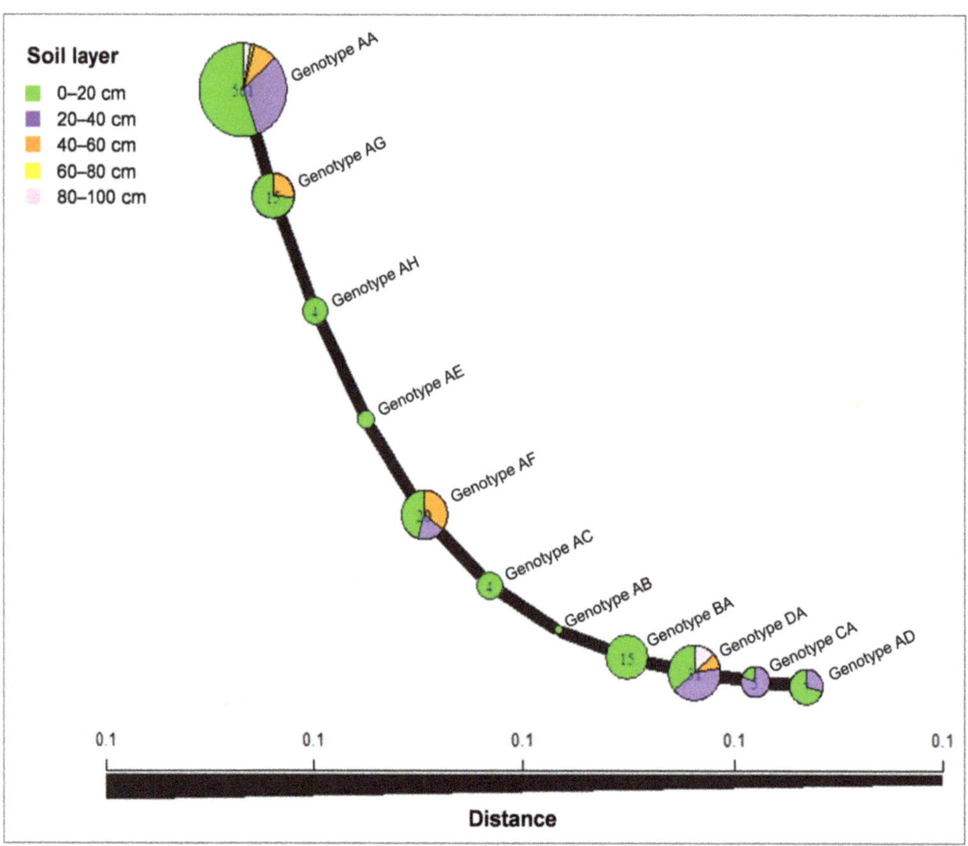

Figure 7. Minimum spanning network constructed using Bruvo's distances showing that the *C. hongkongensis* isolates were grouped into 11 genotypes based on *tef1* and *tub2* sequences. The size of a node is proportional to the number of represented *tef1-tub2* genotypes.

4. Discussion

In this study, more than 1000 *Calonectria* isolates were obtained from five soil layers at 100 sampling points in one *Eucalyptus* plantation. All of the isolates were identified based on DNA sequence comparisons of multiple gene regions. Six *Calonectria* species were identified, namely, *C. aconidialis*, *C. chinensis*, *C. hongkongensis*, *C. ilicicola*, and *C. kyotensis* in the *C. kyotensis* species complex, and *C. orientalis* in the *C. brassicae* species complex. *Calonectria hongkongensis* (64.1% of all of the isolates) was the dominant species, followed by *C. aconidialis* (24.1% of all of the isolates). To our knowledge, this is the first report of *C. orientalis* in China. The species diversity and distribution characteristics of the six species in different soil layers were clarified. The results showed that the number of sampling points from which *Calonectria* was obtained, and the number of *Calonectria* isolates obtained decreased with increasing depth of the soil. The majority of isolates (84.3% of all the isolates) were obtained from soil layers of 0–20 and 20–40 cm. The diversity of the five species in the *C. kyotensis* species complex decreased with increasing soil depth. For each species in the *C. kyotensis* species complex, in most cases, the number of genotypes decreased with increasing soil depth, and the 0–20 cm soil layer contained all of the genotypes of each species.

Five species, namely, *C. aconidialis*, *C. chinensis*, *C. hongkongensis*, *C. ilicicola*, and *C. kyotensis*, in the *C. kyotensis* species complex were isolated from the soil of the *Eucalyptus* plantation in this study. These five species have been frequently isolated from soils in several other regions in southern China, especially from soils in *Eucalyptus* plantations [9,11,14,49]. *Calonectria ilicicola* is considered as a soil-borne fungal pathogen that has been isolated from a number of diseased plant species in China [21,61]. This study presents the first record of *C. ilicicola* isolated from soil in a *Eucalyptus* plantation. Results of this and previous studies suggest that all five of the species in the *C. kyotensis* species complex are potentially widely distributed in *Eucalyptus* plantation soils in other regions of southern China [9,11,14].

This study is the first report of *C. orientalis* in China, and this species is the first *Calonectria* species in the *C. brassicae* species complex found in China. *Calonectria orientalis* has been isolated from soil in Indonesia [29]. Some other species in the *C. brassicae* species complex have also been frequently isolated from soils. With the exception of *C. orientalis*, the other species in the *C. brassicae* species complex isolated from soils were all from Ecuador and Brazil in South America [5,10,29–31,56]. Most of the *Calonectria* species in the *C. brassicae* species complex have only been isolated from South America but not from Asia [5] and *C. orientalis*, in this study, was only isolated from one of the 100 sampling points. These results suggest that *C. orientalis* is not widely distributed in China.

For the five species in the *C. kyotensis* species complex, the results of this study indicate that the diversity of the five species decreased with increasing soil depth, and the number of sampling points containing *Calonectria* and the number of *Calonectria* isolates obtained also decreased with soil depth. Most isolates were obtained from the 0–20 and 20–40 cm soil layers. In most cases, the number of genotypes decreased with increasing soil depth for each species, and the 0–20 cm soil layer contained all of the genotypes of each species. These results suggest that 0–20 cm is the best soil depth for *Calonectria* isolation and for examining the species and genotype diversity of *Calonectria* in soils in *Eucalyptus* plantations in southern China. In several previous studies specialized in the research on *Calonectria* species diversity, soil samples were also exclusively collected from the surface layer, all from the 0–20 cm layer [9–11,13,14,36]. These studies have characterized the diversity of *Calonectria* species well. Results of a number of other studies indicated that microbial diversity and richness are typically affected by the soil depth [62–67], and shallower layers usually have a higher level of microbial diversity [62,63,66–68]. This pattern is consistent with the results of the present study. A possible reason for the vertical distribution of soil microbes is the harsher environment in deeper soil layers, where the soil density is higher, oxygen concentrations are lower, and carbon and nutrients are less available [69]. For *Calonectria*, which includes some important pathogens of various agricultural, horticultural, and forestry crops worldwide, as well as for other genera of fungi in forests, no systematic research has been conducted to examine the species diversity and distribution characteristics in different soil layers. This study showed that the deeper soil layers had comparatively fewer but still contained many *Calonectria*. It remains unknown whether the *Calonectria* were originally distributed in deeper soil layers or whether the fungi in deeper soil layers migrated from surface layers, perhaps through the infiltration of rainwater. Studies on the population diversity differences among different soil layers should be conducted to address this question. Furthermore, the *Calonectria* distributed in deeper soil layers increase the challenge of controlling the diseases caused by these fungi.

This study examined the species diversity and distribution characteristics of *Calonectria* in five soil layers in a *Eucalyptus* plantation in southern China. Six species were isolated from soils in a relatively small *Eucalyptus* plantation, indicating that the diversity of *Calonectria* species in these soils in southern China is relatively high. This study also revealed that the species diversity and number of genotypes of each *Calonectria* species decreased with increasing soil depth, a pattern that helps us to understand the distribution characteristics of *Calonectria* species in different layers of soil. For some *Calonectria* species, there were relatively large numbers of isolates obtained from different soil layers, especially for *C. hongkongensis* and *C. aconidialis* in the 0–20, 20–40, and 40–60 cm soil layers. The genetic structures and population biology of these species in the different soil layers are unknown, but additional studies may increase our understanding of the distribution characteristics and dissemination patterns of *Calonectria* species.

Supplementary Materials: The following are available online at https://www.mdpi.com/article/10.3390/jof7100857/s1, Table S1: Number of sampling points containing each *Calonectria* species in each soil layer, Table S2: All 1037 isolates obtained and sequenced in this study, Figure S1: Number of sampling points that yielded *Calonectria* in each of the five soil layers, Figure S2: Phylogenetic tree of *Calonectria* species based on maximum likelihood (ML) analyses of the *tef1* gene sequences, Figure S3: Phylogenetic tree of *Calonectria* species based on ML analyses of the *tub2* gene sequences, Figure S4: Phylogenetic tree of *Calonectria* species based on ML analyses of the *cmdA* gene sequences, Figure S5: Phylogenetic tree of *Calonectria* species based on ML analyses of the *his3* gene sequences.

Author Contributions: S.C. conceived and designed the experiments. L.L. and S.C. collected the samples. L.L. performed the laboratory work. L.L., W.W. and S.C. analyzed the data. S.C. and L.L. wrote the paper. All authors have read and agreed to the published version of the manuscript.

Funding: This study was initiated through the bilateral agreement between the Governments of South Africa and China and supported by The National Key R&D Program of China (China-South Africa Forestry Joint Research Centre Project; project No. 2018YFE0120900), the National Ten-thousand Talents Program (Project No. W03070115) and the GuangDong Top Young Talents Program (Project No. 20171172).

Institutional Review Board Statement: Not applicable.

Informed Consent Statement: Not applicable.

Data Availability Statement: Data is contained within the article and Supplementary Materials.

Acknowledgments: We thank JiaLong Han, LanSen Sun, Ying Liu, and XueYing Liang for their assistance in collecting samples. We thank QianLi Liu for sequencing the *tub2* gene region of some isolates in Table 4. We thank FeiFei Liu for analyzing the genotype data. We thank GuoQing Li, WenWen Li, and QuanChao Wang who provided assistance in laboratory work and checking the data. We thank LetPub (www.letpub.com; accessed date: 27 August 2021 and 6 October 2021) for providing linguistic assistance during the preparation of this manuscript.

Conflicts of Interest: The authors declare no conflict of interest.

References

1. Crous, P.W. *Taxonomy and pathology of Cylindrocladium (Calonectria) and Allied Genera*; APS Press: St. Paul, MN, USA, 2002.
2. Lombard, L.; Crous, P.W.; Wingfield, B.D.; Wingfield, M.J. Species concepts in *Calonectria* (*Cylindrocladium*). *Stud. Mycol.* **2010**, *66*, 1–14. [CrossRef]
3. Vitale, A.; Crous, P.W.; Lombard, L.; Polizzi, G. *Calonectria* diseases on ornamental plants in Europe and the Mediterranean basin: An overview. *J. Plant Pathol.* **2013**, *95*, 463–476.
4. Daughtrey, M.L. Boxwood blight: Threat to ornamentals. *Annu. Rev. Phytopathol.* **2019**, *57*, 189–209. [CrossRef]
5. Liu, Q.L.; Li, J.Q.; Wingfield, M.J.; Duong, T.A.; Wingfield, B.D.; Crous, P.W.; Chen, S.F. Reconsideration of species boundaries and proposed DNA barcodes for *Calonectria*. *Stud. Mycol.* **2020**, *97*, 100106. [CrossRef] [PubMed]
6. Li, J.Q.; Barnes, I.; Liu, F.F.; Wingfield, M.J.; Chen, S.F. Global genetic diversity and mating type distribution of *Calonectria pauciramosa*: An important wide host-range plant pathogen. *Plant Dis.* **2021**. [CrossRef]
7. Lombard, L.; Rodas, C.A.; Crous, P.W.; Wingfield, B.D.; Wingfield, M.J. *Calonectria* (*Cylindrocladium*) species associated with dying *Pinus* cuttings. *Persoonia* **2009**, *23*, 41–47. [CrossRef] [PubMed]

8. Lombard, L.; Zhou, X.D.; Crous, P.W.; Wingfield, B.D.; Wingfield, M.J. *Calonectria* species associated with cutting rot of *Eucalyptus*. *Persoonia* **2010**, *24*, 1–11. [CrossRef] [PubMed]

9. Lombard, L.; Chen, S.F.; Mou, X.; Zhou, X.D.; Crous, P.W.; Wingfield, M.J. New species, hyper-diversity and potential importance of *Calonectria* spp. from *Eucalyptus* in South China. *Stud. Mycol.* **2015**, *80*, 151–188. [CrossRef]

10. Alfenas, R.F.; Lombard, L.; Pereira, O.L.; Alfenas, A.C.; Crous, P.W. Diversity and potential impact of *Calonectria* species in *Eucalyptus* plantations in Brazil. *Stud. Mycol.* **2015**, *80*, 89–130. [CrossRef]

11. Li, J.Q.; Wingfield, M.J.; Liu, Q.L.; Barnes, I.; Roux, J.; Lombard, L.; Crous, P.W.; Chen, S.F. *Calonectria* species isolated from *Eucalyptus* plantations and nurseries in South China. *IMA Fungus* **2017**, *8*, 259–286. [CrossRef]

12. Pham, N.; Barnes, I.; Chen, S.F.; Liu, F.F.; Dang, Q.; Pham, T.; Lombard, L.; Crous, P.W.; Wingfield, M.J. Ten new species of *Calonectria* from Indonesia and Vietnam. *Mycologia* **2019**, *111*, 78–102. [CrossRef]

13. Wang, Q.C.; Liu, Q.L.; Chen, S.F. Novel species of *Calonectria* isolated from soil near *Eucalyptus* plantations in southern China. *Mycologia* **2019**, *111*, 1028–1040. [CrossRef]

14. Wu, W.X.; Chen, S.F. Species diversity, mating strategy, and pathogenicity of *Calonectria* species from diseased leaves and soils in the *Eucalyptus* plantation in southern China. *J. Fungi* **2021**, *7*, 73. [CrossRef]

15. Old, K.M.; Wingfield, M.J.; Yuan, Z.Q. *A Manual of Diseases of Eucalypts in South-East Asia*; Center for International Forestry Research: Jakarta, Indonesia, 2003.

16. Rodas, C.A.; Lombard, L.; Gryzenhout, M.; Slippers, B.; Wingfield, M.J. *Cylindrocladium* blight of *Eucalyptus grandis* in Colombia. *Australas. Plant Pathol.* **2005**, *34*, 143–149. [CrossRef]

17. Lechat, C.; Crous, P.W.; Groenewald, J.Z. The enigma of *Calonectria* species occurring on leaves of *Ilex aquifolium* in Europe. *IMA Fungus* **2010**, *1*, 101–108. [CrossRef]

18. Chen, S.F.; Lombard, L.; Roux, J.; Xie, Y.J.; Wingfield, M.J.; Zhou, X.D. Novel species of *Calonectria* associated with *Eucalyptus* leaf blight in Southeast China. *Persoonia* **2011**, *26*, 1–12. [CrossRef]

19. Gehesquière, B.; Crouch, J.A.; Marra, R.E.; van Poucke, K.; Rys, F.; Maes, M.; Gobin, B.; Höfte, M.; Heungens, K. Characterization and taxonomic reassessment of the box blight pathogen *Calonectria pseudonaviculata*, introducing *Calonectria henricotiae* sp. nov. *Plant Pathol.* **2016**, *65*, 37–52. [CrossRef]

20. Lopes, U.P.; Alfenas, R.F.; Zambolim, L.; Crous, P.W.; Costa, H.; Pereira, O.L. A new species of *Calonectria* causing rot on ripe strawberry fruit in Brazil. *Australas. Plant Pathol.* **2017**, *47*, 1–11. [CrossRef]

21. Gai, Y.; Deng, Q.; Chen, X.; Guan, M.; Xiao, X.; Xu, D., Deng, M.; Pan, R. Phylogenetic diversity of *Calonectria ilicicola* causing Cylindrocladium black rot of peanut and red crown rot of soybean in southern China. *J. Gen. Plant Pathol.* **2017**, *83*, 273–282. [CrossRef]

22. Freitas, R.G.; Alfenas, R.F.; Guimarães, L.M.S.; Badel, J.L.; Alfenas, A.C. Genetic diversity and aggressiveness of *Calonectria pteridis* in *Eucalyptus* spp. *Plant Pathol.* **2019**, *68*, 869–877. [CrossRef]

23. Wang, Q.C.; Chen, S.F. *Calonectria pentaseptata* causes severe leaf disease on cultivated *Eucalyptus* in Leizhou Peninsula of southern China. *Plant Dis.* **2020**, *104*, 493–509. [CrossRef] [PubMed]

24. Bell, D.K.; Sobers, E.K. A peg, pod, and root necrosis of peanuts caused by a species of *Calonectria*. *Phytopathology* **1966**, *56*, 1361–1364.

25. Johnston, S.A.; Beute, M.K. Histopathology of Cylindrocladium black rot of peanut. *Phytopathology* **1975**, *64*, 649–653. [CrossRef]

26. Pataky, J.K.; Beute, M.K.; Wynne, J.C.; Carlson, G.A. Peanut yield, market quality and value reductions due to Cylindrocladium black rot. *Peanut Sci.* **1983**, *10*, 62–66. [CrossRef]

27. Berner, D.K.; Berggren, G.T.; Snow, J.P.; White, E.P. Distribution and management of red crown rot of soybean in Louisiana. *Appl. Agric. Res.* **1988**, *3*, 160–166.

28. Yamamoto, R.; Nakagawa, A.; Shimada, S.; Komatsu, S.; Kanematsu, S. Histopathology of red crown rot of soybean. *J. Gen. Plant Pathol.* **2016**, *83*, 23–32. [CrossRef]

29. Lombard, L.; Crous, P.W.; Wingfield, B.D.; Wingfield, M.J. Phylogeny and systematics of the genus *Calonectria*. *Stud. Mycol.* **2010**, *66*, 31–69. [CrossRef] [PubMed]

30. Lombard, L.; Wingfield, M.J.; Alfenas, A.C.; Crous, P.W. The forgotten *Calonectria* collection: Pouring old wine into new bags. *Stud. Mycol.* **2016**, *85*, 159–198. [CrossRef]

31. Marin-Felix, Y.; Groenewald, J.Z.; Cai, L.; Chen, Q.; Marincowitz, S.; Barnes, I.; Bensch, K.; Braun, U.; Camporesi, E.; Damm, U.; et al. Genera of phytopathogenic fungi: GOPHY 1. *Stud. Mycol.* **2017**, *86*, 99–216. [CrossRef]

32. Crous, P.W.; Luangsa-ard, J.J.; Wingfield, M.J.; Carnegie, A.J.; Hernández-Restrepo, M.; Lombard, L.; Roux, J.; Barreto, R.W.; Baseia, I.G.; Cano-Lira, J.F.; et al. Fungal Planet description sheets: 785–867. *Persoonia* **2018**, *41*, 238–417. [CrossRef]

33. Crous, P.W.; Carnegie, A.J.; Wingfield, M.J.; Sharma, R.; Mughini, G.; Noordeloos, M.E.; Santini, A.; Shouche, Y.S.; Bezerra, J.D.P.; Dima, B.; et al. Fungal Planet description sheets: 868–950. *Persoonia* **2019**, *42*, 291–473. [CrossRef]

34. Crous, P.W.; Hernández-Restrepo, M.; Schumacher, R.K.; Cowan, D.A.; Maggs-Kölling, G.; Marais, E.; Wingfield, M.J.; Yilmaz, N.; Adan, O.C.G.; Akulov, A.; et al. New and Interesting Fungi. 4. *Fungal Syst. Evol.* **2021**, *7*, 255–343. [CrossRef]

35. Crous, P.W.; Cowan, D.A.; Maggs-Kölling, G.; Yilmaz, N.; Thangavel, R.; Wingfield, M.J.; Noordeloos, M.E.; Dima, B.; Brandrud, T.E.; Jansen, G.M.; et al. Fungal Planet description sheets: 1182–1283. *Persoonia* **2021**, *46*, 313–528.

36. Liu, Q.L.; Chen, S.F. Two novel species of *Calonectria* isolated from soil in a natural forest in China. *MycoKeys* **2017**, *26*, 25–60. [CrossRef]

37. Phipps, P.M.; Beute, M.K.; Barker, K.R. An elutriation method for quantitative isolation of *Cylindrocladium crotalariae* microsclerotia from peanut field soil. *Phytopathology* **1976**, *66*, 1255–1259. [CrossRef]
38. Thies, W.F.; Patton, R.F. The biology of *Cylindrocladium scoparium* in Wisconsin forest tree nurseries. *Phytopathology* **1970**, *60*, 1662–1668. [CrossRef]
39. Sobers, E.K.; Littrell, R.H. Pathogenicity of three species of *Cylindrocladium* to select hosts. *Plant Dis. Report.* **1974**, *58*, 1017–1019.
40. Anderson, P.J. *Rose Canker and Its Control*; Massachusetts Agricultural Experiment Station Bulletin: Boston, MA, USA, 1918; Volume 183, pp. 11–46.
41. Dumas, M.T.; Greifenhagen, S.; Halicki-Hayden, G.; Meyer, T.R. Effect of seedbed steaming on *Cylindrocladium floridanum*, soil microbes and the development of white pine seedlings. *Phytoprotection* **1998**, *79*, 35–43. [CrossRef]
42. Xu, Y.; Du, A.; Wang, Z.; Zhu, W.; Li, C.; Wu, L. Effects of different rotation periods of *Eucalyptus* plantations on soil physiochemical properties, enzyme activities, microbial biomass and microbial community structure and diversity. *For. Ecol. Manag.* **2020**, *456*, 117683. [CrossRef]
43. Cooperative Research Group on Chinese Soil Taxonomy. *Chinese Soil Taxonomy*; Science Press: Beijing, China, 2001.
44. van Burik, J.A.; Schreckhise, R.W.; White, T.C.; Bowden, R.A.; Myerson, D. Comparison of six extraction techniques for isolation of DNA from filamentous fungi. *Med. Mycol.* **1998**, *36*, 299–303. [CrossRef]
45. Kumar, S.; Stecher, G.; Tamura, K. MEGA7: Molecular Evolutionary Genetics Analysis version 7.0 for bigger datasets. *Mol. Biol. Evol.* **2016**, *33*, 1870–1874. [CrossRef]
46. Graça, R.N.; Alfenas, A.C.; Maffia, L.A.; Titon, M.; Alfenas, R.F.; Lau, D.; Rocabado, J.M.A. Factors influencing infection of eucalypts by *Cylindrocladium pteridis*. *Plant Pathol.* **2009**, *58*, 971–981. [CrossRef]
47. Paradis, E.; Claude, J.; Strimmer, K. APE: Analyses of phylogenetics and evolution in R language. *Bioinformatics* **2004**, *20*, 289–290. [CrossRef] [PubMed]
48. Kamvar, Z.N.; Tabima, J.F.; Grünwald, N.J. Poppr: An R package for genetic analysis of populations with clonal, partially clonal, and/or sexual reproduction. *PeerJ* **2014**, *2*, e281. [CrossRef] [PubMed]
49. Crous, P.W.; Groenewald, J.Z.; Risede, J.M.; Simoneau, P.; Hywel-Jones, N.L. *Calonectria* species and their *Cylindrocladium* anamorphs: Species with sphaeropedunculate vesicles. *Stud. Mycol.* **2004**, *50*, 415–430.
50. Kang, J.C.; Crous, P.W.; Schoch, C.L. Species concepts in the *Cylindrocladium floridanum* and *Cy. spathiphylli* complexes (*Hypocreaceae*) based on multi-allelic sequence data, sexual compatibility and morphology. *Syst. Appl. Microbiol.* **2001**, *24*, 206–217. [CrossRef] [PubMed]
51. Victor, D.; Crous, P.W.; Janse, B.J.H.; Wingfield, M.J. Genetic variation in *Cylindrocladium floridanum* and other morphologically similar *Cylindrocladium* species. *Syst. Appl. Microbiol.* **1997**, *20*, 268–285. [CrossRef]
52. Boedijn, K.B.; Reitsma, J. Notes on the genus *Cylindrocladium* (Fungi: *Mucedineae*). *Reinwardtia* **1950**, *1*, 51–60.
53. Terashita, T. A new species of *Calonectria* and its conidial state. *Trans. Mycol. Soc. Japan* **1968**, *8*, 124–129.
54. Tubaki, K. Studies on the Japanese *Hyphomycetes*. V. Leaf & stem group with a discussion of the classification of *Hyphomycetes* and their perfect stages. *J. Hattori Bot. Lab.* **1958**, *20*, 142–244.
55. El-Gholl, N.E.; Alfieri, S.A., Jr.; Barnard, E.L. Description and pathogenicity of *Calonectria clavata* sp. nov. *Mycotaxon* **1993**, *48*, 201–216.
56. Crous, P.W.; Groenewald, J.Z.; Risède, J.M.; Simoneau, P.; Hyde, K.D. *Calonectria* species and their *Cylindrocladium* anamorphs: Species with clavate vesicles. *Stud. Mycol.* **2006**, *55*, 213–226. [CrossRef]
57. Crous, P.W.; Wingfield, M.J.; Alfenas, A.C. Additions to *Calonectria*. *Mycotaxon* **1993**, *46*, 217–234.
58. Decock, C.; Crous, P.W. *Curvicladium* gen. nov., a new hyphomycete genus from French Guiana. *Mycologia* **1998**, *90*, 276–281. [CrossRef]
59. Lombard, L.; van der Merwe, N.A.; Groenewald, J.Z.; Crous, P.W. Generic concepts in *Nectriaceae*. *Stud. Mycol.* **2015**, *80*, 189–245. [CrossRef]
60. Cunningham, C.W. Can three incongruence tests predict when data should be combined? *Mol. Biol. Evol.* **1997**, *14*, 733–740. [CrossRef] [PubMed]
61. Pei, W.H.; Cao, J.F.; Yang, M.Y.; Zhao, Z.J.; Xue, S.M. First report of black rot of *medicago sativa* caused by *Cylindrocladium parasiticum* (teleomorph *Calonectria ilicicola*) in Yunnan province, China. *Plant Dis.* **2015**, *99*, 890. [CrossRef]
62. Fierer, N.; Schimel, J.P.; Holden, P.A. Variations in microbial community composition through two soil depth profiles. *Soil Biol. Biochem.* **2003**, *35*, 167–176. [CrossRef]
63. Jumpponen, A.; Jones, K.L.; Blair, J. Vertical distribution of fungal communities in tallgrass prairie soil. *Mycologia* **2010**, *102*, 1027–1041. [CrossRef] [PubMed]
64. de Araujo Pereira, A.P.; Santana, M.C.; Bonfim, J.A.; de Lourdes Mescolotti, D.; Cardoso, E.J.B.N. Digging deeper to study the distribution of mycorrhizal arbuscular fungi along the soil profile in pure and mixed *Eucalyptus grandis* and *Acacia mangium* plantations. *Appl. Soil Ecol.* **2018**, *128*, 1–11. [CrossRef]
65. Grishkan, I.; Lázaro, R.; Kidron, G.J. Vertical divergence of cultivable microfungal communities through biocrusted and bare soil profiles at the Tabernas Desert, Spain. *Geomicrobiol. J.* **2020**, *37*, 534–549. [CrossRef]
66. Upton, R.N.; Sielaff, A.C.; Hofmockel, K.S.; Xu, X.; Polley, H.W.; Wilsey, B.J. Soil depth and grassland origin cooperatively shape microbial community co-occurrence and function. *Ecosphere* **2020**, *11*, e02973. [CrossRef]

67. Frey, B.; Walthert, L.; Perez-Mon, C.; Stierli, B.; Köchli, R.; Dharmarajah, A.; Brunner, I. Deep soil layers of drought-exposed forests harbor poorly known bacterial and fungal communities. *Front. Microbiol.* **2021**, *12*, 674160. [CrossRef]
68. Na, X.; Ma, S.; Ma, C.; Liu, Z.; Xu, P.; Zhu, H.; Liang, W.; Kardol, P. *Lycium barbarum* L. (goji berry) monocropping causes microbial diversity loss and induces *Fusarium* spp. enrichment at distinct soil layers. *Appl. Soil Ecol.* **2021**, *168*, 104107. [CrossRef]
69. Lennon, J.T. Microbial life deep underfoot. *mBio* **2020**, *11*, e03201-19. [CrossRef] [PubMed]

Journal of
Fungi

Article

Four New Species of *Hemileccinum* (Xerocomoideae, Boletaceae) from Southwestern China

Mei-Xiang Li [1,2,3], Gang Wu [1,2] and Zhu L. Yang [1,2,*]

[1] CAS Key Laboratory for Plant Diversity and Biogeography of East Asia, Kunming Institute of Botany, Chinese Academy of Sciences, Kunming 650201, China; limeixiang@mail.kib.ac.cn (M.-X.L.); wugang@mail.kib.ac.cn (G.W.)

[2] Yunnan Key Laboratory for Fungal Diversity and Green Development, Kunming Institute of Botany, Chinese Academy of Sciences, Kunming 650201, China

[3] College of Life Sciences, University of Chinese Academy of Sciences, Beijing 100049, China

* Correspondence: fungi@mail.kib.ac.cn

Abstract: The genus *Hemileccinum* belongs to the subfamily Xerocomoideae of the family Boletaceae. In this study, phylogenetic inferences of *Hemileccinum* based on sequences of a single-locus (ITS) and a multi-locus (nrLSU, *tef1-α*, *rpb1*, *rpb2*) were conducted. Four new species, namely *H. abidum*, *H. brevisporum*, *H. ferrugineipes* and *H. parvum* were delimited and proposed based on morphological and molecular evidence. Descriptions and line-drawings of them were presented, as well as their comparisons to allied taxa. Our study shed new light on the recognition of the genus. The pileipellis of the species in this genus should mostly be regarded as (sub)epithelium to hyphoepithelium, because the pileipellis of most studied species here is composed of short inflated cells in the inner layer (subpellis) and filamentous hyphae in outer layer (suprapellis). The basidiospores of the studied species, including the type species, *H. impolitum*, have a warty surface.

Keywords: boletes; taxonomy; morphology; molecular phylogeny

Citation: Li, M.-X.; Wu, G.; Yang, Z.L. Four New Species of *Hemileccinum* (Xerocomoideae, Boletaceae) from Southwestern China. *J. Fungi* **2021**, *7*, 823. https://doi.org/10.3390/jof7100823

Academic Editors: Anush Kosakyan, Rodica Catana and Alona Biketova

Received: 27 August 2021
Accepted: 28 September 2021
Published: 30 September 2021

Publisher's Note: MDPI stays neutral with regard to jurisdictional claims in published maps and institutional affiliations.

1. Introduction

The genus *Hemileccinum* Šutara was created based on the species *H. impolitum* (Fr.) Šutara as the type, and *H. depilatum* (Redeuilh) Šutara [1]. These two species were both originally placed in the genus *Boletus* L. [2] and later were transferred to the genus *Leccinum* because of the lateral stipe stratum of the leccinoid type which is predominantly anticlinally arranged, breaking up into characteristic fascicles of hyphae ending in elements of the caulohymenium during growth of the stipe [1,3,4]. However, molecular phylogenetic analyses indicated that these two species are very distant from the species of both *Leccinum* and *Boletus*, but similar to the species of *Xerocomus*; thus they were accordingly transferred to *Xerocomus* [5,6]. Based on the previous molecular evidence and his own further morphological observations, Šutara established the genus *Hemileccinum* to accommodate these two species. He emphasized that *Hemileccinum* was diagnosed by the anatomical structure of the peripheral stipe layers having a lateral stipe stratum of the leccinoid type, which distinguished this genus not only from all the other species belonging to the *Xerocomus* s.l. but also from those in the genus *Boletus* [1]. Wu et al. confirmed the monophyly of *Hemileccinum* and found an additional diagnosed character of this genus, namely the irregularly warty basidiospores under SEM [7,8]. Meanwhile, the genus *Corneroboletus* N.K. Zeng & Zhu L. Yang was confirmed as a synonym of *Hemileccinum* due to the similar basidiospore ornamentation and the closely phylogenetic relationship [8].

As ectomycorrhizal fungi, species in the genus *Hemileccinum* are widely distributed in temperate, subtropical and tropical regions, and play an important role in forest ecology [1,8–14]. However, the species diversity of *Hemileccinum* was relatively poorly known in the world until recent studies suggested existence of other potentially new specie of the genus [8,10]. Until now, only 10 species of the genus have been reported in the world according to the database of INDEX FUNGORUM (accessed date: 27 August 2021). Among them, *H. impolitum* and *H. depilatum* are from Europe [1], *H. subglabripes* (Peck) Halling, *H. rubropunctum* (Peck) Halling and *H. hortonii* (A.H. Sm. & Thiers) M. Kuo & B. Ortiz are from North America [8,9]. In Asia, *H. rugosum* G. Wu & Zhu L. Yang is described from China, and *H. indecorum* (Massee) G. Wu & Zhu L. Yang is from tropical China, Singapore, Thailand [8,15,16].

During past fungal investigations in southwestern China, we encountered four potential new species of *Hemileccinum*. Our aim in this study is to clarify their molecular phylogenetic positions and to delimit them based on morphological data and molecular evidence.

2. Materials and Methods

2.1. Sample Collection and Morphological Study

In total, seventeen collections were examined in this study, which were collected from the Yunnan Province of southwestern China during the years 2007–2017 (Figure 1). The macroscopic characters of the specimens were described based on fresh basidiomata, and the dried specimens were deposited in the Cryptogamic Herbarium of the Kunming Institute of Botany, Chinese Academy of Sciences (KUN-HKAS). Color codes of the form "5C4" indicate the plate, row, and color block from Kornerup and Wanscher [17]. For microscopic observation, a ZEISS Axiostar Plus microscope (Oberkochen, Germany) was used and the dried specimens were revived in 5% KOH solution or distilled water. Moreover, Melzer's reagent was applied to test color reactions of the tissue fragments to the solution. Microscopic studies follow Li et al. and Zhou et al. [18,19]. In the descriptions of basidiospores, the abbreviation [n/m/p] means n basidiospores measured from m basidiomata of p collections. The range notation (a)b–c(d) stands for the dimensions of basidiospores in which b–c contains a minimum of 90% of the measured values while a and d in the brackets stand for the extreme values. Q is used to imply "length/width ratio" of a basidiospore in side view; Q_m means average Q of all basidiospores $\pm$ sample standard deviation. To observe basidiospore ornamentations, a ZEISS Sigma 300 scanning electron microscope (SEM) (Oberkochen, Germany) was used. Genera are abbreviated as follows: *H.* for *Hemileccinum*, *Ca.* for *Castanopsis*, *C.* for *Castanea*, *L.* for *Lithocarpus*, *P.* for *Pinus*, *Q.* for *Quercus* and *Rug.* for *Rugiboletus*.

2.2. Molecular Procedures and Phylogenetic Analyses

Total genomic DNA was obtained with the Ezup Column Fungi Genomic DNA Purification Kit (Sangon Biotech, Shanghai, China) according to the manual from material dried with silica gel. A total of five nuclear loci were sequenced, including the internal transcribed spacer (ITS), the large subunit of nuclear ribosomal RNA gene (nrLSU), the polymerase II subunit one (*rpb1*) gene, the second largest subunit of RNA polymerase II (*rpb2*), and the translation elongation factor 1-α gene (*tef1-α*). The primer pairs of ITS1/ITS4 [20,21], LROR/LR5 [22,23] were used for amplifying ITS, nrLSU, respectively. The primer pairs used for amplifying the *rpb1*, *rpb2*, *tef1-α*, followed those in Wu et al. [7]. PCR was performed in a total volume of 25 µL containing 1 µL forward primer, 1 µL reverse primer, 9.5 µL nuclease-free H_2O, 12.5 µL BlasTaq^{TM} 2×PCR MasterMix (abm, Richmond, VA, Canada) and 1 µL DNA template. PCR protocol was as follows: pre-denaturation at 95 °C for 5 min, followed by 35 cycles of denaturation at 95 °C for 60 s, 52 °C for 60 s, and 72 °C for 80 s, and then a final elongation at 72 °C for 8 min was included. The PCR products were purified with a Gel Extraction and PCR Purification Combo Kit (Spincolumn) (Bioteke, Beijing, China), and then sequenced by ABI-3730-XL DNA Analyzer (Applied Biosystems, Foster City, CA, USA) by using the same primer pairs as in the PCR amplification for sequencing.

Figure 1. Fresh basidiomata of *Hemileccinum* species. (**a–c**) *H. albidum* ((**a**,**b**) Type, KUN-HKAS81120, (**c**) KUN-HKAS87225); (**d–f**) *H. brevisporum* ((**d**) KUN-HKAS67896, (**e**) KUN-HKAS59445, (**f**) Type, KUN-HKAS89150); (**g–i**) *H. ferrugineipes* ((**g**,**h**) Type, KUN-HKAS115554, (**i**) KUN-HKAS75054); (**j–l**) *H. parvum* ((**j–l**) Type, KUN-HKAS115553) Bars = 30 mm.

2.3. Phylogenetic Analyses

We used BLAST to compare the obtained sequences of the newly collected materials with those in the GenBank database. The BLAST results were used to predict the phylogenetic relationship between the newly collected specimens and known species and indicated that the new materials were genetically similar to the other species of *Hemileccinum*. In this study, two datasets were produced, the ITS dataset, and the combined nrLSU, *tef1-α*, *rpb1* and *rpb2* dataset. The ITS sequences of *Hemileccinum* species from China were used to infer relationships of Chinese species with those from Europe, North America and East Asia. In the analysis of ITS dataset, *Phylloporus rubrosquamosus* N.K. Zeng, Zhu L. Yang & L.P. Tang, *Phylloporus rubeolus* N.K. Zeng, Zhu L. Yang & L.P. Tang, *Hourangia cheoi* (W.F. Chiu) Xue T. Zhu & Zhu L. Yang and *Hourangia pumila* (M.A. Neves & Halling) Xue T. Zhu, Halling & Zhu L. Yang were chosen as outgroup [24–26]. The combined dataset was mainly used to infer phylogenetic relationships and systematic positions of the Chinese species. In the multigene phylogenetic analysis, including all known genera in the subfamily Xerocomoideae were included. We screened the relevant sequences deposited in GenBank, which were mainly submitted by Wu et al. [7,8], Gelardi et al. [27], Zhu et al. [26], Zeng et al. [24], Neves et al. [28]. We collected a total of 13 ingroup species of 8 known genera within Xerocomoideae and 2 outgroup species outside Xerocomoideae but in the Boletaceae. Detailed information of the voucher specimens is given in Table 1.

The sequences were assembled with SeqMan implemented in Lasergene v7.1 (DNASTAR Inc., Madison, WI, USA), and then aligned by using MAFFT v7.310 [29]. The software Bioedit v7.2.5 [30] was used to check aligned matrices. To assess any potential conflicts in the gene tree topologies for these five loci, single-locus matrices were analyzed using maximum likelihood (ML) in RAxML v8.0.20 [31]. Sequences of the loci without conflicts were then concatenated using Phyutility 2.2 [32,33] for further phylogenetic analyses. The best-fitted substitution model for each gene was determined through MrModeltest v2.4 [34] by using Akaike Information Criterion (AIC). GTR + I + G was inferred as the best-fit model for the nrLSU, *tef1-α*, *rpb1* and ITS selected according to the AIC in MrModeltest v2.4 [34]. SYM + I + G was selected as the best model for *rpb2*. For the ultimate phylogenetic analyses, Maximum Likelihood (ML) analysis and Bayesian Inference were conducted by RAxML v8.0.20 [31] and MRBAYES v3.2.7 [35], respectively. The parameters of RAxML were set as defaults with 500 bootstrap replicates, except the substitution model which was set as GTRGAMMAI.

BI analyses were conducted with two independent runs of one cold and three heated chains. Runs were performed for 2 million generations, and trees sampled every 100 generations. The convergence was determined with the average standard deviation of split frequencies (<0.01) Chain convergence was determined using Tracer v1.5 to confirm sufficiently large ESS values (>200). The sampled trees were subsequently summarized by using the "sump" and "sumt" commands with a 25% burn-in [31,35]. The Bayesian posterior probabilities (BPP) of internodes were estimated based on the majority rule consensus with the remaining trees.

Table 1. Specimens used in phylogenetic analysis and their GenBank accession numbers. The newly generated sequences are shown in bold.

Species	Voucher	Locality	GenBank Accession Number					References
			ITS	nrLSU	rpb2	rpb1	tef1-α	
Hemileccinum rugosum	KUN-HKAS84355	China	–	KT990578	KT990413	KT990931	KT990774	[8]
Hemileccinum rugosum	KUN-HKAS84970	China	–	KT990577	KT990412	–	KT990773	[8]
Hemileccinum rugosum	KUN-HKAS50284	China	–	KT990576	KT990411	–	KT990772	[8]
Hemileccinum subglabripes	MICH:KUO-07230802	USA	–	MK601738	MK766300	–	MK721092	[10]
Hemileccinum subglabripes	MICH:KUO-07070702	USA	–	MK601737	MK766299	–	MK721091	[10]
Hemileccinum subglabripes	MICH:KUO-08301402	USA	–	MK601739	MK766301	–	MK721093	[10]
Hemileccinum subglabripes	72206	USA	–	KF030303	–	KF030374	KF030404	[36]
Hemileccinum subglabripes	294169	USA	MN128237	–	–	–	–	from GenBank
Hemileccinum subglabripes	3660	–	KM248936	–	–	–	–	from GenBank
Hemileccinum depilatum	2137333	USA	AY127032	–	–	–	–	from GenBank
Hemileccinum depilatum	AF2845	Belgium	–	–	MG212633	–	MG212591	[37]
Hemileccinum depilatum	Bd1	–	–	AF139712	–	–	–	[5]
Hemileccinum impolitum	Bim 1	Germany	–	AF139715	–	KF030375	JQ327034	[36]
Hemileccinum impolitum	47698	Portugal	AJ419187	–	–	–	–	[38]
Hemileccinum impolitum	BI57407	Thailand	KM235997	–	–	–	–	from GenBank
Hemileccinum impolitum	BI57408	Thailand	KM235998	–	–	–	–	from GenBank
Hemileccinum impolitum	17173	USA	JF907783	–	–	–	–	[39]
Hemileccinum impolitum	KUN-HKAS84869	Germany	–	KT990575	KT990410	KT990930	KT990771	[8]
Hemileccinum indecorum	KUN-HKAS63126	China	–	KF112440	–	–	–	[7]
Hemileccinum indecorum	OR0863	Thailand	–	–	MH614772	–	MH614726	[16]
Hemileccinum rubropunctum	JLF56666	USA	MH190826	–	–	–	–	from GenBank
Hemileccinum rubropunctum	MES256	USA	FJ480428	–	–	–	–	[40]
Hemileccinum rubropunctum	NY-792788REH-8501	USA	–	MK601768	MK766327	–	MK721122	[10]
Hemileccinum rubropunctum	NY-01193924REH-9597	USA	–	MK601769	MK766328	–	MK721123	[10]
Hemileccinum sp.	KUN-HKAS53421	China	–	KF112432	KF112751	KF112565	KF112235	[7]
Hemileccinum hortonii	MICH KUO-07050706	USA	–	MK601821	MK766377	–	MK721175	[10]
Hemileccinum albidum	KUN-HKAS87225	China	**MZ923777**	**MZ923774**	**MZ936317**	**MZ936334**	**MZ936351**	**This study**
Hemileccinum albidum	KUN-HKAS83355	China	**MZ923778**	**MZ923775**	**MZ936321**	**MZ936340**	**MZ936357**	**This study**
Hemileccinum albidum (T)	KUN-HKAS81120	China	**MZ923782**	**MZ923766**	**MZ936320**	**MZ936339**	**MZ936352**	**This study**
Hemileccinum albidum	KUN-HKAS50503	China	**MZ923781**	**MZ923767**	**MZ936319**	**MZ936335**	**MZ936355**	**This study**
Hemileccinum albidum	KUN-HKAS50350	China	**MZ923779**	**MZ923768**	**MZ936323**	**MZ936342**	**MZ936359**	**This study**
Hemileccinum albidum	KUN-HKAS84554	China	**MZ923780**	–	**MZ936318**	**MZ936336**	**MZ936358**	**This study**
Hemileccinum albidum	KUN-HKAS85753	China	**MZ923786**	–	**MZ936325**	**MZ936337**	**MZ936353**	**This study**

Table 1. *Cont.*

Species	Voucher	Locality	GenBank Accession Number					References
			ITS	nrLSU	rpb2	rpb1	tef1-α	
Hemileccinum albidum	KUN-HKAS87105	China	-	MZ923769	MZ936327	MZ936338	MZ936356	This study
Hemileccinum albidum	KUN-HKAS83333	China	MZ923784	-	MZ936326	MZ936344	MZ936361	This study
Hemileccinum albidum	KUN-HKAS83400	China	MZ923783	MZ923770	MZ936324	MZ936341	MZ936354	This study
Hemileccinum albidum	KUN-HKAS115749	China	MZ923785	-	MZ936322	MZ936343	MZ936360	This study
Hemileccinum brevisporum (T)	KUN-HKAS89150	China	MZ923788	MZ923764	MZ936328	MZ936345	MZ936362	This study
Hemileccinum brevisporum	KUN-HKAS59445	China	-	KT990579	KT990414	KT990932	KT990775	[8]
Hemileccinum brevisporum	KUN-HKAS67896	China	MZ923787	MZ923765	MZ936329	MZ936346	MZ936363	This study
Hemileccinum ferrugineipes (T)	KUN-HKAS115554	China	MZ923792	MZ923773	MZ936330	MZ936350	MZ973011	This study
Hemileccinum ferrugineipes	KUN-HKAS75054	China	-	KF112377	KF112749	KF112563	KF112234	[7]
Hemileccinum ferrugineipes	KUN-HKAS93310	China	MZ923791	-	MZ936331	MZ936347	MZ973012	This study
Hemileccinum parvum	KUN-HKAS99764	China	MZ923789	MZ923771	MZ936332	MZ936349	MZ973009	This study
Hemileccinum parvum (T)	KUN-HKAS115553	China	MZ923790	MZ923772	MZ936333	MZ936348	MZ973010	This study
Heimioporus sp.	KUN-HKAS53451	China	-	KF112345	KF112805	KF112616	KF112226	[7]
Heimioporus aff. *japonicus*	KUN-HKAS52236	China	-	KF112346	KF112807	KF112617	KF112227	[7]
Heimioporus japonicas	KUN-HKAS52237	China	-	KF112347	KF112806	KF112618	KF112228	[7]
Aureoboletus tenuis	KUN-HKAS75104	China	-	KT990518	KT990359	KT990897	KT990722	[8]
Aureoboletus thibetanus	KUN-HKAS76655	China	-	KF112420	KF112752	KF112626	KF112236	[7]
Pulchroboletus roseoalbidus	AMB 12757	Italy	-	NG_060126	-	-	KJ729512	[27]
Alessioporus ichnusanus	AMB 12756	Italy	-	NG_057044	-	-	KJ729513	[27]
Phylloporus rubrosquamosus	KUN-HKAS52552	China	-	KF112391	KF112780	-	KF112289	[7]
Phylloporus rubrosquamosus	KUN-HKAS54559	China	NR120124	NG_042668	-	-	JQ967175	[24,25]
Phylloporus rubeolus	KUN-HKAS52573	China	JQ967259	JQ967216	-	-	JQ967172	[24,25]
Xerocomus fraternus	KUN-HKAS55328	China	-	KT990681	KT990497	-	KT990869	[8]
Xerocomus velutinus	KUN-HKAS68135	China	-	KT990673	-	KT991011	KT990861	[8]
Hourangia cheoi	Yang 5153	China	KP136997	KP136947	KP136975	KP136966	KP136924	[26]
Hourangia pumila	REH8063	Indonesia	JQ003626	NG_060636	-	-	-	[28]
Boletellus indistinctus	KUN-HKAS77623	China	-	KT990531	KT990371	-	KT990733	[8]
Boletellus indistinctus	KUN-HKAS80681	China	-	KT990532	KT990368	KT990903	KT990734	[8]
Leccinum variicolor	KUN-HKAS57758	China	-	KF112445	KF112725	KF112591	KF112251	[7]
Leccinum aff. *scabrum*	KUN-HKAS57266	China	-	KF112442	KF112722	KF112590	KF112248	[7]
Leccinum monticola	KUN-HKAS76669	China	-	KF112443	KF112723	KF112592	KF112249	[7]
Leccinellum cremeum	KUN-HKAS90639	China	-	-	KT990420	KT990936	KT990781	[8]
Leccinellum sp.	KUN-HKAS53410	China	-	KT990585	KT990421	KT990937	-	[8]

3. Results

3.1. Molecular Phylogenetic Analysis

A total of 79 sequences, including 16 for ITS, 12 for nrLSU, 17 for *tef1-α*, 17 for *rpb1*, and 17 for *rpb2* were newly generated in the present study and aligned with sequences downloaded from GenBank and previous studies. Sequences retrieved from GenBank and obtained in this study were listed in Table 1. ML and BI analyses of the ITS dataset resulted in almost identical topologies and thus only the tree inferred from ML analysis was displayed (Figure 2). Our phylogenetic analyses indicated that *Hemileccinum* formed a monophyletic group with evident support (MLB/BPP = 100%/1.0). Eight phylogenetic species of the genus *Hemileccinum* were retrieved, and four of them could be new to science.

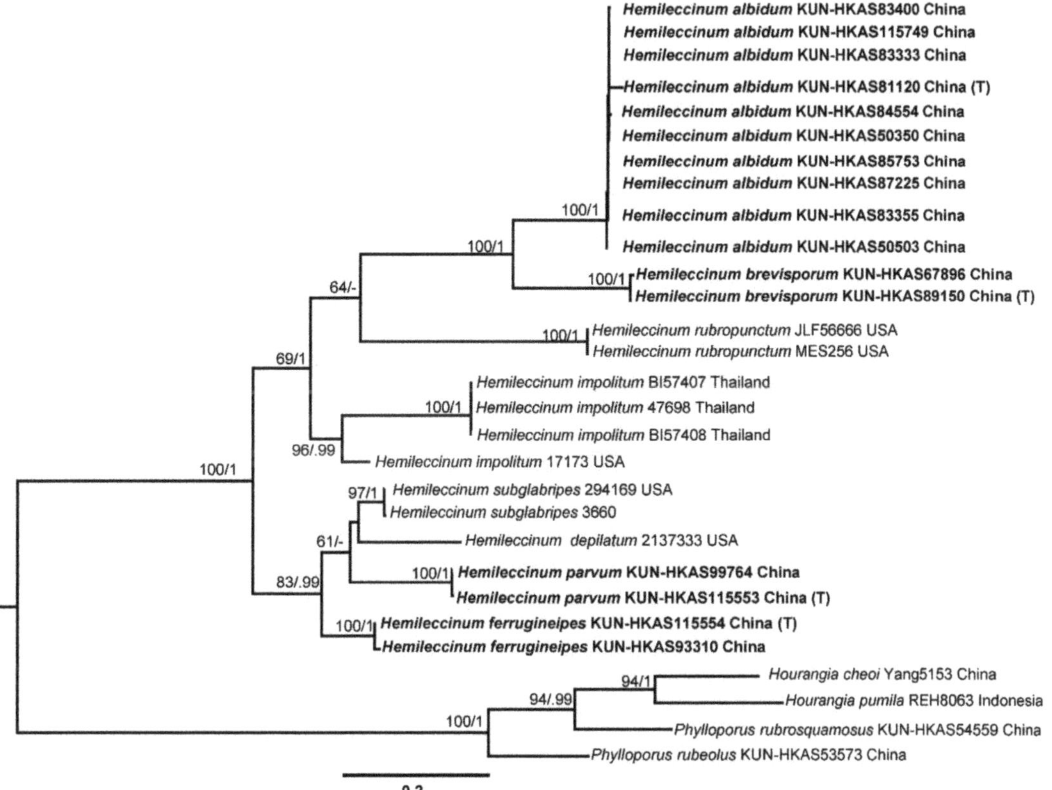

Figure 2. Maximum-Likelihood phylogenetic tree generated from ITS dataset. Bootstrap values (BP) ≥ 50% from ML analysis and Bayesian posterior probabilities (PP) ≥ 0.90 are shown on the branches. Newly described species are indicated in bold and their type specimens are marked with (T).

According to the four single-locus phylogenetic analyses, no strongly supported (>70% of ML) conflict of topologies was observed. Therefore, sequences of the four DNA loci were concatenated for the final analysis. ML and BI analyses of the concatenated data set resulted in almost identical topologies and thus only the tree inferred from ML analysis was displayed (Figure 3). Our molecular phylogenetic analysis indicated that *Hemileccinum* is a monophyletic genus with high statistic supports (BP = 98%, PP = 1). Thirteen phylogenetic species of the genus *Hemileccinum* were retrieved, and four of them could be new to science. By further morphological examinations of the related specimens of those four potential new species, we verified their taxonomic statuses of new species. For detailed information of each species, see below.

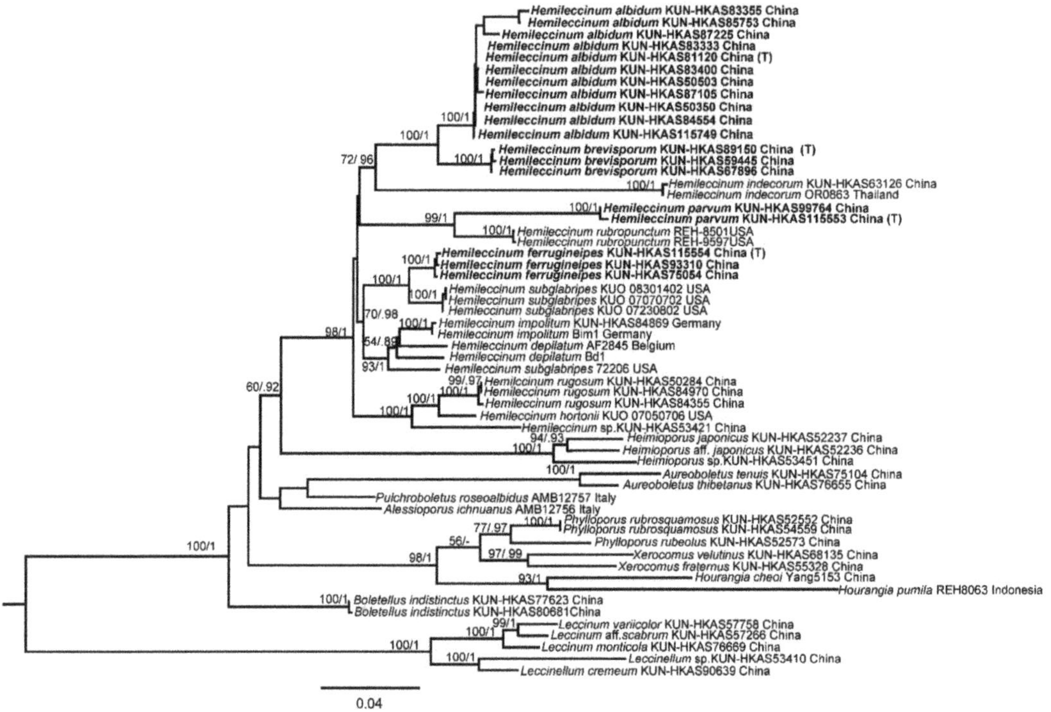

Figure 3. Maximum-Likelihood analysis of *Hemileccinum* with nrLSU, *tef1-α*, *rpb1* and *rpb2* sequence data. Bootstrap values (BP) ≥ 50% from ML analysis and Bayesian posterior probabilities (PP) ≥ 0.90 are shown on the branches. Newly described species are indicated in bold and their type specimens are marked with (T).

3.2. Taxonomy

Hemileccinum albidum Mei-Xiang Li, Zhu L. Yang & G. Wu, **sp. nov.**, Figure 1a–c, Figures 4a–c and 5.

MycoBank no: 840704

Etymology: The epithet '*albidum*' refers to the somewhat white stipe of this species.

Type: CHINA. Yunnan Province: Jingdong County, Ailao Mt., alt. 2490 m, associated with *Castanopsis ceratacantha*, Ca. *rufescens*, *Lithocarpus xylocarpus*, *Quercus griffithii*, 22 July 2013, Jiao Qin 682 (KUN-HKAS81120).

Diagnosis: *Hemileccinum albidum* is distinguished by the combination characters of the even pileus, and the whitish, nearly smooth stipe, with only small, granular scales at the base.

Description: *Basidioma* stipitate-pileate, small to medium-sized. *Pileus* 3–9 cm diam, hemispherical to applanate, finely rugose, then smooth, finely subtomentose, dry or slightly viscid when wet; surface of *Pileus* grey-brown when young, then chrome yellow (5A8), pompeian yellow (5C7) to ochraceous (2D2–5) or golden brown (5D7), somewhat paler along the pileus edge; context white (1A1), yellowish (2A4–5) to brownish (5B5–8, 6C6–8), unchanging on exposure. Hymenophore depressed around the apex of the stipe; hymenophoral surface yellowish (2A4–5) to yellow to sulphur yellow (4A5–4A6) or olivaceous yellow (2A6–7), unchanging when bruised; pores roundish, 0.5–1.0(1.5)/mm; tubes up to 10 mm long, concolorous with the hymenophoral surface, unchanging when bruised.

Stipe 5–16 cm long, 1.0–2.5 cm wide, subcylindrical; surface whitish (8A1), cream (1A2) to pale yellow-brown (2A3) or pinkish (13A2) to purplish (11B3–5), fibrillose, sometimes covered with small pale granular scales; context unchanging in color when cut. *Basal mycelium* white (1A1).

Basidia 25–38 × 10–14 µm, clavate, 4-spored, sterigmata 4–6 µm long. *Basidiospores* [120/3/3] (10)11–12.5 × (4.0)4.5–5.5 µm, [Q = (2.00)2.18–2.66(2.75), Q_m = 2.36 ± 0.12], subfusiform in side view with distinct suprahilar depression, subfusoid in ventral view, brownish yellow, inamyloid, with tiny warts and pinholes on the surface under SEM. *Hymenophoral trama* nearly phylloporoid with hyphae of the lateral strata touching or almost touching each other with hyphae diverging from the central strand to the subhymenium. *Cheilocystidia* 41–50 × 8–11 µm, lanceolate to clavate or ventricose, thin-walled, colorless. *Pleurocysitidia* 46–56 × 8–13 µm, ventricose-subfusiform, with long beak, thin-walled. *Pileipellis* an hyphoepithelium 170–230 µm thick, composed of moniliform hyphal segments 5–37 µm wide, thin-walled, with narrowly cylindrical to shortly cystidioid terminal cells 10–75 × 3–20 µm. *Pileal trama* composed of interwoven hyphae 5–34 µm wide. *Stipitipellis* ca. 130 µm thick, hymeniform, terminal cells broadly clavate, 20–43 × 10–22 µm, sometimes connected with narrow, filamentous hyphae at the outer layer. *Caulocystidia* abundant, 26–43 × 7–12 µm, thin-walled. *Stipe trama* composed of parallel hyphae, 3.5–12.0 µm wide. *Clamp connections* absent in all tissues.

Habitat and distribution: Scattered in subtropical forests dominated by plants of the family Fagaceae (*Castanopsis ceratacantha*, Ca. *rufescens*, Ca. *calathiformis*, *Lithocarpus xylocarpus*, *L. hancei*, *L. mairei* and *Quercus griffithii*); on acidic, loamy, humid soils; moderately common in southwestern China; fruiting in June to August in southwestern China (Yunnan Province) between 1968 and 2490 m altitude.

Additional specimens examined: CHINA. Yunnan Province: Jingdong County, Ailao Mt., alt. 2490 m, associated with *Castanopsis ceratacantha*, Ca. *rufescens*, *Lithocarpus xylocarpus* and *Quercus griffithii*, 21 July 2006, Zhu-Liang Yang 4706 (KUN-HKAS50503); same location, 20 July 2006, Yan-Chun Li 596 (KUN-HKAS50350); same location, 23 July 2013, Bang Feng 1359 (KUN-HKAS115749); Longling County, Zhenan Town, alt. 1968 m, associated with *Castanopsis calathiformis* and *Lithocarpus hancei*, 11 July 2014, Xiao-Bin Liu 459 (KUN-HKAS87105); same location, 22 July 2014, Xiao-Bin Liu 673 (KUN-HKAS87225); same location, 25 August 2014, Chen Yan 155 (KUN-HKAS85753); Longling County, Xueshan Village, alt. 2000 m, associated with *Castanopsis ceratacantha*, *Lithocarpus mairei* and *Quercus griffithii*, 19 June 2014, Jiao Qin 916 (KUN-HKAS83333); same location, 21 June 2014, Jiao Qin 938 (KUN-HKAS83355); same location, 31 July 2014, Jiao Qin 983 (KUN-HKAS83400); same location, 14 June 2014, Li-Hong Han 258 (KUN-HKAS84554).

Notes: *Hemileccinum albidum* is distinguished by combination characters of the even pileus and whitish stipe surface covered with concolorous, small granular scales. Phylogenetically, *H. albidum* is closely related to *H. brevisporum*. However, the former species differs in its whitish stipe and larger basidiospores (11.0–12.5 × 4.5–5.5 µm). Morphologically, the size, pileus color and shape of *H. albidum* are similar to those of the European *H. depilatum*. However, the latter is different from the former by its wrinkled or hammered pileus and the pileipellis composed of hyphae of spherical and shortly cylindrical, terminal cells 16.5–44.0 × 8.5–30.0 µm [4]. Ecologically, *H. albidum* occurs under trees of Fagaceae in subtropical regions while *H. depilatum* is distributed in hardwoods, especially with trees of *Ulmus* and *Carpinus* in temperate regions [41,42] (Appendix A).

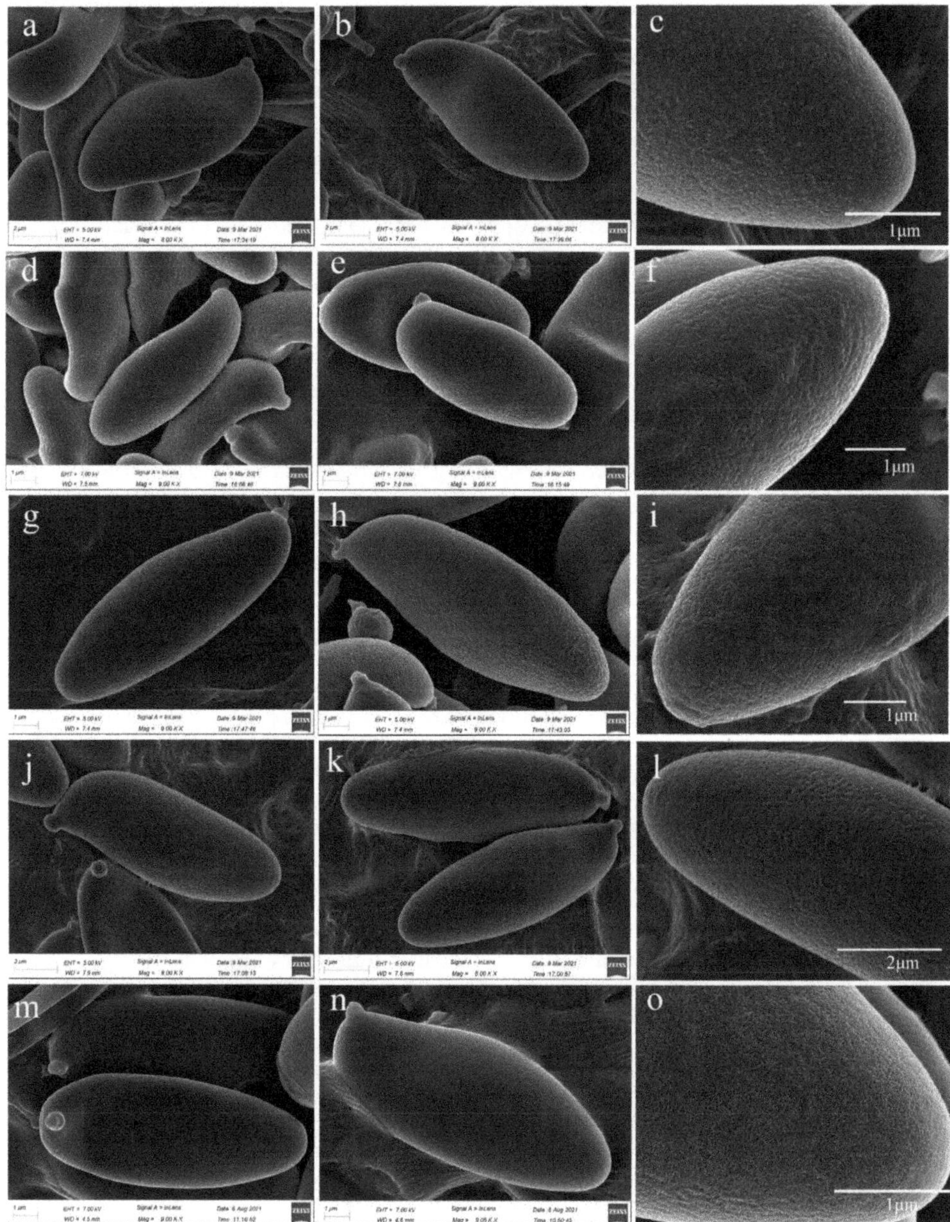

Figure 4. Basidiospores of *Hemileccinum albidum*, *H. brevisporum*, *H. ferrugineipes*, *H. parvum* and *H. impolitum* under SEM. (**a–c**) *H. albidum* (Type, KUN-HKAS81120); (**d–f**) *H. brevisporum* (Type, KUN-HKAS89150); (**g–i**) *H. ferrugineipes* (Type, KUN-HKAS115554,); (**j–l**) *H. parvum* (Type, KUN-HKAS115553); (**m–o**) *H. impolitum* (KUN-HKAS84869).

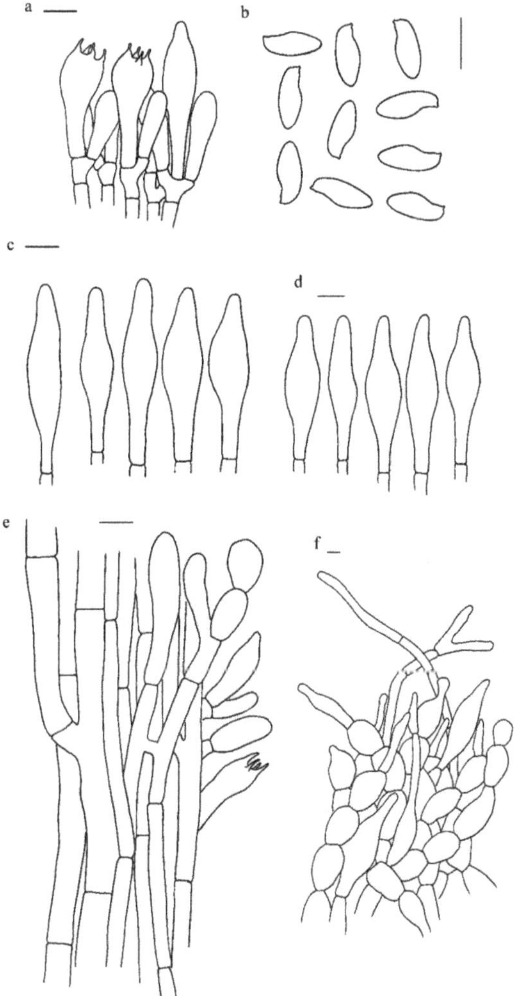

Figure 5. Microscopic features of *H. albidum* (Type, KUN-HKAS81120). (**a**). Hymenium and subhymenium; (**b**). Basidiospores; (**c**). Cheilocystidia; (**d**). Pleurocystidia; (**e**). Stipitipellis; (**f**). Pileipellis. *Bars*: a = 20 μm, b = 30 μm, c–e = 20 μm, f = 10 μm.

Hemileccinum brevisporum Mei-Xiang Li, Zhu L. Yang & G. Wu, **sp. nov.**, Figure 1d–f, Figures 4d–f and 6.

MycoBank no: 840701

Etymology: The epithet '*brevisporum*' refers to the short basidiospores.

Type: CHINA. Yunnan Province: Menghai County, alt. 1700 m, associated with *Castanopsis calathiformis*, Ca. *indica* and *Lithocarpus truncates*, 1 July 2014, Kuan Zhao 487 (KUN-HKAS89150).

Diagnosis: Differs from other *Hemileccinum* species by the combined characters of the dense fine-grained scales on the stipe surface, the shorter basidiospores measuring 9–11 × 4–5 μm and small basidia measuring 18.5–27.0 × 8–11 μm.

Description: *Basidioma* stipitate-pileate, fleshy, small to medium-sized. *Pileus* 9 cm diam, glabrous to slightly subtomentose, dry, convex to planate, pale yellow-brown (2A3) to pale red-brown (7A5); context yellowish (3A5–3A6), unchanging when bruised. Hymenophoral surface and tubes concolorous, flash yellow (3A8) to dull yellow (3B3–3B4), unchanging when bruised, pores roundish, 1.0–1.5/mm, tubes 11 mm long, unchanging when injured. *Stipe* 13–15 cm long, 2.0–2.3 cm wide, subcyclindrical, surface of stipe cream (2A2–3A2) to yellowish (2A4–2A5) at upper part, pale yellow-brown to yellow-brown (6C8) at lower part, covered with small yellowish brown (5D8) dotted scales, context of stipe cream to pale yellow (1A2–1A3), unchanging when bruised. *Basal mycelium* white to cream (2A2–3A2).

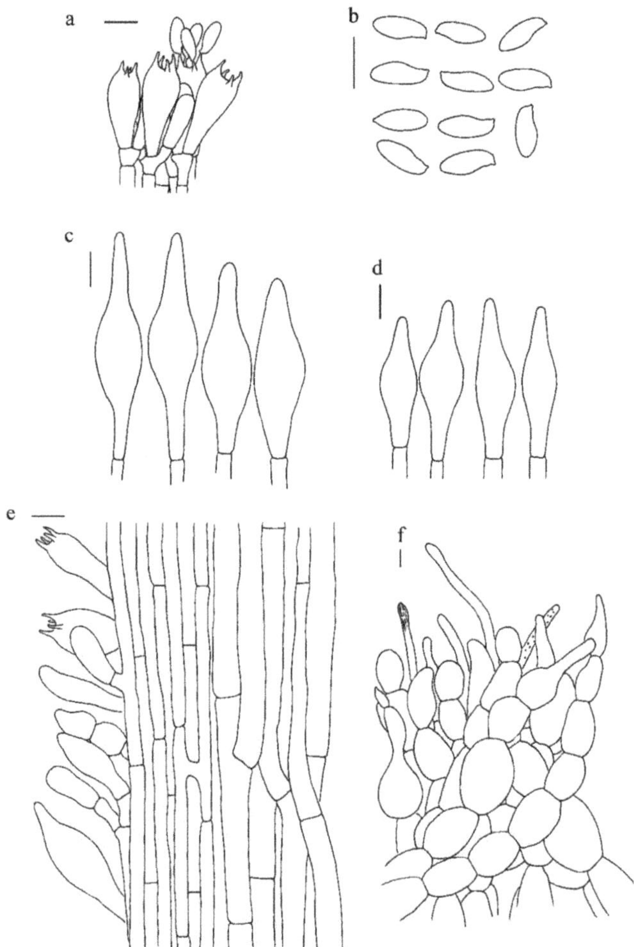

Figure 6. Microscopic features of *H. brevisporum* (Type, KUN-HKAS89150). (**a**). Hymenium and subhymenium; (**b**). Basidiospores; (**c**). Pleurocystidia; (**d**). Cheilocystidia; (**e**). Stipitipellis; (**f**). Pileipellis. *Bars*: a = 20 μm, b = 30 μm, c–e = 20 μm, f = 10 μm.

Basidia 18.5–27.0 × 8–11 μm, clavate, hyaline in 5% KOH, 4-spored. *Basidiospores* [80/2/2], 9–11 × 4–5 μm, [Q = (2.22)2.35–2.50(2.75), Q_m = 2.37 ± 0.15], subfusiform and inequilateral in side view with distinct suprahilar depression, subfusoid in ventral view, yellowish to brownish, smooth under light microscopy, but with tiny warts on the surface under SEM. *Hymenophoral trama* nearly phylloporoid with hyphae of the lateral strata touching or almost touching each other with hyphae diverging from the central strand to the subhymenium; hyphae subcylindrical to cylindrical, 3.5–14.0 μm wide. *Cheilocystidia* 37–50 × 11–13 μm, ventricose-subfusiform, with long beak, thin-walled. *Pleurocystidia* 48–67 × 12–16 μm, ventricose subfusiform, with long beak, thin-walled. *Pileipellis* an hyphoepithelium 150–210 μm thick, composed of moniliform hyphal segments 5–35 μm wide, thin-walled, with narrowly cylindrical to shortly cystidioid terminal cells 6–53 × 4–20 μm. *Pileal trama* composed of interwoven hyphae 5–37 μm wide. *Stipitipellis* ca. 100 μm thick, hymeniform, terminal cells broadly clavate, 13–30 × 7.0–12.5 μm, sometimes connected with narrow, filamentous hyphae at the outer layer. *Caulobasidia* abundant, 18.5–28.0 × 9.0–12.0 μm, thin-walled. *Stipe trama* composed of parallel hyphae, 4–12 μm wide. *Clamp connections* absent.

Habitat and distribution: Scattered in subtropical forests dominated by the families Fagaceae (*Castanopsis calathiformis*, Ca. *indica*, Ca. *orthacantha*, *Lithocarpus hancei*, *L. mairei* and *Quercus griffithii*) and Pinaceae (*Pinus yunnanensis* or *P. armandii*); on acidic or slightly alkaline, loamy soils; rather rare; fruiting in July to August in southwestern to northwestern Yunnan between 1700 and 2120 m altitude.

Additional specimens examined: CHINA. Yunnan Province: Longling County, alt. 2010 m, associated with *Castanopsis calathiformis*, *Lithocarpus hancei* and *Pinus yunnanensis*, 9 July 2009, Yan-Chun Li 1698 (KUN-HKAS59445); Jianchuan County, Laojunshan town, alt. 2120 m, associated with *Castanopsis orthacantha*, *Lithocarpus mairei*, *Quercus griffithii* and *Pinus armandii*, 9 August 2010, Qing Cai 334 (KUN-HKAS67896).

Notes: *Hemileccinum brevisporum* is morphologically similar to *H. impolitum* because of the ornamentation in the stipe and the slightly subtomentose pileus surface [1,41,42]. However, *H. impolitum*, originally described from Europe, differs from *H. brevisporum*, by its much stockier stipe, and larger basidiospores (12–15 × 4–6 μm). Ecologically, *H. brevisporum* occurs under trees of Fagaceae and Pinaceae in subtropical regions while *H. impolitum* is distributed in hardwood or floodplain forests, especially with trees of *Quercus* and *Fagus* in temperate regions [41,42] (Appendix A).

Hemileccinum ferrugineipes Mei-Xiang Li, Zhu L. Yang & G. Wu, **sp. nov.**, Figure 1g–i, Figures 4g–i and 7.

MycoBank no: 840700

Etymology: The epithet '*ferrugineipes*' refers to the reddish brown stipe of this species.

Type: CHINA. Yunnan Province: Pu'er City, Simao District, Taiyanghe Nature Reserve, alt. 1200 m, associated with *Castanopsis ferox*, Ca. *calathiformis*, *Cyclobalanopsis xanthotricha*, *Quercus fabri*, *Q. variabilis* and *Lithocarpus glabra*, 24 June 2016, Jian-Wei Liu 584 (KUN-HKAS115554).

Diagnosis: Differs from other *Hemileccinum* species by the combined characters of rugose pileus, creamy yellow stipe surface when young becoming reddish when mature, and densely scaled surface of the stipe.

Description: *Basidioma* stipitate-pileate, small to medium-sized. *Pileus* 3–10 cm diam, clavate to planate, surface rugose, slightly subtomentose, dry, yellowish brown (5E5), olive brown (4E5–6) to dull brown (5E8–5F8), context cream to yellowish (2A4–5), unchanging when bruised. Hymenophoral surface and tubes concolorous, yellow (1A2–1A3) to ochreous (5B7–5C7), unchanging when bruised, pores roundish, 1.5–2.5/mm; tubes 4–6 mm long, unchanging when injured. *Stipe* 4–10 cm long, 1–2 cm wide, subcylindrical, surface yellowish to yellow at upper part, lower part pale red-brown of stipe pileus; covered with longitudinal striations and densely dotted scales, context cream (1A2) to yellowish, unchanging when bruised. *Basal mycelium* cream.

Basidia 23–35 × 9–13 μm, clavate, hyaline in 5% KOH, 4-spored. *Basidiospores* [80/2/2], 11.0–12.5 × 4–5 μm, [Q = (2.30)2.40–2.78(3.00), Q_m = 2.63 ± 0.19], subfusiform in side view with distinct suprahilar depression, ellipsoid to somewhat oblong in face view, yellowish to brownish, smooth under light microscopy, but with tiny warts under SEM. *Hymenophoral trama* nearly phylloporoid with hyphae of the lateral strata touching or almost touching each other with hyphae diverging from the central strand to the subhymenium; hyphae subcylindrical to cylindrical, 4–13 μm wide. *Cheilocystidia* 36–63 × 7–11 μm, ventricose-subfusiform, with long beak, thin-walled. *Pleurocystidia* 37–62 × 8–12 μm, ventricose subfusiform, with long beak, thin-walled. *Pileipellis* an hyphoepithelium 170–270 μm thick, composed of moniliform hyphal segments 5–42 μm wide, thin-walled; always with narrowly cylindrical to shortly cystidioid terminal cells, 20–127 × 4–12 μm. *Pileal trama* composed of interwoven hyphae 6–38 μm wide. *Stipitipellis* ca.30–40 μm thick, hymeniform, terminal cells broadly clavate, 15.0–32.0 × 6.5–15.5 μm. *Caulobasidia* abundant, 28–44 × 9–12 μm, thin-walled. *Stipe trama* composed of parallel hyphae, 5–13 μm wide. *Clamp connections* absent.

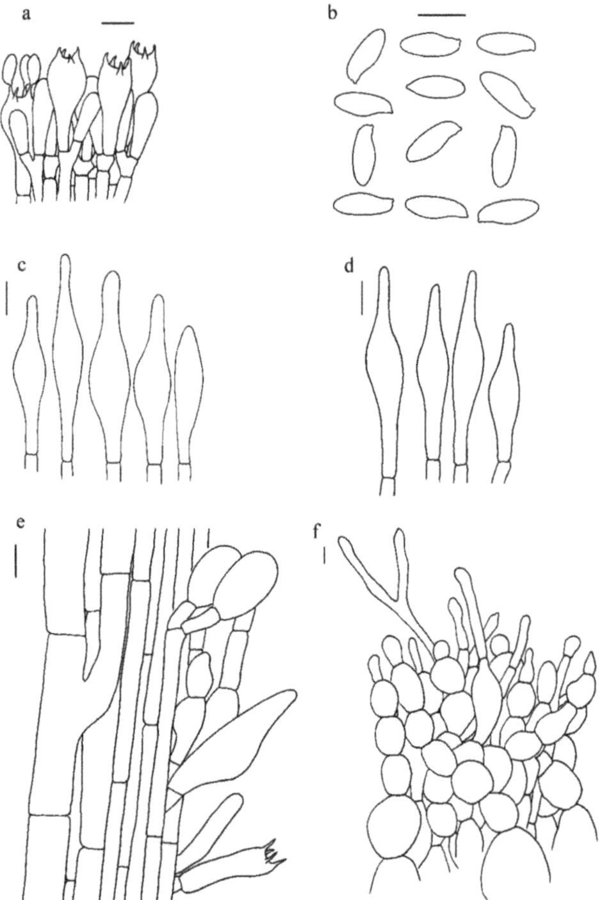

Figure 7. Microscopic features of *H. ferrugineipes* (Type, KUN-HKAS115554). (**a**). Hymenium and subhymenium; (**b**). Basidiospores; (**c**). Pleurocystidia; (**d**). Cheilocystidia; (**e**). Stipitipellis; (**f**). Pileipellis. *Bars*: a = 20 μm, b = 30 μm, c–e = 20 μm, f = 10 μm.

Habitat and distribution: Scattered in subtropical forests dominated by plants of the family Fagaceae (*Castanopsis ferox*, Ca. *calathiformis*, Ca. *hystrix*, *Cyclobalanopsis xanthotricha*, *Quercus fabri*, *Q. variabilis* and *Lithocarpus glabra*); on acidic or slightly alkaline, loamy soils; rather rare; fruiting in June to August in southwestern to northwestern Yunnan between 1200 and 1690 m altitude.

Additional specimens examined: CHINA. Yunnan Province: Baoshan City, Longyang District, alt. 1690 m, associated with *Castanopsis calathiformis*, *Quercus fabri* and *Lithocarpus glabra*, 30 July 2017, Pan-Meng Wang 350 (KUN-HKAS93310); Lanping County, alt. 1400 m, associated with *Castanopsis hystrix*, *Quercus fabri* and *Lithocarpus glabra*, 16 August 2011, Gang Wu 759 (KUN-HKAS75054).

Notes: Hemileccinum ferrugineipes is characterized by its rugose pileus and small, dense, dotted scales on the reddish-brown stipe. Phylogenetically, the American species *H. subglabripes* is close to *H. ferrugineipes*, but differs from it by its fairly long and slender, nearly smooth stipe [9,10,43–45]. Morphologically, *H. ferrugineipes* is similar to *Rugiboletus extremiorientalis* (Lj.N.Vassiljeva) G. Wu & Zhu L. Yang and *H. hortonii* in the rugose pileus and dense scales on the stipe [44–47]. However, *H. ferrugineipes* differs from *Rug. extremiorientalis*, originally described from subtropical Yunnan, China, by its reddish slightly densely scaled surface of the stipe and hyphoepithelium pileipellis. *H. ferrugineipes* differs from *H. hortonii*, originally described from Illinois, USA, by its tightly wrinkled pileus and the stockier stipe [9,44,45]. Ecologically, *H. ferrugineipes* occurs under trees of Fagaceae in subtropical regions; *H. hortonii* is scattered or in groups on the ground under mixed deciduous woods, occasionally under conifers; *H. hortonii* is rather rare and might be found in eastern North America, west to Michigan [44,45] (Appendix A).

Hemileccinum parvum Mei-Xiang Li, Zhu L. Yang & G. Wu, **sp. nov.**, Figure 1j–l, Figures 4j–l and 8.

MycoBank no: 840703

Etymology: The epithet *'parvum'* refers to the small basidioma.

Type: CHINA. Yunnan Province: Wenshan City, Malipo County, alt. 1200 m, associated with *Castanea henryi*, *C. mollissima*, *Lithocarpus bonnetii* and *Quercus marlipoensis*, 30 July 2017, 532624MF-201-Wu 2299 (KUN-HKAS115553).

Diagnosis: Differs from other *Hemileccinum* species by the combined characters of the small basidioma, and the rugose surface of pileus, the pale yellow context staining pale blue very slowly when bruised.

Description: Basidioma stipitate-pileate, small. *Pileus* 3.3–3.6 cm diam, rugose, slightly subtomentose, hemispherical, brownish (5B5–8, 6C6–8) at the central part, becoming paler towards the margin (brownish or yellowish); context pale yellow (1A2–1A3), staining pale blue very slowly when bruised at some spots, 4–5 mm thick. Hymenophoral surface and tubes concolorous, light yellow (3B4–3B5), pores roundish, 1.5–2.0/mm, unchanging when bruised; tubes 4–5 mm long, sinuate near the stipe. *Stipe* 6.0–9.7 cm long, 0.4–0.9 cm wide, clavate, central, solid, pale yellow (2A2–2A4) at the upper part and becoming paler downwards, surface ornamented with coarsely small squamules; context light yellow (3B4–3B5), unchanging when bruised. *Basal mycelium* white (1A1).

Basidia 20.5–32.0 × 8.0–10.5 μm, clavate, 4-spored; sterigmata up to 4–5 μm long. *Basidiospores* [80/2/2], 12–14 × 4.5–5.0 μm, [Q = (2.40)2.50–2.80(2.88), Q_m = 2.69 ± 0.11], subfusiform and inequilateral in side view with distinct suprahilar depression, subfusoid in ventral view, yellowish to brownish, inamyloid, smooth under light microscopy, but with tiny warts on the surface under SEM. *Hymenophoral trama* phylloporoid with hyphae of the lateral strata touching or almost touching each other with hyphae diverging from the central strand to the subhymenium; hyphae subcylindrical to cylindrical, 4–12 μm wide. *Cheilocystidia* 41–50 × 8–11 μm, lanceolate to clavate or ventricose, thin-walled, colorless. *Pleurocysitidia* 45–65 × 9–11 μm, ventricose-subfusiform, with long beak, thin-walled. *Pileipellis* an hyphoepithelium 160–240 μm thick, composed of moniliform hyphal segments 6–30 μm wide, thin-walled, with narrowly cylindrical to shortly cystidioid terminal cells 10–87 × 5–17 μm. *Pileal trama* composed of interwoven hyphae 6–30 μm wide. *Stipitipellis* ca. 100 μm thick, hymeniform, terminal cells broadly clavate, 20.0–43.0 × 10.0–21.5 μm, sometimes connected with narrow, filamentous hyphae at the outer layer. *Caulocystidia* abundant, 24.5–60.0 × 10.5–19.0 μm, thin-walled. *Stipe trama* composed of parallel hyphae, 3.5–12.0 μm wide. *Clamp connections* absent in all tissues.

Habitat and distribution: Scattered in subtropical forests dominated by plants of the family Fagaceae (*Castanea henryi*, *C. mollissima*, *Lithocarpus bonnetii* and *Quercus marlipoensis*);

on acidic, wet, fertile soils; rather rare; fruiting in July in southeastern Yunnan between 1200 and 1300 m altitude.

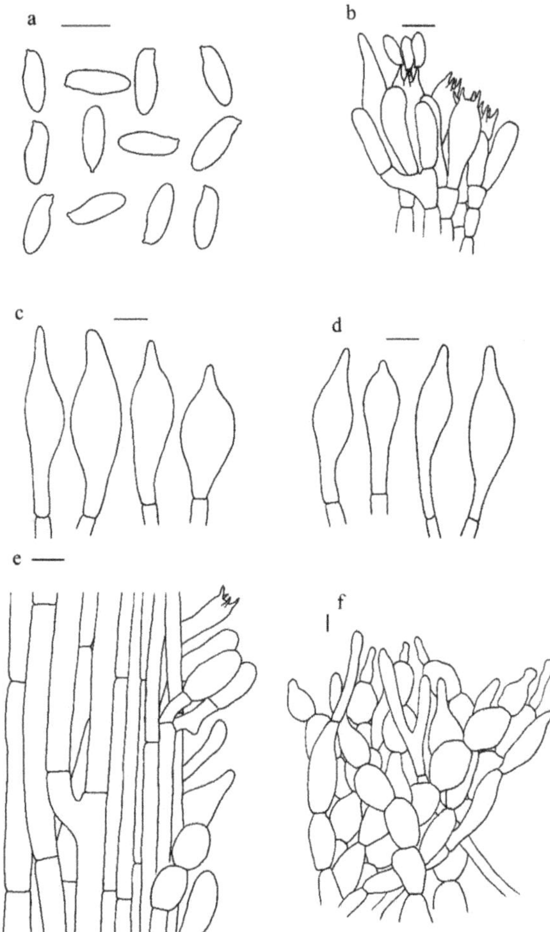

Figure 8. Microscopic features of *H. parvum* (Type, KUN-HKAS115553) (**a**). Basidiospores; (**b**). Hymenium and subhymenium; (**c**). Pleurocystidia; (**d**). Cheilocystidia; (**e**). Stipitipellis; (**f**). Pileipellis. *Bars*: a = 30 μm, b–e = 20 μm, f = 10 μm.

Additional specimens examined: Yunnan Province: Wenshan City, Malipo County, alt. 1300 m, associated with *Castanea henryi*, *C. mollissima*, *Lithocarpus bonnetii* and *Quercus marlipoensis*, 27 July 2016, Gang Wu 1645 (KUN-HKAS99764).

Notes: *Hemileccinum parvum* is morphologically similar to *H. subglabripes* because of the slightly wrinkled pileus and the slender stipe [9,48], However, *H. subglabripes*, originally described from the USA, differs from *H. parvum* by the nearly smooth stipe of the latter covered with branny particles on the stem which are pale and easily overlooked, and the larger basidioma. Our data show that *H. parvum* is phylogenetically close to *H. rubropunctum*, but the latter differs by its longer stipe and the red scales on it [10,44,45]. Ecologically, *H. parvum* occurs under trees of Fagaceae in subtropical southeastern Yunnan; *H. subglabripes* inhabits mixed deciduous trees, sometimes under spruce in eastern and particularly northern North America; and *H. rubropunctum* grows in mixed woods with oak or chestnut in northeastern North America [44,45] (Appendix A).

4. Discussion

The genus *Hemileccinum* Šutara is geographically widely distributed, but its species diversity is poorly known. In Asia, only two species have been previously reported with molecular evidence. One is *H. indecorum* from tropical areas, and the other is *H. rugosum* from subtropical Yunnan, China [8,15]. In this study, four new species in China were recognized and delimited. They are well-supported by molecular phylogenetic and morphological evidence. The host specificity, altitude and edaphic factors seem to be important for determining the distribution of different species of *Hemileccinum*. Our newly described species are distributed in the broad-leaved and mixed forests in southwestern China. *Hemileccinum albidum* and *H. brevisporum* are distributed on high altitudes: between 1700 and 2500 m a.s.l., while *H. ferrugineipes*: 1200–1700 m a.s.l., and *H. parvum*: 1200-1300 m a.s.l. *Hemileccinum albidum, H. ferrugineipes* and *H. parvum* are found in subtropical forests and associated with plants of the family Fagaceae (*Castanopsis ceratacantha*, Ca. *rufescens*, Ca. *ferox*, Ca. *hystrix*, Ca. *calathiformis*; *Castanea henryi, C. mollissima; Cyclobalanopsis xanthotricha; Lithocarpus xylocarpus, L. hancei, L. mairei, L. glabra, L. bonnetii; Quercus griffithii, Q. fabri, Q. variabilis, Q. marlipoensis*), growing mostly on acidic soils. However, *H. ferrugineipes* can also be found in slightly alkaline habitats. *Hemileccinum brevisporum* is found in subtropical broad-leaved and mixed forests, growing with members of Fagaceae (*Castanopsis calathiformis*, Ca. *indica*, Ca. *orthacantha; Lithocarpus hancei, L. mairei; Quercus griffithii*) and Pinaceae (*Pinus yunnanensis* or *P. armandii*) on acidic or slightly alkaline soils. The species we described here are hardly seen in the wild mushroom market, thus their edibility is unknown yet. However, referring to the edibility of the European/American species of *Hemileccinum* [41–45,49], the newly described species could also be edible, but we need more investigations to confirm this.

Overall, the proposed new species are significantly different from the Asian species *H. indecorum* because the viscid pileus and stipe of the latter species are densely covered with whitish to dirty white, small conical to subconical to irregular squamules [15]. They also quite differ from the European species *H. impolitum*, which has a relatively bald pileal surface and a collapsed trichoderm pileipellis when mature [1]. Šutara reported that the basidiospores of *H. impolitum* are smooth [1]. Our re-examination of European material of *H. impolitum* under SEM indicated that there are irregular warts on the surface of basidiospores as with those of other species in *Hemileccinum* (Figure 4m–o). Accordingly, *H. depilatum* (Redeuilh) Šutara should also have a warty basidiospore surface.

The description of the new species also sheds new light on the recognition of the genus. The pileipellis of the species in this genus should mostly be regarded as (sub)epithelium to hyphoepithelium, because the pileipellis of most studied species here are composed of short inflated cells in the inner layer (subpellis) and filamentous hyphae in outer layer (suprapellis), with *H. indecorum* standing at one extremity with whitish to dirty white, small conical to subconical to irregularly shaped squamules on the pileus surface [15] and *H. impolitum* located at the other extremity with collapsed trichoderm pileipellis when mature [1]. The lateral stipe stratum of *H. impolitum* in this genus was diagnosed as the leccinoid type, predominantly anticlinally arranged hyphae ending in elements of the caulohymenium [1,3,4]. However, on the basis of the observation on our new species, this feature is not present in all species of *Hemileccinum*. The structure of the lateral stipe stratum is traceable in our species.

Based on the current study, we increased the species diversity of the genus *Hemileccinum* from Asia and reconstructed a comprehensive phylogenetic tree which included almost all known species of this genus. However, probably due to the limitations of species sampling or insufficient genetic variation of the DNA loci we used, the deep phylogenetic relationships within the genus remain unresolved. In future work, more species with detailed morphological observations and phylogenomic analysis will provide new evidence for these relationships.

Author Contributions: Conceptualization: Z.L.Y., G.W. and M.-X.L.; molecular experiments and data analysis: M.-X.L.; original draft writing: M.-X.L.; draft review and editing: Z.L.Y. and G.W. All authors have read and agreed to the published version of the manuscript.

Funding: This work was supported by Yunnan Ten-Thousand-Talents Plan-Yunling Scholar Project, Yunnan Ten-Thousand-Talents Plan-Young & Elite Talents Project, and the National Natural Science Foundation of China (No. 31970015).

Institutional Review Board Statement: Not applicable.

Informed Consent Statement: Not applicable.

Data Availability Statement: Publicly available datasets were analyzed in this study. This data can be found here: https://www.ncbi.nlm.nih.gov/; http://www.mycobank.org/; http://purl.org/phylo/treebase/phylows/study/TB2:S28729, accessed on 18 September 2021.

Acknowledgments: The authors are very grateful to their colleagues at Kunming Institute of Botany, Chinese Academy of Sciences, including Kuan Zhao, Jiao Qin, Bang Feng, Yan-Chun Li, Chen Yan, Qing Cai, Li-Hong Han, and Xiao-Bin Liu, and Pan-Meng Wang and Jian-Wei Liu for providing specimens, and Yang-Yang Cui and Geng-Shen Wang for help in morphological observation and phylogenetic analysis, and Zhi-Jia Gu for arranging the scanning electron microscopy.

Conflicts of Interest: The authors declare that there are no conflict of interest.

Appendix A

Key to known species of *Hemileccinum* in the world	
1. Pileus surface pale brown, brown to reddish brown, more or less even	2
1′. Pileus surface orange to reddish orange, distinctively wrinkled	9
2. Pileus surface slightly subtomentose, without small conical to subconical to irregularly shaped squamules, known in subtropical and temperate regions in the Northern Hemisphere ...	3
2′. Pileus surface with whitish to dirty white, small conical to subconical to irregularly shaped squamules, known from tropical southeast Asia ...	*H. indecorum*
3. Basidioma varied in size, yellowish context without color change when bruised ...	4
3′. Basidioma usually small in size ($\leq$ 4 cm in daim), the pale yellow context staining pale blue very slowly when bruised, known from subtropical areas ...	*H. parvum*
4. Stipe surface covered with obvious ornaments ...	5
4′. Stipe surface covered with small scales, not obvious	6
5. Stipe surface ornamented with distinctly reddish brown longitudinal streaks, known from East Asia	*H. ferrugineipes*
5′. Stipe surface often covered with red dense fine-grained scales, known from North America ...	*H. rubropunctum*
6. Basidiospores longer in length (> 11 μm), and in other parts of the world ...	7
6′. Basidiospores shorter in length ($\leq$ 11 μm), and distributed in subtropical forests in southwestern China. ...	*H. brevisporum*
7. Stipe stout (> 2.5 cm in diam.); pileipellis an trichoderm, collapsed when mature, restricted to Europe	*H. impolitum*
7′. Stipe slender ($\leq$ 2.5 cm in diam.); pileipellis an hyphoepithelium	8
8. Stipe surface whitish, covered with small, granular scales only at the base, distributed in subtropical China	*H. albidum*
8′. Stipe surface yellowish, nearly smooth, covered with indistinctive tiny scales, known from eastern North America	*H. subglabripes*
9. Basidiospores smaller (10–12 × 4–5 μm), distributed in subtropical and tropical China	*H. rugosum*
9′. Basidiospores larger ($\geq$ 12 μm in length), distributed in temperate regions ...	10
10. Basidiospores narrower (12.0–15.0 × 3.5–4.5 μm), known from North America ...	*H. hortonii*
10′. Basidiospores wider (12.0–15.0 × 5–6 μm), known from Europe ...	*H. depilatum*

References

1. Šutara, J. *Xerocomus s. l.* in the light of the present state of knowledge. *Czech Mycol.* **2008**, *60*, 29–62. [CrossRef]
2. Fries, E.M. *Epicrisis Systematis Mycologici, seu Synopsis Hymenomycetum*; Typographia Academica: Munich, Germany, 1839; p. 421.
3. Bertault, R. Amanites du Maroc (*Troisième contribution*). *Bull. Société Mycol. Fr.* **1980**, *96*, 271–287.
4. Šutara, J. The delimitation of the genus *Leccinum*. *Czech Mycol.* **1989**, *43*, 1–12.
5. Binder, M.; Besl, H. 28S rDNA sequence data and chemotaxonomical analyses on the generic concept of *Leccinum* (Boletales). *Micologia* **2000**, 71–82.
6. Binder, M.; Hibbett, D.S. Molecular systematics and biological diversification of Boletales. *Mycology* **2006**, *98*, 971–981. [CrossRef]

7. Wu, G.; Feng, B.; Xu, J.; Zhu, X.-T.; Li, Y.-C.; Zeng, N.-K.; Hosen, I.; Yang, Z.L. Molecular phylogenetic analyses redefine seven major clades and reveal 22 new generic clades in the fungal family Boletaceae. *Fungal Divers.* **2014**, *69*, 93–115. [CrossRef]

8. Wu, G.; Li, Y.-C.; Zhu, X.-T.; Zhao, K.; Han, L.-H.; Cui, Y.-Y.; Li, F.; Xu, J.-P.; Yang, Z.L. One hundred noteworthy boletes from China. *Fungal Divers.* **2016**, *81*, 25–188. [CrossRef]

9. Halling, R.E.; Fechner, N.; Nuhn, M.; Osmundson, T.; Soytong, K.; Arora, D.; Binder, M.; Hibbett, D. Evolutionary relationships of *Heimioporus* and *Boletellus* (Boletales), with an emphasis on Australian taxa including new species and new combinations in *Aureoboletus, Hemileccinum* and *Xerocomus. Aust. Syst. Bot.* **2015**, *28*, 1–22. [CrossRef]

10. Kuo, M.; Ortiz-Santana, B. Revision of leccinoid fungi, with emphasis on North American taxa, based on molecular and morphological data. *Mycology* **2020**, *112*, 197–211. [CrossRef]

11. Singer, R. *The Agaricales in Modern Taxonomy*, 4th ed.; Koeltz Scientific Books: Koenigstein, Germany, 1986; pp. 1–981.

12. Agerer, R. Fungal relationships and structural identity of their ectomycorrhizae. *Mycol. Prog.* **2006**, *5*, 67–107. [CrossRef]

13. Tedersoo, L.; May, T.; Smith, M.E. Ectomycorrhizal lifestyle in fungi: Global diversity, distribution, and evolution of phylogenetic lineages. *Mycorrhiza* **2009**, *20*, 217–263. [CrossRef] [PubMed]

14. Ryberg, M.; Matheny, P.B. Asynchronous origins of ectomycorrhizal clades of Agaricales. *Proc. R. Soc. B Boil. Sci.* **2011**, *279*, 2003–2011. [CrossRef]

15. Zeng, N.-K.; Cai, Q.; Yang, Z.L. *Corneroboletus*, a new genus to accommodate the southeastern Asian *Boletus indecorus. Mycology* **2012**, *104*, 1420–1432. [CrossRef]

16. Vadthanarat, S.; Lumyong, S.; Raspé, O. *Cacaoporus*, a new Boletaceae genus, with two new species from Thailand. *MycoKeys* **2019**, *54*, 1–29. [CrossRef]

17. Kornerup, A.; Wanscher, J.H. *Methuen Handbook of Colour*, 3rd ed.; Eyre Methuen: London, UK, 1967; pp. 1–252.

18. Li, Y.C.; Yang, Z.L.; Tolgor, B. Phylogenetic and biogeographic relationships of *Chroogomphus* species as inferred from molecular and morphological data. *Fungal Divers.* **2009**, *38*, 85–104.

19. Zhou, M.; Dai, Y.-C.; Vlasák, J.; Yuan, Y. Molecular Phylogeny and Global Diversity of the Genus *Haploporus* (Polyporales, Basidiomycota). *J. Fungi* **2021**, *7*, 96. [CrossRef]

20. White, T.J.; Bruns, T.; Lee, S.; Taylor, J. Amplification and Direct Sequencing of Fungal Ribosomal RNA Genes for Phylogenetics. In *PCR Protocols: A Guide to Methods and Applications*; Innis, M., Gelfand, D., Sninsky, J., White, T., Eds.; Academic Press Inc.: New York, NY, USA, 1990; p. 315.

21. Gardes, M.; Bruns, T.D. ITS primers with enhanced specificity for *basidiomycetes*—Application to the identification of mycorrhizae and rusts. *Mol. Ecol.* **1993**, *2*, 113–118. [CrossRef] [PubMed]

22. James, T.Y.; Kauff, F.; Schoch, C.L.; Matheny, P.B.; Hofstetter, V.; Cox, C.; Celio, G.; Gueidan, C.; Fraker, E.; Miadlikowska, J.; et al. Reconstructing the early evolution of Fungi using a six-gene phylogeny. *Nature* **2006**, *443*, 818–822. [CrossRef] [PubMed]

23. Vilgalys, R.; Hester, M. Rapid genetic identification and mapping of enzymatically amplified ribosomal DNA from several *Cryptococcus* species. *J. Bacteriol.* **1990**, *172*, 4238–4246. [CrossRef] [PubMed]

24. Zeng, N.-K.; Tang, L.-P.; Li, Y.-C.; Tolgor, B.; Zhu, X.-T.; Zhao, Q.; Yang, Z.L. The genus *Phylloporus* (Boletaceae, Boletales) from China: Morphological and multilocus DNA sequence analyses. *Fungal Divers.* **2012**, *58*, 73–101. [CrossRef]

25. Schoch, C.L.; Robbertse, B.; Robert, V.; Vu, D.; Cardinali, G.; Irinyi, L.; Meyer, W.; Nilsson, R.H.; Hughes, K.; Miller, A.N.; et al. Finding needles in haystacks: Linking scientific names, reference specimens and molecular data for Fungi. *Database* **2014**, *2014*, bau061. [CrossRef] [PubMed]

26. Zhu, X.-T.; Wu, G.; Zhao, K.; Halling, R.E.; Yang, Z.L. *Hourangia*, a new genus of Boletaceae to accommodate *Xerocomus cheoi* and its allied species. *Mycol. Prog.* **2015**, *14*, 1–10. [CrossRef]

27. Gelardi, M.; Simonini, G.; Ercole, E.; Vizzini, A. Alessioporusand Pulchroboletus (Boletaceae, Boletineae), two novel genera for *Xerocomus ichnusanus* and *X. roseoalbidusfrom* the European Mediterranean basin: Molecular and morphological evidence. *Mycology* **2014**, *106*, 1168–1187. [CrossRef] [PubMed]

28. Neves, M.A.; Binder, M.; Halling, R.; Hibbett, D.; Soytong, K. The phylogeny of selected *Phylloporus* species, inferred from NUC-LSU and ITS sequences, and descriptions of new species from the Old World. *Fungal Divers.* **2012**, *55*, 109–123. [CrossRef]

29. Katoh, K.; Standley, D.M. MAFFT Multiple Sequence Alignment Software Version 7: Improvements in Performance and Usability. *Mol. Biol. Evol.* **2013**, *30*, 772–780. [CrossRef] [PubMed]

30. Hall, T.A. BioEdit: A user-friendly biological sequence alignment editor and analyses program for Windows 95/98/NT. *Nucleic Acids Symp. Ser.* **1999**, *41*, 95–98.

31. Stamatakis, A. RAxML-VI-HPC: Maximum likelihood-based phylogenetic analyses with thousands of taxa and mixed models. *Bioinformatics* **2006**, *22*, 2688–2690. [CrossRef]

32. Stamatakis, A.; Hoover, P.; Rougemont, J. A Rapid Bootstrap Algorithm for the RAxML Web Servers. *Syst. Biol.* **2008**, *57*, 758–771. [CrossRef]

33. Smith, S.A.; Dunn, C. Phyutility: A phyloinformatics tool for trees, alignments and molecular data. *Bioinformatics* **2008**, *24*, 715–716. [CrossRef]

34. Nylander, J.A.A. *MrModeltest v2. Program Distributed by the Author*; Evolutionary Biology Centre, Uppsala University: Uppsala, Sweden, 2004.

35. Ronquist, F.; Huelsenbeck, J.P. MrBayes 3: Bayesian phylogenetic inference under mixed models. *Bioinformatics* **2003**, *19*, 1572–1574. [CrossRef]

36. Nuhn, M.E.; Binder, M.; Taylor, A.F.; Halling, R.E.; Hibbett, D.S. Phylogenetic overview of the Boletineae. *Fungal Biol.* **2013**, *117*, 479–511. [CrossRef]

37. Vadthanarat, S.; Raspé, O.; Lumyong, S. Phylogenetic affinities of the sequestrate genus Rhodactina (Boletaceae), with a new species, *R. rostratispora* from Thailand. *MycoKeys* **2018**, *29*, 63–80. [CrossRef] [PubMed]

38. Martín, M.; Raidl, S. The taxonomic position of *Rhizopogon melanogastroides* (Boletales). *Mycotaxon* **2002**, *84*, 221–228.

39. Osmundson, T.W.; Robert, V.A.; Schoch, C.L.; Baker, L.J.; Smith, A.; Robich, G.; Mizzan, L.; Garbelotto, M.M. Filling Gaps in Biodiversity Knowledge for Macrofungi: Contributions and Assessment of an Herbarium Collection DNA Barcode Sequencing Project. *PLoS ONE* **2013**, *8*, e62419. [CrossRef] [PubMed]

40. Smith, M.E.; Pfister, D.H. Tuberculate ectomycorrhizae of angiosperms: The interaction betweenBoletus rubropunctus (Boletaceae) and *Quercusspecies* (Fagaceae) in the United States and Mexico. *Am. J. Bot.* **2009**, *96*, 1665–1675. [CrossRef]

41. Breitenbach, J.; Kränzlin, F. *Fungi of Switzerland 3*; Boletales and Agaricales Mykologia: Luzern, Switzerland, 1991; pp. 1–361.

42. Courtecuisse, R.; Duhem, B. *Mushroom and Toadstools of Britain and Europe*; Harper Collins Publishers: New York, NY, USA, 1995; p. 432.

43. Peck, C.H. *Boleti of the United States*; Bulletin of the New York State Museum: New York, NY, USA, 1889; Volume 2, p. 112.

44. Bessette, A.E.; Bessette, A.R.; Fisher, D.W. *Mushroom of Northeastern North America*; Syracause University Press: New York, NY, USA, 1997; p. 328.

45. Phillips, R. *Mushrooms and Other Fungi of North America*; Firefly Books: New York, NY, USA, 2010; p. 263.

46. Wu, G.; Zhao, K.; Li, Y.-C.; Zeng, N.-K.; Feng, B.; Halling, R.E.; Yang, Z.L. Four new genera of the fungal family Boletaceae. *Fungal Divers.* **2015**, *81*, 1–24. [CrossRef]

47. Smith, A.H.; Thiers, H.D. *The Boletes of Michigan*; University of Michigan Press: Ann Arbor, MI, USA, 1971; p. 428.

48. Peck, C.H. Report of the state botanist. *Ann. Rep. N. Y. St. Mus. Nat. Hist.* **1896**, *50*, 77–159.

49. Kuo, M. Retrieved from the Mushroom Expert. Com 2020. Available online: http://www.mushroomexpert.com/ (accessed on 30 September 2021).

Journal of
Fungi

MDPI

Article

Lack of Phylogenetic Differences in Ectomycorrhizal Fungi among Distinct Mediterranean Pine Forest Habitats

Irene Adamo [1,2,*], Carles Castaño [3], José Antonio Bonet [1,2], Carlos Colinas [2,4], Juan Martínez de Aragón [1,4] and Josu G. Alday [1,2]

[1] Joint Research Unit CTFC-AGROTECNIO-CERCA, Av. Alcalde Rovira Roure 191, E25198 Lleida, Spain; jantonio.bonet@exchange.ctfc.es (J.A.B.); mtzda@ctfc.es (J.M.d.A.); josucham@gmail.com (J.G.A.)

[2] Department of Crop and Forest Sciences, University of Lleida, Av. Alcalde Rovira Roure 191, E25198 Lleida, Spain; carlos.colinas@udl.cat

[3] Department of Forest Mycology and Plant Pathology, Swedish University of Agricultural Sciences, SE-75007 Uppsala, Sweden; carles.castanyo@c.ctfc.cat

[4] Forest Science and Technology Centre of Catalonia, Ctra. Sant Llorenç de Morunys km 2, E25280 Solsona, Spain

* Correspondence: irene.adamo@udl.cat

Citation: Adamo, I.; Castaño, C.; Bonet, J.A.; Colinas, C.; Martínez de Aragón, J.; Alday, J.G. Lack of Phylogenetic Differences in Ectomycorrhizal Fungi among Distinct Mediterranean Pine Forest Habitats. *J. Fungi* **2021**, *7*, 793. https://doi.org/10.3390/jof7100793

Academic Editors: Anush Kosakyan, Rodica Catana and Alona Biketova

Received: 5 July 2021
Accepted: 17 September 2021
Published: 24 September 2021

Publisher's Note: MDPI stays neutral with regard to jurisdictional claims in published maps and institutional affiliations.

Abstract: Understanding whether the occurrences of ectomycorrhizal species in a given tree host are phylogenetically determined can help in assessing different conservational needs for each fungal species. In this study, we characterized ectomycorrhizal phylogenetic composition and phylogenetic structure in 42 plots with five different Mediterranean pine forests: i.e., pure forests dominated by *P. nigra*, *P. halepensis*, and *P. sylvestris*, and mixed forests of *P. nigra-P. halepensis* and *P. nigra-P. sylvestris*, and tested whether the phylogenetic structure of ectomycorrhizal communities differs among these. We found that ectomycorrhizal communities were not different among pine tree hosts neither in phylogenetic composition nor in structure and phylogenetic diversity. Moreover, we detected a weak abiotic filtering effect (4%), with pH being the only significant variable influencing the phylogenetic ectomycorrhizal community, while the phylogenetic structure was slightly influenced by the shared effect of stand structure, soil, and geographic distance. However, the phylogenetic community similarity increased at lower pH values, supporting that fewer, closely related species were found at lower pH values. Also, no phylogenetic signal was detected among exploration types, although short and contact were the most abundant types in these forest ecosystems. Our results demonstrate that pH but not tree host, acts as a strong abiotic filter on ectomycorrhizal phylogenetic communities in Mediterranean pine forests at a local scale. Finally, our study shed light on dominant ectomycorrhizal foraging strategies in drought-prone ecosystems such as Mediterranean forests.

Keywords: DNA metabarcoding; phylogenetic structure; habitat filtering

1. Introduction

Ectomycorrhizal fungi are essential organisms in forests, as they form symbiotic relations with trees providing them nutrients in exchange for photosynthetic carbon [1–3]. Some ectomycorrhizal fungi are host specific [3–6] and are influenced by tree species as well as by soil abiotic factors such as pH and nutrient availability [7–10]. Therefore, host effect and abiotic soil parameters are often fundamental drivers of ectomycorrhizal community assembly [11–16]. Moreover, previous studies showed that ectomycorrhizal taxonomic community composition does not significantly change between Mediterranean congeneric pine species [17]. Nevertheless, how ectomycorrhizal fungi are phylogenetically structured among Mediterranean pine host species and whether at both taxonomic and phylogenetic level respond to similar abiotic factors has not been assessed yet. Previous studies showed that ectomycorrhizal responses to climate warming are modulated by host plant performance and nutrient availability [18–20]. Therefore, it is crucial to disentangle whether these drivers influence ectomycorrhizal phylogenetic composition and structure, to better understand forest ecosystem functioning [21].

Phylogenetic analyses are useful tools to estimate the relative importance of evolutionary and ecological forces structuring communities [12,22,23]. In this regard, phylogenetic indices have been implemented to calculate the phylogenetic relatedness of an observed community and compare the value to expectations of community assembly under neutral processes from a regional species pool [24]. Therefore, these indices enable us to characterize whether communities are more phylogenetically related (phylogenetic clustering) or less phylogenetically related (phylogenetic overdispersion) than expected by chance [22–24]. In general, habitat filtering is the dominant assembly process when closely related species that share similar traits are selected to coexist within the community (i.e., phylogenetic clustering). In contrast, competition processes occur when distantly related species with dissimilar traits are selected to co-occur within a community (i.e., phylogenetic overdispersion [22], while the random phylogenetic structure is detected when none of the above processes are inferred [22–25]. For example [26], observed phylogenetic clustering of Agaricomycotina communities (including mycorrhizal and saprotrophs) and observed that xeric oak-dominated forests acted as a filter for these communities. Likewise [27], found phylogenetically clustered arbuscular mycorrhizal communities along an altitudinal gradient and observed that environment was the primary ecological factor structuring these communities, either via changes in host plant or fungal niches. Although the ecological processes filtering communities have recently received criticism [28], investigating the communities' phylogenetic responses to the environment in different ecosystems is fundamental to understand the mechanisms that structure communities [29,30]. However, how ectomycorrhizal communities are phylogenetic structured in Mediterranean pine forests has not been studied yet.

The description of phylogenetic relations between ectomycorrhizal fungi might help to understand the evolutionary ecology of traits, species, and entire communities [31]. In this regard, exploration types of ectomycorrhizal fungi represent an important group of functional traits, which are defined according to the hyphal morphology, i.e., long distance, medium distance, medium distance fringe, short distance, or contact exploration types [32]. The hyphal morphology determines access to distinct nutrient sources, for example, nitrogen (N) [33–35]. Ectomycorrhizal species with short, contact, and medium smooth distance exploration types may preferentially use soluble inorganic forms of N close to the host roots due to the lack the enzymes to access organic N forms [33,36]. Conversely, some fungi have enzymes (i.e., fenton peroxidase) to access insoluble N substrates such as organic substrates and they usually show medium mat and long-distance exploration types [35–37]. However, long and medium fringe exploration types might demand higher carbon cost on the host than shorter distance exploration types [18,32], therefore species with shorter exploration types may be favored under stressful conditions [18,38]. In this regard, several studies have addressed ectomycorrhizal exploration types' responses to environmental drivers [39–42], however, the phylogenetic pattern of the trait in Mediterranean ecosystems has rarely been assessed. Thus, understanding the phylogenetic relationships between ectomycorrhizal species and the evolution of hyphal morphologies in the current climate change context might shed light on the future impacts on Mediterranean ecosystem functioning.

In this study, we aim to characterize the ectomycorrhizal phylogenetic composition and phylogenetic structure in 42 plots of five different Mediterranean pine forests: i.e., pure forests dominated by *P. nigra*, *P. halepensis*, and *P. sylvestris*, and mixed forests of *P. nigra-P. halepensis* and *P. nigra-P. sylvestris*. In line with the above premises, we hypothesized that:

- Considering that *P. halepensis*, *P. nigra*, and *P. sylvestris* are phylogenetically closely related [43–45], we expect that ectomycorrhizal phylogenetic composition, structure, and diversity will not be different among them due to co-evolutionary processes [46].
- Previous studies have identified that ectomycorrhizal taxonomic composition is influenced by soil parameters followed by geographical distance [27,47] Thus, we hypothesized that soil physico-chemistry will act as the main habitat filter on ectomycorrhizal

phylogenetic composition [48]. Finally, among abiotic filters pH, P, and CN ratio strongly influenced ectomycorrhizal taxonomic community composition [17]. Here, we tested if these filters would act similarly over ectomycorrhizal phylogenetic composition.

- In Mediterranean ecosystems, soil N might not be limited due to warmer temperatures which may enhance N mineralization by increasing decomposition of the organic matter [17,49]. Therefore, short exploration types could uptake nutrients close to the host roots. Here, we expected that short and contact exploration types will be dominant, thus, both traits will be overrepresented and dispersed across the ectomycorrhizal phylogenetic tree in comparison with medium and long-term exploration types.

2. Materials and Methods

2.1. Sites Selection

The study was conducted in the mountainous pre-Pyrenees region of Catalonia, North-eastern Spain (Figure S1) in a set of long-term monitoring plots in which fungal fruiting has been recorded for ~20 years [50]. The region is under the influence of the Mediterranean climate, with a summer drought period from June to August and mean annual temperatures from 6 to 9 °C with most of the precipitation occurring in spring and autumn [51]. The 42 pine forests were randomly selected from the 579 sites included in the Forest Ecological Inventory of Catalonia carried out by Centre de Recerca Ecològica i Aplicacions Forestals (CREAF, Barcelona, Spain, 1992), trying to preserve even-aged forest. From the total 42 forests, 32 correspond to pure pine forests, with 14 plots corresponding to *P. nigra* and *P. sylvestris* species and 4 to *P. halepensis*, whilst 10 plots were mixed plots (7 mixed plots of *P. sylvestris* and *P. nigra* species and 3 plots dominated by *P. nigra* and *P. halepensis*). The main features of the study plots are summarized in Table 1 and Table S1.

Table 1. Table summarizing the main features of the study plots: BA (Basal area), Number of trees per hectare, Altitude, Slope, pH, CN ratio and P (Phosphorus). Ps: *P. sylvestris*, Pn: *P. nigra*, Ph: *P. halepensis*, Ps-Pn: *P. sylvestris-nigra*, Pn-Ph: *P. nigra-halepensis*.

Forest Type	Range	BA, m^2 ha^{-1}	N. of Tree Per Hectare	Altitude, m a.s.l	Slope, %	pH	CN Ratio	P
Ps	Min.	18.0	681	854	4	4.8	6.9	2
(14)	Mean	29.8	1362	1197	22	7.2	12.4	5.8
	Max.	41.5	1517	1615	37	8.3	19.5	9
Pn	Min.	16.1	638	397	5	8.0	4.0	3
(14)	Mean	27.7	1692	763	16	8.2	14.4	5.0
	Max.	39.1	2838	1040	32	8.4	21.3	9
Ph	Min.	24.0	1006	520	10	8.2	12.5	3
(4)	Mean	28.8	2093	612	16	8.3	13.6	4.8
	Max.	33.6	3088	661	34	8.4	14.8	6
Ps–Pn	Min.	11.5	477	1030	8	6.6	12.1	2
(7)	Mean	23.5	1161	1085	24	7.7	14.5	3.3
	Max.	31.8	2870	1148	31	8.3	19.8	5
Pn–Ph	Min.	17.6	1229	390	9	8.2	11.1	2
(3)	Mean	19.7	1806	469	12	8.3	13.2	4.0
	Max.	20.9	2761	577	13	8.4	15.4	5

2.2. Soil Sampling

Soils were sampled during the Autumn season (October and November) in 2009. In each of the selected forest stands, a 10 m × 10 m plot was established in the center for long-term monitoring of fungal fruiting [51]. In each plot, we took four soil subsamples, i.e., one per plot-side [52], with a rectangular steel drill (a 30 cm depth and a 6 × 4.5 cm width). The four soil subsamples were pooled in the field and around 1 kg of the mixed sample was placed on ice and taken to the laboratory for fungal DNA extraction. A similar procedure was followed for soil samples to determine soil physico-chemical parameters.

2.3. Soil Analysis

Soil samples were analyzed using the methodology described in [53]. Each sample was air-dried, sieved ($\leq$2 mm mesh), and soil texture (clay, sand, and lime proportions) was analyzed using the Bouyoucos—method [54]. Soil pH and electrical conductivity (EC) using a conductivity meter in a 1:2.5 soil:deionized water slurry [55]. Total nitrogen concentration using the Kjeldahl method [56]. Moreover, available phosphorus concentration using the Olsen method [57]; total organic matter and total carbon concentration using the Walkley-Black method [58]. Finally, exchangeable cations such as sodium (Na), potassium (K+) and magnesium (Mg2+) with atomic absorption spectroscopy after extraction with 1 N ammonium acetate (pH 7; [55–59]).

2.4. Fungal Community and Bioinformatic Analysis

Fungal DNA was extracted from 0.5 g of homogenized soil using the NucleoSpin® NSP soil kit (Macherey-Nagel, Duren, Germany) following the manufacturer's protocol. Fungal internal transcribed spacer 2 (ITS2) region was amplified in a 2720 Thermal Cycler (Life Technologies, Carlsbad, CA, USA) using the primers gITS7 [60], ITS4, and ITS4A [61,62]. We optimized the number of PCR cycles in each sample aiming for weak to medium PCR bands at the agarose gels, which was achieved in most of the samples by using 21–26 cycles. The final concentrations in the PCR reactions, PCR conditions, DNA purification and sequencing, and bioinformatics analyses were as explained by Adamo et al. (2021). Sequence data are archived at NCBI's Sequence Read Archive under accession number PRJNA641823 (www.ncbi.nlm.nih.gov/sra, accessed on 25 June 2020).

2.5. Taxonomic and Functional Identification

We taxonomically identified the 600 most abundant OTUs, which represented 93% of the total sequences. We selected the most abundant sequence from each OTU for taxonomic identification, using PROTAX software [63] implemented in PlutoF, using a 50% probability of correct classification (called "plausible identifications") [63]. These identifications were confirmed and some of them improved using massBLASTer in PlutoF against the UNITE [64]. Taxonomic identities at species level were assigned based on >98.5% similarity with database references, or to other lower levels using the next criteria: genus on >97%, family on >95%, order on >92%, and phylum on >90% similarity. OTUs were assigned to the following functional guilds: (a) root-associated basidiomycetes, (b) root-associated ascomycetes, (c) molds, (d) yeasts, (e) litter-associated basidiomycetes, (f) litter-associated ascomycetes, (g) pathogens, (h) moss-associated fungi, (i) soil saprotrophs (saprotrophic taxa commonly found in N-rich mineral soils), (j) unknown function, based on the UNITE database, DEEMY (www.deemy.de) or FUNGuild [65]. ECM species were assigned to exploration types according to the DEEMY database [66,67].

2.6. Phylogenetic and Statistical Analyses

The ghost-tree approach [68], which allows sequence data to be integrated into a single tree, was used to reconstruct the fungal phylogenetic tree. Foundation phylogeny at the family level was derived by (Treebase ID S20837) [69], following methodology and was based on the sequences of six genes 18sS rRNA, 28S rRNA, 5.8S rRNA, translation elongation factor 1-α (tef1α), and RNA polymerase II (two subunits: RPB1 and RPB2) [70].

Statistical analyses were implemented in the R software environment (version 3.6.1, R Development Core Team 2019). The *ape* package was used to load and manipulate the phylogenetic tree in newick format [71], while the *phyloseq* package was used to import and handle OTU counts [72], taxonomic assignments, and associated phylogenetic tree. The *philR* package was used to analyze compositional data using the phylogenetic tree information [73]. The *picante* package was used to calculate the ectomycorrhizal phylogenetic structure indices (NRI and NTI) and Faith's phylogenetic diversity [74,75]. The *vegan* package was used for the multivariate analyses [76].

For all compositional analyses, the ectomycorrhizal species abundance matrix was previously filtered to exclude the taxa that were not seen in at least 10% of samples to eliminate random noise. We analyzed ectomycorrhizal phylogenetic community composition using philR which enables us to transform compositional data into an orthogonal unconstrained space with phylogenetic and evolutionary interpretation [73]. First, PhilR Isometric Log Ratio transformations were built from the phylogenetic tree utilizing a weighted reference [73], then a Euclidean distance matrix was built from the philR transformed data. After that, Redundancy analysis, (RDA function *"rda"*) was used to visualize ectomycorrhizal phylogenetic compositional differences between tree host species. Moreover, ectomycorrhizal phylogenetic differences between tree host species were tested using permutational multivariate analyses of variance (PMAV, function *"adonis"*) on the Euclidean dissimilarity matrix based on the philR transformed data. To test the phylogenetic structure of ectomycorrhizal communities between tree host species were calculated the standardized effect size of mean pairwise distances and mean nearest taxon distances using *ses.mpd* (Standardized effect of mean pairwise distances in communities) and *ses.mntd* (Standardized effect of nearest taxon index in communities) functions from *picante*. In each stand type, we compared the MPD and MNTD values with the MPD and MNTD distributions of random communities in order to identify whether communities were more over-dispersed or under-dispersed than expected by chance. We used the *independentswap* null model, which randomizes community data matrix with the independent swap algorithm maintaining species occurrence frequency and sample species richness, to construct from 9999 randomly assembled communities [77]. After calculating SES.MPD and SES.MNTD, the values were multiplied by -1 as these values are equivalent to -1 times NRI (net relatedness index) and NTI (nearest taxon index), respectively. Importantly, an increase in the NRI value indicates increasing phylogenetic clustering (or decreasing overall relatedness) of a set of species relative to the source pool [25]. On the other hand, the nearest taxon index (NTI) is a standardized measure of the mean phylogenetic distance to the nearest taxon in each sample/community [25]. The NRI measures the standardized effect size of the mean phylogenetic distance (MPD), which estimates the average phylogenetic relatedness between all possible pairs of taxa in a community. The NTI calculates the mean nearest phylogenetic neighbor among the individuals in a community. The ectomycorrhizal phylogenetic diversity comparisons between tree host species were done using Faith's PD phylogenetic diversity index with the function *pd*. Moreover, to assess the phylogenetic relationships among species change across space, we computed multiple-site phylogenetic turnover, nestedness, and phylo-beta diversity (Sorensen similarity index) per tree host species using *"phylo.beta.multi"* function in the betapart package [78].

Second, variation partitioning (function *"varpart"*) was used to test the relative importance as variation sources of geographical distances, soil parameters, and stand structure in ectomycorrhizal phylogenetic composition (philR transformed data) and structure (NRI, NTI). To avoid multicollinearity before variation partitioning analysis highly correlated environmental variables were removed ($r > 0.7$). The geographical distances included were previously evaluated using principal coordinates of neighbors' matrices spatial eigenvectors (PCNM, *pcnm* function) based on UTM coordinates of the sampled stands with Euclidean distances. Thus, significant spatial eigenvectors were forward selected and the selected spatial eigenvectors were used as explanatory variables in the variation partitioning, together with soil (Sand content, K, Mg, organic matter, Na, N, P, water pH, and CN ratio) and stand structural variables (Tree species, Altitude, Slope, Trees per hectare, and Basal Area). The significance of each partition was tested using multivariate ANOVAs. Moreover, to evaluate the effect of pH, CN, and P on the ectomycorrhizal phylogenetic composition we conducted a redundancy analysis (*rda* function). In addition, the *sm.density.compare* (*n* of permutations = 999) from the *sm* package was used to randomly assign pH values between the five tree hosts and estimate how different the densities were using a permutational test of density equality [79]. Lastly, to visualize if the ectomycorrhizal phylogenetic communities were clustering across the pH gradient, we performed a hierarchical cluster analysis on the ectomycorrhizal phylogenetic compositional data based on the Euclidian distance matrix using the function *hclust* in the *stats* package.

Finally, a binary data matrix was compiled with ectomycorrhizal exploration traits (contact, short, medium smooth, medium mat, medium-fringe, and long). Then, we calculated the trait *ses.mpd* and *ses.mntd* using the *independentswap* null model to assess trait structure following the same methodology for communities. Finally, to test for a phylogenetic signal to exploration types, K' Blomberg statistics were calculated for the presence of the traits using the function *MultiPhylosignal* in the *picante* package [74,80]. Moreover, the traits were visualized on the phylogenetic tree by plotting the exploration types at the tips of the phylogenetic tree following [74].

3. Results

3.1. Ectomycorrhizal Phylogenetic Description

The hybrid phylogenetic tree of ectomycorrhizal fungi was consistent with Mikryukov et al. (2020) (Figure 1). The families Sebacinaceae, Clavulinaceae, and Hydnaceae clearly formed a monophyletic group, while Bankeraceae Thelephoraceae, Russulaceae, and Albatrellaceae formed two distinct clades (Figure 1). Moreover, two other family groups were identified, one including Atheliaceae, Sclerodermataceae, Boletaceae, Gomphidiaceae, and Suillaceae, and the other including Tricholomataceae, Amanitaceae, Hydnangiaceae, Cortinariaceae, Hymenogastraceae, and Inocybaceae (Figure 1). Finally, the most abundant species in each tree host were indicated in Table S2.

3.2. Ectomycorrhizal Phylogenetic Composition, Structure, and Diversity

There were no significant differences in the ectomycorrhizal phylogenetic composition among tree host species (r^2 = 0.10, $F_{(4,41)}$ = 1.12, p = 0.281). The RDA and the sd-ellipses based on the philR Euclidean distance matrix clearly showed that all forest types were overlapping at the ordination center (Figure 2). Redundancy analyses resulted in two main axes that explained together 22% of the variance. However, *P. halepensis-nigra*, *P. halepensis*, and *P. nigra* communities were less spread (homogeneous), while, *P. sylvestris* and *P. sylvestris-nigra* communities were more overdispersed in the ordination space (heterogeneous). Regarding ectomycorrhizal phylogenetic structure, no significant difference was detected for NRI ($F_{(4,41)}$ = 0.26, p = 0.901) between tree host species. Positive mean values of NRI were detected in *P. halepensis* (0.51 ± 0.09), indicating ectomycorrhizal higher phylogenetic clustering. *P. nigra-halepensis* (0.14 ± 0.83), *P. sylvestris-nigra* (0.06 ± 0.25) and *P. sylvestris* (0.05 ± 0.27), and *P. nigra* (−0.02 ± 0.26) showed dispersion of NRI values positive and negative around 0 (Figure 3a). However, we detected significant differences in NTI values ($F_{(4,41)}$ = 2.96, p = 0.031) between tree host species. Mean positive NTI values were detected across all tree host species, except in *P. sylvestris-nigra* (−0.46 ± 0.37), indicating ectomycorrhizal phylogenetic clustering in *P. halepensis*, *P. nigra-halepenesis*, while *P. nigra*, *P. sylvestris* were not clearly defined, with values around 0, and a marginal phylogenetic overdispersion was detected in *P. sylvestris-nigra* (Figure 3b).

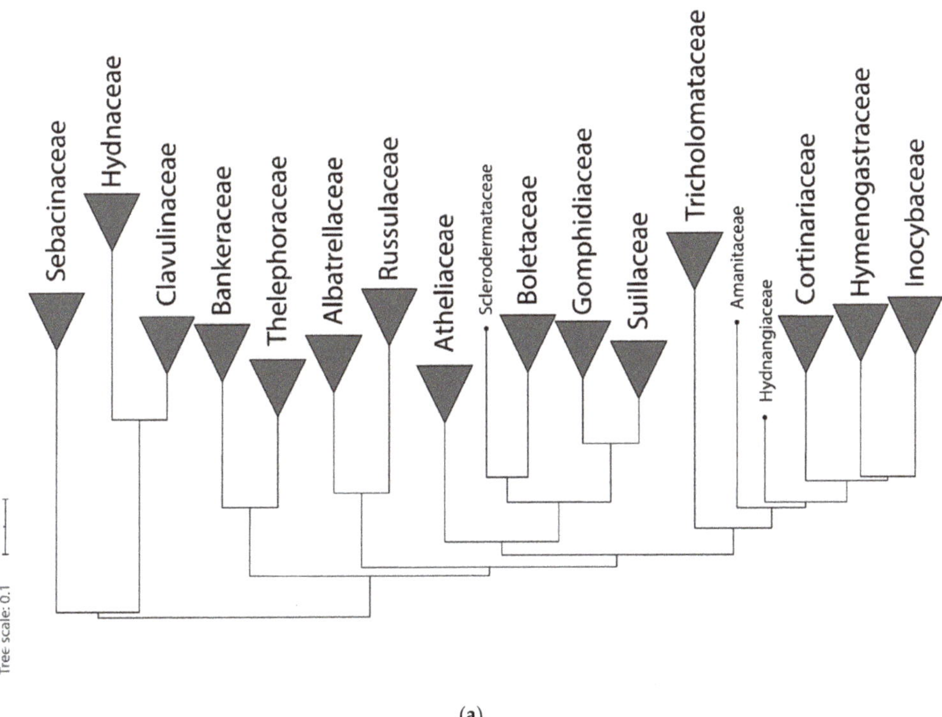

(a)

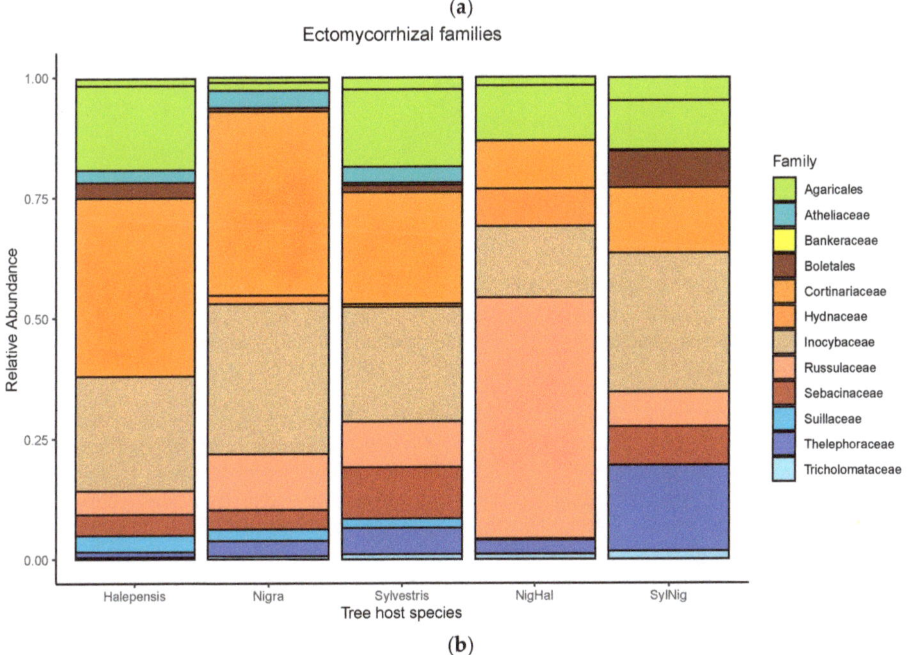

(b)

Figure 1. (a) The hybrid phylogenetic tree of ectomycorrhizal families based on the foundation phylogeny derived by Zhao et al. (2017), based on the sequences of six genes 18sS rRNA, 28S rRNA, 5.8S rRNA, translation elongation factor 1-α (tef1α) and RNA polymerase II. (b) Relative abundance of the most abundant ectomycorrhizal families.

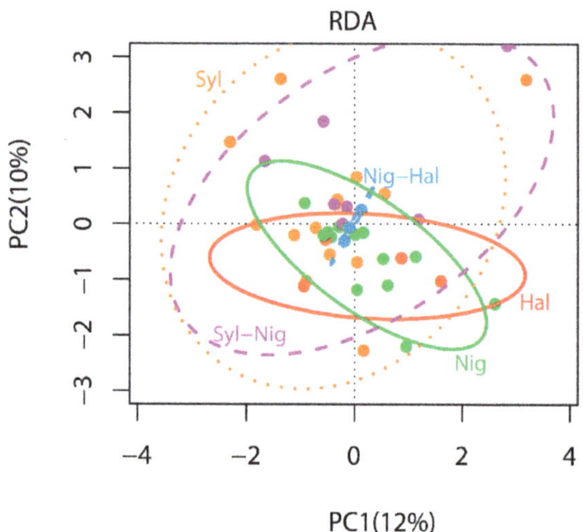

Figure 2. philR RDA ordination based on Euclidean distance matrix displaying ectomycorrhizal phylogenetic community composition of *P. halepensis*, *P. nigra-halepensis*, *P. nigra*, *P. sylvestris-nigra* and *P. sylvestris* forest and the sd ellipses of each forest.

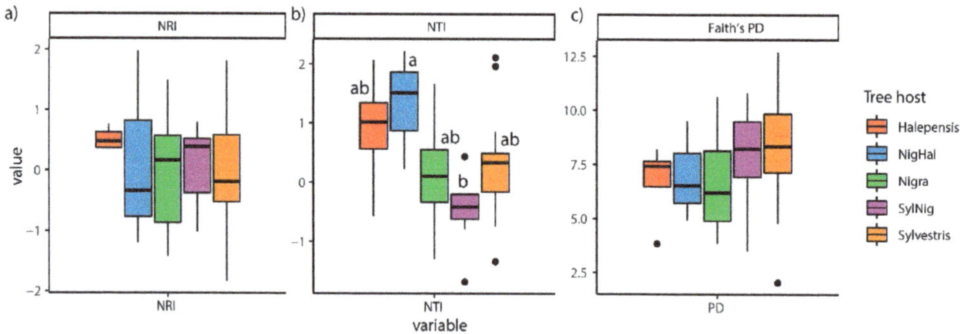

Figure 3. Boxplots displaying (**a**) Net Relatedness Index (NRI) values (**b**) Nearest Taxon Index (NTI) (**c**) Faith's PD values between tree host species (Halepensis: *P. halepensis*, NigHal: *P. nigra-halepensis*, Nigra: *P. nigra*, SylNig: *P. sylvestris-nigra* and Sylvestris: *P. sylvestris*). Means were compared using ANOVA and Tukey's HSD tests, with letters denoting significant differences between host species.

The ectomycorrhizal phylogenetic diversity analysis showed no significant differences between tree host species ($F_{(4,41)} = 0.92$, $p = 0.458$, Figure 3c), with PD mean values ranging from 6.6 of *P. nigra* and 8.3 *P. sylvestris* (Figure 3c). In addition, analysis of multiple-site phylogenetic similarities showed that total beta diversity values were similar across host tree species (Table S1), although, species turnover resulted strongly higher than species nestedness across the tree host species and with similar values, except for *P. nigra-halepensis* (Phylo beta.sim: 0.40; Phylo beta.sne: 0.14; Table S1).

3.3. Main Drivers of Ectomycorrhizal Phylogenetic Composition and Structure

When testing the relative importance of geographic distance, soil parameters, and stand structure on ectomycorrhizal phylogenetic composition, soil accounted for the greatest proportion of the total variance (4%) followed by geographic distance, however, these fractions were not significant ($p > 0.05$, Figure 4a). Moreover, stand structure, soil, and geographic distance shared 4% of the total variance. Conversely, when the phylogenetic structure was analyzed, stand structure, soil, and geographical distance shared an 8% proportion of variation, while stand structure accounted for 4% of the total variance ($p > 0.05$) (Figure 4b). Finally, the phylogenetic structure was marginally influenced by soil (2%) and not by geographic distance.

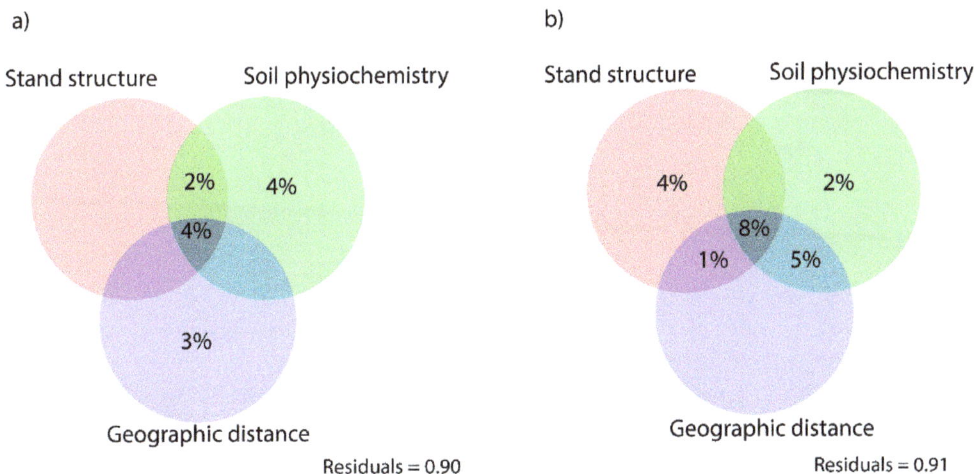

Figure 4. Variance partitioning analyses for (**a**) ectomycorrhizal phylogenetic composition and (**b**) ectomycorrhizal phylogenetic structure (NRI, NTI indices) in response to stand structure, soil physiochemistry and geographic distance. Values show the fraction of variation explained by each group of parameters, as well as the shared contribution of each combination of them.

pH was the only significant soil predictor influencing phylogenetic composition (Variance = 0.55, F = 2.76, $p = 0.002$). Thus, the distribution of pH values was significantly different across tree hosts ($p = 0.041$). Moreover, when pH densities were compared only *P. sylvestris* and *P. nigra* differed significantly ($p = 0.009$) showing a larger left tail towards lower pH values (Figure S2). The hierarchical clustering of the ectomycorrhizal phylogenetic composition showed that communities were clustered into two main groups (Figure S3), Here, the group composed of *P. halepensis*, *P. sylvestris*, and *P. sylvestris-nigra* communities clustered at lower pH values (<7), while *P. nigra* and *P. nigra-halepensis* communities only occurred at higher pH values (>7) (Figure S3).

3.4. Trait Evolution of the Exploration Types

When the exploration traits were visualized on the phylogenetic tree, 59 OTUs out of 184 had short exploration types, up to 53 had contact exploration types, while 39 OTUs and 25 OTUs had medium fringe and medium smooth exploration types. Conversely, medium mat and long exploration types were the least abundant with 9 and 8 OTUs, respectively. Finally, we did not find any phylogenetic signal for any exploration type ($0.25 < K < 0.77$, $p > 0.05$), as exploration types were dispersed across the phylogenetic tree (Figure 5).

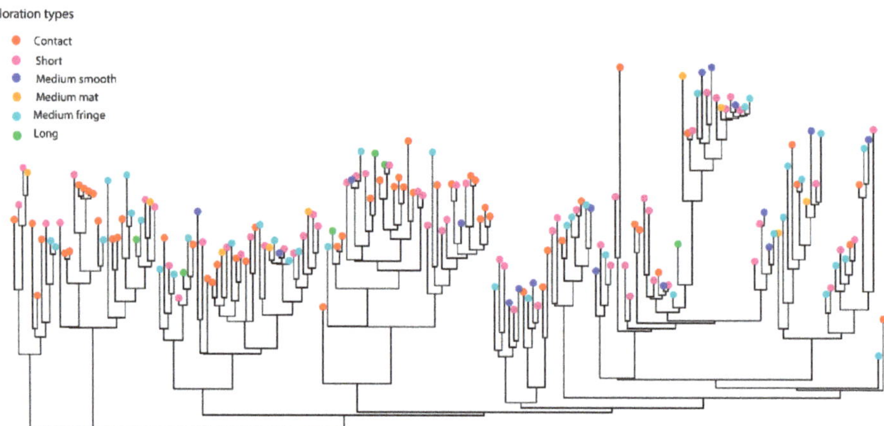

Figure 5. The hybrid phylogenetic tree displaying the distribution of the exploration types (Contact, short, medium smooth, medium fringe and long).

4. Discussion

The results of our phylogenetic study on ectomycorrhizal communities in Mediterranean pine forests showed that phylogenetic composition, structure, and diversity were similar among habitats with distinct pine tree hosts. However, significant differences were found in nearest taxon index values between *P. nigra-halepensis* and *P. sylvestris-nigra*, probably not directly caused by differences in tree hosts but due to higher differences in the local abiotic conditions in *P. sylvestris-nigra* than in *P. halepensis-nigra* sites. Moreover, we detected a weak abiotic filtering effect on the ectomycorrhizal phylogenetic compositional variation, being pH the only variable among soil variables that significantly influence the ectomycorrhizal phylogenetic community. This finding suggests that pH acts as a strong abiotic filter on the ectomycorrhizal community at both phylogenetic and taxonomic levels [17]. In contrast, ectomycorrhizal phylogenetic structure variation was marginally influenced only by the shared effect of stand structure, soil, and geographic distance. Therefore, the phylogenetic structure may be indirectly influenced by other processes (i.e., competition; [30]) not directly tested in this study. Finally, we identified that short and contact exploration types were the most abundant in these forest ecosystems. Conversely, long exploration types were the least abundant, although there was no phylogenetic signal since exploration types were dispersed across the phylogenetic tree.

4.1. Ectomycorrhizal Phylogenetic Description

Our study allowed us to investigate the phylogenetic relationships between 256 OTUs using a multiple gene tree at family level as a foundation tree which allows us to build a better-supported tree (Figure 1a) [70]. Also, we were able to identify monophyletic groups of families, such as Sebacinaceae, Clavulinaceae, and Hydnaceae, and Atheliaceae, Sclerodermataceae, Boletaceae, Gomphidiaceae, and Suillaceae, however, this last clade formed a paraphyletic group with Russulaceae and Albatrellaceae. Moreover, two other family groups were identified, one including Bankeraceae, Thelephoraceae, Sclerodermataceae, Boletaceae, Gomphidiaceae, and Suillaceae, and the other including Tricholomataceae, Amanitaceae, Hydnangiaceae, Cortinariaceae, Hymenogastraceae, and Inocybaceae. Therefore, the resolved phylogenetic tree resulted in a strong backbone for the downstream analyses as the level of resolution allows us to perform reliable phylogenetic diversity analyses [81]. Finally, disentangling the ectomycorrhizal phylogenetic community structure in our study region, where the current climate change may lead to changes in ecosystems functioning, is crucial to predict the impacts on ectomycorrhizal taxonomic and phylogenetic community composition and diversity [82].

4.2. Ectomycorrhizal Phylogenetic Composition, Structure, and Diversity

Our results demonstrated that ectomycorrhizal phylogenetic community and diversity were not significantly different among pine tree host species or in NRI values, although there were differences in NTI values between *P. sylvestris-nigra* and *P. nigra-halepensis*. These results are in accordance with previous taxonomical studies on ectomycorrhizal communities between congeneric tree hosts [9,83], in which a lack of phylogenetic differences was observed. Similarly, ectomycorrhizal community composition was not different between phylogenetically related pines in China [84]. In contrast, several studies reported taxonomical differences between ectomycorrhizal communities between hosts of different families or genera [7,85]. Thus, it seems that at both taxonomic and phylogenetic levels, ectomycorrhizal communities are not varying significantly among phylogenetically close related tree hosts [14].

Similar phylogenetic studies detected phylogenetic clustering of Agaromycotina communities in xeric oak-dominated forests and concluded that oak acted as the main habitat filter [26]. Here, our results showed an opposite trend, with no significant differences in ectomycorrhizal phylogenetic structure and diversity between pine tree hosts. However, we observed significant differences in phylogenetic dispersion among habitats with distinct pines hosts. For example, ectomycorrhizal species in *P. halepensis-nigra* forest resulted in phylogenetic clusters, while in *P. sylvestris-nigra* were slightly more overdispersed at the tip of the phylogeny (Figure 3a), probably due to the low number of *P. nigra-halepensis* sites which may have caused underestimation of the differences between phylogenetic taxa. In this regard, the three *P. nigra-halepensis* sites showed similar soil properties (i.e., values range from pH: 8.18–8.38, CN: 11–151, P: 2–5, N: 0.12–0.16), which may have resulted in the occurrence of closely related species that are adapted to these similar abiotic conditions. These results may imply that Mediterranean pine host tree species are weak habitat filters for ectomycorrhizal fungi, probably due to a lack of host specificity among congeneric hosts. Thus, our results are in agreement with the hypothesis that the lack of phylogenetic composition, structure, and diversity between pine host species may be partially explained by possible conserved symbiosis between *Pinus* and ectomycorrhizal fungi [86].

Finally, we found high and similar turnover values in all the tree host species forest, while nestedness was significantly lower, except in the case of *P. nigra-halepensis* forest. It seems that both environmental filtering by soil and dispersal limitation may, to a certain extent, promote species replacement among sites [87]. However, in *P. nigra-halepensis* forest higher nestedness might indicate local species loss probably due to its soil site conditions that resulted in the occurrence of a locally adapted subset of species.

4.3. Main Drivers of Ectomycorrhizal Phylogenetic Composition and Structure

In this study, we observed that soil parameters influenced ectomycorrhizal phylogenetic composition, while phylogenetic structure variation was primarily influenced by the shared effect of the three environmental filters. However, these three fractions were not significant and explained a residual amount of variation, thus, the second hypothesis is not accepted. Although previous studies have identified that soil parameters are the main drivers of taxonomic ectomycorrhizal community variation in Mediterranean pine forest [17], here, soil parameters were marginally important in driving ectomycorrhizal phylogenetic composition. This may imply that at the phylogenetic level, the lack of strong abiotic gradients results in the occurrence of non-closely related species which are adapted to heterogeneous but not specific environmental conditions [88].

Soil properties have been widely described as a strong abiotic filter on taxonomic fungal communities at different spatial scales [14,62,89,90]. In contrast, we observed a weak abiotic filtering effect of soil physico-chemistry on phylogenetic community composition, with pH resulting in the only significant variable. The importance of pH as an influential variable over ectomycorrhizal community composition at local and regional scales has been widely described [87,90,91]. However, in view of our results, it seems that pH acts as an abiotic filter at both taxonomic and phylogenetic levels [87,92,93]. In addition, our results showed that a left tail of *P. sylvestris* and *P. sylvestris-nigra* in the distribution of the pH values resulting in a wider niche for species adapted to low pH values (Figure S2). In this regard, phylogenetic fungal communities were more clustered at lower values of pH (<7), thus, it seems that lower pH values might result in the occurrence of only adapted fungal species that can grow and maintain cellular function in acidic environments [94], causing higher phylogenetic similarity [87,95,96].

Regarding phylogenetic structure, stand structure alone explained a proportion of its variation although this was not significant. However, stand structure, soil physico-chemistry, and geographic distance shared an important proportion of variance. In Mediterranean ecosystems, the influence of stand structural variables on fungi has already been assessed over mushroom yields with important effects [81]. Although weakly, differences in stand structural variables may result in the occurrence of different less phylogenetically related fungal species that are better adapted to certain local forest conditions. In view of our results, we argue that ectomycorrhizal phylogenetic structure is more importantly influenced by the combined effect of all environmental variables. Hence, phylogenetic relatedness between species decreases with increasing geographic distance, differences in stand structure, and soil conditions. Finally, we hypothesize that there may be other processes influencing the ectomycorrhizal phylogenetic community, such as competition for space and resources [30]. that were not directly tested, therefore further studies are needed to further disentangle whether other processes influence phylogenetic structure in these ecosystems.

4.4. Trait Evolution of the Exploration Types

We observed that 51% of the ectomycorrhizal species had short and contact exploration types, 21% and 13% of the species had medium fringe and medium exploration types, respectively, and only 4% of the species had long exploration types. Similarly, [38] found that in *P. pinaster*-dominated Mediterranean forests, long distance were the least abundant exploration types, while short and contact types were dominating the community. Moreover, our results showed that traits were dispersed across the phylogenetic tree (Figure 5). Thus, the third hypothesis is accepted. In this regard, the dispersion of traits across the phylogenetic tree suggests that even more distant related species showed the same exploration type, resulting in a random trait pattern with a lack of phylogenetic signal. In addition, [38] found that mycorrhizal species with long distance exploration types were less abundant under drier conditions, whereas short-distance and contact type species increased. Recent studies suggest that drier conditions may favor short-contact types [18,38]. Similarly, based on our results, we argue that dispersion of short and contact exploration types might be an adaptation to the Mediterranean stress conditions where the limiting factor is water and not nutrients. Therefore, having medium mat and long exploration types might be a disadvantage due to their higher C demand on the host [18,32,38]. At the same time, in northern and temperate ecosystems, soil N is a limiting nutrient [81], and previous studies have shown that species with long medium mat exploration occurs in soils where N is limiting and patchily distributed [37,42], while short exploration types are more efficient in up-taking soluble inorganic N [36]. However, despite these observed trends and since exploration types of mycorrhizae represent a distinct set of fungal traits, the use of exploration types to study fungal trait responses to environmental changes can be often misleading, and further research should be addressed. In any case, previous work in this area showed a lack of N effect on mycorrhizal communities in Mediterranean pine forests [17], therefore, as N is not limiting it can be easily captured by ectomycorrhizal fungi close to the host with no need of investing in high biomass exploration types.

Finally, we acknowledge the accuracy of the ITS2 region in species identification and resolution [97], but also its limitation in phylogenetic applications due to its high variability [70,98]. However, the use of a backbone phylogenetic tree at family level constructed from multiple gene sequences provides a sufficient taxonomic resolution, thus can be an accurate predictor of phylogenetic diversity metrics [99]. Moreover, it is known that the identification of basidiomycetes through ITS2 amplification is more efficient than in other taxa (i.e., Ascomycetes) [100]. Therefore, future studies aiming to disentangle fungal phylogenetic patterns in community structure should include a robust backbone phylogenetic tree and at least the whole ITS region.

5. Conclusions

In this study, we found no differences neither in ectomycorrhizal phylogenetic community composition nor structure and diversity, indicating that ectomycorrhizal communities at both phylogenetic and taxonomic levels do not change among phylogenetically closely related tree hosts. Moreover, soil parameters only had a marginal filtering effect on ectomycorrhizal phylogenetic variation as pH resulted in the only significant driver of the phylogenetic community. In this regard, our results showed that pH acts as the broadest abiotic filter of ectomycorrhizal communities at a local scale.

Conversely, the ectomycorrhizal phylogenetic structure was marginally influenced by the combined effect of soil, stand structure, and geographic distance, indicating that phylogenetic structure is mainly influenced by their combined effect.

Finally, short and contact distance were the dominant exploration types, as they may be favored under drought stress conditions but also under high nutrient availability. Our results shed light on the drivers of ectomycorrhizal phylogenetic community variation in Mediterranean pine forests, being fundamental to get a better insight on the drivers of community assembly and ecosystem functioning. Nevertheless, further research on ectomycorrhizal phylogenetic communities is needed to better understand how changes in deterministic processes will affect ectomycorrhizal communities and forest ecosystems' functioning.

Supplementary Materials: The following are available online at https://www.mdpi.com/article/ 10.3390/jof7100793/s1, Figure S1: Catalonia map displaying the location of the 42 plots, Figure S2: Density curves of pH values of the five tree hosts from the permutational test of density equality, Figure S3: Hierarchical clustering of ectomycorrhizal phylogenetic compositional data based on a Euclidean distance matrix, Table S1: Table summarizing the main texture and moisture properties of the study plots, Table S2: Most abundant ectomycorrhizal species detected in pure and mixed stands of *Pinus* spp., Table S3: Phylogenetic species turnover, nestedness and total beta diversity values across host tree species stands.

Author Contributions: C.C. (Carlos Colinas), J.A.B., J.M.d.A. and J.G.A. planned and designed the fungal research. J.M.d.A. sampled the soils and together with C.C. (Carlos Colinas), measured the environmental variables. C.C. (Carles Castaño) performed the lab works and the bioinformatic analyses. I.A. analysed the data and wrote the manuscript, with inputs of J.G.A. and C.C. (Carles Castaño). All authors provided inputs on the last version. All authors have read and agreed to the published version of the manuscript.

Funding: This project has received funding from the European Union's H2020 research and innovation programme under Marie Sklodowska-Curie grant agreement No 801586. This work was partially supported by the Spanish Ministry of Science, Innovation and Universities, grants RTI2018-099315-A-I00. I.A. was supported by a H2020-Marie Slodowska Curie Action Cofund fellowship (801596), J.G.A. was supported by Ramon y Cajal fellowship (RYC-2016-20528) and J.A.B. benefitted from a Serra-Húnter Fellowship provided by the Generalitat of Catalunya.

Institutional Review Board Statement: Not applicable.

Informed Consent Statement: Not applicable.

Data Availability Statement: Available upon reasonable request.

Conflicts of Interest: The authors declare that they have no known competing financial interest or personal relationships that could have appeared to influence the work reported in this paper.

References

1. Allen, M.F. Mycorrhizal Fungi: Highways for Water and Nutrients in Arid Soils. *Vadose Zone J.* **2007**, *6*, 291–297. [CrossRef]
2. Van Der Heijden, M.G.A.; Bardgett, R.D.; Van Straalen, N.M. The unseen majority: Soil microbes as drivers of plant diversity and productivity in terrestrial ecosystems. *Ecol. Lett.* **2008**, *11*, 296–310. [CrossRef] [PubMed]
3. Smith, S.E.; Read, D.J. *Mycorrhizal Symbiosis*, 3rd ed.; Academic Press: San Diego, CA, USA, 2008.
4. Ishida, T.A.; Nara, K.; Hogetsu, T. Host effects on ectomycorrhizal fungal communities: Insight from eight host species in mixed conifer-broadleaf forests. *New Phytol.* **2007**, *174*, 430–440. [CrossRef] [PubMed]
5. Urbanová, M.; Šnajdr, J.; Baldrian, P. Composition of fungal and bacterial communities in forest litter and soil is largely determined by dominant trees. *Soil Biol. Biochem.* **2015**, *84*, 53–64. [CrossRef]
6. Van der Linde, S.; Suz, L.M.; Orme, C.D.L.; Cox, F.; Andreae, H.; Asi, E.; Atkinson, B.; Benham, S.; Carroll, C.; Cools, N.; et al. Environment and host as large-scale controls of ectomycorrhizal fungi. *Nature* **2018**, *558*, 243–248. [CrossRef] [PubMed]
7. Suz, L.M.; Kallow, S.; Reed, K.; Bidartondo, M.; Barsoum, N. Pine mycorrhizal communities in pure and mixed pine-oak forests: Abiotic environment trumps neighboring oak host effects. *For. Ecol. Manag.* **2017**, *406*, 370–380. [CrossRef]
8. Glassman, S.I.; Peay, K.; Talbot, J.M.; Smith, D.P.; Chung, J.A.; Taylor, J.W.; Vilgalys, R.; Bruns, T.D. A continental view of pine-associated ectomycorrhizal fungal spore banks: A quiescent functional guild with a strong biogeographic pattern. *New Phytol.* **2015**, *205*, 1619–1631. [CrossRef]
9. Erlandson, S.R.; Savage, J.A.; Cavender-Bares, J.M.; Peay, K.G. Soil moisture and chemistry influence diversity of ectomycorrhizal fungal communities associating with willow along an hydrologic gradient. *FEMS Microbiol. Ecol.* **2016**, *92*, fiv148. [CrossRef]
10. Kyaschenko, J.; Clemmensen, K.E.; Karltun, E.; Lindahl, B.D. Below-ground organic matter accumulation along a boreal forest fertility gradient relates to guild interaction within fungal communities. *Ecol. Lett.* **2017**, *20*, 1546–1555. [CrossRef]
11. Dumbrell, A.J.; Nelson, M.; Helgason, T.; Dytham, C.; Fitter, A.H. Relative roles of niche and neutral processes in structuring a soil microbial community. *ISME J.* **2009**, *4*, 337–345. [CrossRef]
12. Peay, K.G.; Garbelotto, M.; Bruns, T.D. Evidence of dispersal limitation in soil microorganisms: Isolation reduces species richness on mycorrhizal tree islands. *Ecology* **2010**, *91*, 3631–3640. [CrossRef] [PubMed]
13. Powell, J.R.; Bennett, A.E. Unpredictable assembly of arbuscular mycorrhizal fungal communities. *Pedobiologia* **2016**, *59*, 11–15. [CrossRef]
14. Tedersoo, L.; Bahram, M.; Põlme, S.; Kõljalg, U.; Yorou, N.S.; Wijesundera, R.L.C.; Ruiz, L.V.; Vasco-Palacios, A.M.; Thu, P.Q.; Suija, A.; et al. Global diversity and geography of soil fungi. *Science* **2014**, *346*, 1256688. [CrossRef] [PubMed]
15. Boeraeve, M.; Honnay, O.; Jacquemyn, H. Effects of host species, environmental filtering and forest age on community assembly of ectomycorrhizal fungi in fragmented forests. *Fungal Ecol.* **2018**, *36*, 89–98. [CrossRef]
16. Pérez-Izquierdo, L.; Zabal-Aguirre, M.; González-Martínez, S.C.; Buée, M.; Verdú, M.; Rincón, A.; Goberna, M. Plant intraspecific variation modulates nutrient cycling through its below ground rhizospheric microbiome. *J. Ecol.* **2019**, *107*, 1594–1605. [CrossRef]
17. Adamo, I.; Castaño, C.; Bonet, J.A.; Colinas, C.; de Aragón, J.M.; Alday, J.G. Soil physico-chemical properties have a greater effect on soil fungi than host species in Mediterranean pure and mixed pine forests. *Soil Biol. Biochem.* **2021**, *160*, 108320. [CrossRef]
18. Fernandez, C.W.; Kennedy, P.G. Revisiting the "Gadgil effect": Do interguild fungal interactions control carbon cycling in forest soils? *New Phytol.* **2016**, *209*, 1382–1394. [CrossRef]
19. Hartmann, M.; Brunner, I.; Hagedorn, F.; Bardgett, R.D.; Stierli, B.; Herzog, C.; Chen, X.; Zingg, A.; Graf-Pannatier, E.; Rigling, A.; et al. A decade of irrigation transforms the soil microbiome of a semi-arid pine forest. *Mol. Ecol.* **2017**, *26*, 1190–1206. [CrossRef]
20. Mohan, J.E.; Cowden, C.C.; Baas, P.; Dawadi, A.; Frankson, P.T.; Helmick, K.; Hughes, E.; Khan, S.; Lang, A.; Machmuller, M.; et al. Mycorrhizal fungi mediation of terrestrial ecosystem responses to global change: Mini-review. *Fungal Ecol.* **2014**, *10*, 3–19. [CrossRef]
21. Sardans, J.; Peñuelas, J. Plant-soil interactions in Mediterranean forest and shrublands: Impacts of climatic change. *Plant Soil* **2013**, *365*, 1–33. [CrossRef]
22. Webb, C.O.; Ackerly, D.D.; McPeek, M.A.; Donoghue, M.J. Phylogenies and community ecology. *Annu. Rev. Ecol. Syst.* **2002**, *33*, 475–505. [CrossRef]
23. Maherali, H.; Klironomos, J.N. Influence of Phylogeny on Fungal Community Assembly and Ecosystem Functioning. *Science* **2007**, *316*, 1746–1748. [CrossRef]
24. Webb, C.O. Exploring the phylogenetic structure of ecological communities: An example for rain forest trees. *Am. Nat.* **2000**, *156*, 2. [CrossRef]
25. Vamosi, S.M.; Heard, S.B.; Vamosi, J.C.; Webb, C.O. Emerging patterns in the comparative analysis of phylogenetic community structure. *Mol. Ecol.* **2009**, *18*, 572–592. [CrossRef]
26. Edwards, I.P.; Zak, D.R. Phylogenetic similarity and structure of Agaricomycotina communities across a forested landscape. *Mol. Ecol.* **2010**, *19*, 1469–1482. [CrossRef]
27. Egan, C.P.; Callaway, R.M.; Hart, M.M.; Pither, J.; Klironomos, J. Phylogenetic structure of arbuscular mycorrhizal fungal communities along an elevation gradient. *Mycorrhiza* **2016**, *27*, 273–282. [CrossRef] [PubMed]
28. Kraft, N.J.B.; Adler, P.B.; Godoy, O.; James, E.C.; Fuller, S.; Levine, J.M. Community assembly, coexistence and the environmental filtering metaphor. *Funct. Ecol.* **2015**, *29*, 592–599. [CrossRef]
29. Emerson, B.C.; Gillespie, R.G. Phylogenetic analysis of community assembly and structure over space and time. *Trends Ecol. Evol.* **2008**, *23*, 619–630. [CrossRef]

30. Tucker, C.M.; Cadotte, M.W.; Carvalho, S.; Davies, T.J.; Ferrier, S.; Fritz, S.; Grenyer, R.; Helmus, M.; Jin, L.S.; Mooers, A.O.; et al. A guide to phylogenetic metrics for conservation, community ecology and macroecology. *Biol. Rev.* **2016**, *92*, 698–715. [CrossRef]
31. Weber, M.G.; Agrawal, A.A. Phylogeny, ecology, and the coupling of comparative and experimental approaches. *Trends Ecol. Evol.* **2012**, *27*, 394–403. [CrossRef] [PubMed]
32. Agerer, R. Exploration types of ectomycorrhizae. *Mycorrhiza* **2001**, *11*, 107–114. [CrossRef]
33. Hobbie, E.A.; Agerer, R. Nitrogen isotopes in ectomycorrhizal sporocarps correspond to belowground exploration types. *Plant Soil* **2010**, *327*, 71–83. [CrossRef]
34. Pena, R.; Tejedor, J.; Zeller, B.; Dannemann, M.; Polle, A. Interspecific temporal and spatial differences in the acquisition by litter-derived Nitrogen by ectomycorrhizal assemblages. *New Phytol.* **2013**, *199*, 520–528. [CrossRef] [PubMed]
35. Tedersoo, L.; Smith, M.E. Lineages of ectomycorrhizal fungi revisited: Foraging strategies and novel lineages revealed by sequences from belowground. *Fungal Biol. Rev.* **2013**, *27*, 83–99. [CrossRef]
36. Lilleskov, E.A.; Fahey, T.J.; Horton, T.R.; Lovett, G.M. Belowground ectomycorrhizal fungal community change over a nitrogen deposition gradient in alaska. *Ecology* **2002**, *83*, 104–115. [CrossRef]
37. Koide, R.T.; Fernandez, C.; Malcolm, G. Determining place and process: Functional traits of ectomycorrhizal fungi that affect both community structure and ecosystem function. *New Phytol.* **2014**, *201*, 433–439. [CrossRef] [PubMed]
38. Castaño, C.; Lindahl, B.D.; Alday, J.G.; Hagenbo, A.; de Aragón, J.M.; Parladé, J.; Pera, J.; Bonet, J.A. Soil microclimate changes affect soil fungal communities in a Mediterranean pine forest. *New Phytol.* **2018**, *220*, 1211–1221. [CrossRef]
39. Lilleskov, E.A.; Hobbie, E.A.; Horton, T.R. Conservation of ectomycorrhizal fungi: Exploring the linkages between functional and taxonomic responses to anthropogenic N deposition. *Fungal Ecol.* **2011**, *4*, 174–183. [CrossRef]
40. Moeller, H.V.; Peay, K.G.; Fukami, T. Ectomycorrhizal fungal traits reflect environmental conditions along a coastal California edaphic gradient. *FEMS Microbiol. Ecol.* **2014**, *87*, 797–806. [CrossRef]
41. Pena, R.; Lang, C.; Lohaus, G.; Boch, S.; Schall, P.; Schöning, I.; Ammer, C.; Fischer, M.; Polle, A. Phylogenetic and functional traits of ectomycorrhizal assemblages in top soil from different biogeographic regions and forest types. *Mycorrhiza* **2017**, *27*, 233–245. [CrossRef] [PubMed]
42. Defrenne, C.E.; Philpott, T.J.; Guichon, S.H.A.; Roach, W.J.; Pickles, B.J.; Simard, S.W. Shifts in Ectomycorrhizal Fungal Communities and Exploration Types Relate to the Environment and Fine-Root Traits Across Interior Douglas-Fir Forests of Western Canada. *Front. Plant Sci.* **2019**, *10*, 643. [CrossRef]
43. Liston, A.; Robinson, W.A.; Piñero, D.; Alvarez-Buylla, E.R. Phylogenetics ofPinus(Pinaceae) Based on Nuclear Ribosomal DNA Internal Transcribed Spacer Region Sequences. *Mol. Phylogenet. Evol.* **1999**, *11*, 95–109. [CrossRef]
44. Gernandt, D.S.; Geada López, G.; Ortiz García, S.; Liston, A. Phylogeny and classification of Pinus. *Taxon* **2005**, *54*, 29–42. [CrossRef]
45. Saladin, B.; Leslie, A.B.; Wüest, R.O.; Litsios, G.; Conti, E.; Salamin, N.; Zimmermann, N.E. Fossils matter: Improved estimates of divergence times in Pinus reveal older diversification. *BMC Evol. Biol.* **2017**, *17*, 95. [CrossRef]
46. Cavender-Bares, J.; Kozak, K.H.; Fine, P.V.A.; Kembel, S.W. The merging of community ecology and phylogenetic biology. *Ecol. Lett.* **2009**, *12*, 693–715. [CrossRef]
47. Narwani, A.; Matthews, B.; Fox, J.; Venail, P. Using phylogenetics in community assembly and ecosystem functioning research. *Funct. Ecol.* **2015**, *29*, 589–591. [CrossRef]
48. Pérez-Valera, E.; Verdú, M.; Navarro-Cano, J.A.; Goberna, M. Resilience to fire of phylogenetic diversity across biological domains. *Mol. Ecol.* **2018**, *27*, 2896–2908. [CrossRef] [PubMed]
49. Guo, J.; Ling, N.; Chen, Z.; Xue, C.; Li, L.; Liu, L.; Gao, L.; Wang, M.; Ruan, J.; Guo, S.; et al. Soil fungal assemblage complexity is dependent on soil fertility and dominated by deterministic processes. *New Phytol.* **2019**, *226*, 232–243. [CrossRef]
50. Martínez de Aragón, J.; Bonet, J.A.; Fischer, C.R.; Colinas, C. Productivity of ectomycorrhizal and selected edible saprotrophic fungi in pine forests of the pre-Pyrenees mountains, Spain: Predictive equations for forest management of mycological resources. *For. Ecol. Manag.* **2007**, *252*, 239–256. [CrossRef]
51. Alday, J.G.; Martínez De Aragón, J.; De-Miguel, S.; Bonet, J.A. Mushroom biomass and diversity are driven by different spatio-Temporal scales along Mediterranean elevation gradients. *Sci. Rep.* **2017**, *7*, 45824. [CrossRef]
52. Adamo, I.; Piñuela, Y.; Bonet, J.A.; Castaño, C.; de Aragón, J.M.; Parladé, J.; Pera, J.; Alday, J.G. Sampling forest soils to describe fungal diversity and composition. Which is the optimal sampling size in mediterranean pure and mixed pine oak forests? *Fungal Biol.* **2021**, *125*, 469–476. [CrossRef]
53. Alday, J.G.; Marrs, R.H.; Martínez-Ruiz, C. Soil and vegetation development during early succession on restored coal wastes: A six-year permanent plot study. *Plant Soil* **2012**, *352*, 305–320. [CrossRef]
54. Day, P.R. Particle fractionation and particle-size analysis. In *Methods of Soil Analysis Part 1. Agronomy No. 9*; Black, C.A., Ed.; American Society of Agronomy: Madison, WI, USA, 1965.
55. Allen, S.E. *Chemical Analysis of Ecological Materials*; Blackwell's: Oxford, UK, 1989.
56. Bremner, J.M.; Mulvaney, C.S. Nitrogen total. In *Methods of Soil Analysis*, 2nd ed.; Miller, A.L., Keeney, D.R., Eds.; American Society of Agronomy: Madison, WI, USA, 1982; pp. 595–624.
57. Olsen, S.R.; Sommers, L. Phosphorus. In *Methods of Soil Analysis*; Miller, A.L., Keeney, D.R., Eds.; American Society of Agronomy: Madison, WI, USA, 1982; pp. 403–427.
58. Walkley, A. A critical examination of rapid method for determining organic carbon in soils. *Soil Sci.* **1947**, *63*, 251–254. [CrossRef]

59. Anderson, J.M.; Ingram, J.S.I. *Tropical Soil Biology and Fertility: A Handbook of Methods*, 2nd ed.; C.A.B. International: Wallingford, UK, 1993.
60. Ihrmark, K.; Bödeker, I.T.; Cruz-Martinez, K.; Friberg, H.; Kubartova, A.; Schenck, J.; Strid, Y.; Stenlid, J.; Brandström-Durling, M.; Clemmensen, K.; et al. New primers to amplify the fungal ITS2 region—Evaluation by 454-sequencing of artificial and natural communities. *FEMS Microbiol. Ecol.* **2012**, *82*, 666–677. [CrossRef]
61. White, T.J.; Bruns, T.; Lee, S.; Taylor, J. Amplification and direct sequencing of fungal ribosomal RNA genes for phylogenetics. In *PCR Protocols*; Academic Press: San Diego, CA, USA, 1990; pp. 315–322. [CrossRef]
62. Sterkenburg, E.; Clemmensen, K.; Ekblad, A.; Finlay, R.D.; Lindahl, B.D. Contrasting effects of ectomycorrhizal fungi on early and late stage decomposition in a boreal forest. *ISME J.* **2018**, *12*, 2187–2197. [CrossRef] [PubMed]
63. Somervuo, P.; Koskela, S.; Pennanen, J.; Nilsson, R.H.; Ovaskainen, O. Unbiased probabilistic taxonomic classification for DNA barcoding. *Bioinformatics* **2016**, *32*, 2920–2927. [CrossRef] [PubMed]
64. Abarenkov, K.; Nilsson, H.; Larsson, K.; Alexander, I.J.; Eberhardt, U.; Erland, S.; Høiland, K.; Kjøller, R.; Larsson, E.; Pennanen, T.; et al. The UNITE database for molecular identification of fungi—Recent updates and future perspectives. *New Phytol.* **2010**, *186*, 281–285. [CrossRef] [PubMed]
65. Nguyen, N.H.; Song, Z.; Bates, S.T.; Branco, S.; Tedersoo, L.; Menke, J.; Schilling, J.S.; Kennedy, P.G. FUNGuild: An open annotation tool for parsing fungal community datasets by ecological guild. *Fungal Ecol.* **2016**, *20*, 241–248. [CrossRef]
66. Suz, L.M.; Barsoum, N.; Benham, S.; Dietrich, H.-P.; Fetzer, K.D.; Fischer, R.; García, P.; Gehrman, J.; Kristöfel, F.; Manninger, M.; et al. Environmental drivers of ectomycorrhizal communities in Europe's temperate oak forests. *Mol. Ecol.* **2014**, *23*, 5628–5644. [CrossRef] [PubMed]
67. Agerer, R.; Rambold, G. *DEEMY—An Information System for the Characterization and Determination of Ectomycorrhizae*; Ludwig-Maximilians-Universität München: Munchen, Germany; Available online: www.deemy.de (accessed on 4 July 2021).
68. Fouquier, J.; Rideout, J.R.; Bolyen, E.; Chase, J.; Shiffer, A.; McDonald, D.; Knight, R.; Caporaso, J.G.; Kelley, S.T. Ghost-tree: Creating hybrid-gene phylogenetic trees for diversity analyses. *Microbiome* **2016**, *4*, 11. [CrossRef] [PubMed]
69. Zhao, R.-L.; Li, G.-J.; Sanchez-Ramirez, S.; Stata, M.; Yang, Z.-L.; Wu, G.; Dai, Y.-C.; He, S.-H.; Cui, B.; Zhou, J.-L.; et al. A six-gene phylogenetic overview of Basidiomycota and allied phyla with estimated divergence times of higher taxa and a phyloproteomics perspective. *Fungal Divers.* **2017**, *84*, 43–74. [CrossRef]
70. Mikryukov, V.S.; Dulya, O.V.; Modorov, M.V. Phylogenetic signature of fungal response to long-term chemical pollution. *Soil Biol. Biochem.* **2020**, *140*, 107644. [CrossRef]
71. Paradis, E.; Claude, J.; Strimmer, K. APE: Analyses of phylogenetics and evolution in R language. *Bioinformatics* **2004**, *20*, 289–290. [CrossRef]
72. McMurdie, P.J.; Holmes, S. Phyloseq: An R Package for Reproducible Interactive Analysis and Graphics of Microbiome Census Data. *PLoS ONE* **2013**, *8*, e611217. [CrossRef]
73. Silverman, J.D.; Washburne, A.D.; Mukherjee, S.; David, L.A. A phylogenetic transform enhances analysis of compositional microbiota data. *Elife* **2017**, *6*, e21887. [CrossRef] [PubMed]
74. Kembel, S.W.; Cowan, P.D.; Helmus, M.; Cornwell, W.; Morlon, H.; Ackerly, D.; Blomberg, S.; Webb, C. Picante: R tools for integrating phylogenies and ecology. *Bioinformatics* **2010**, *26*, 1463–1564. [CrossRef] [PubMed]
75. Faith, D.P. Conservation evaluation and phylogenetic diversity. *Biol. Conserv.* **1992**, *61*, 1–10. [CrossRef]
76. Oksanen, J.; Blanchet, F.G.; Friendly, M.; Kindt, R.; Legendre, P.; Mcglinn, D.; Minchin, P.R.; O'Hara, R.B.; Simposon, G.L.; Solymos, P. Package "Vegan". 2018. Available online: https://cran.r-project.org (accessed on 4 July 2021).
77. Gotelli, N.J. Null model analysis of species co-occurrence patterns. *Ecology* **2000**, *81*, 2606–2621. [CrossRef]
78. Baselga, A.; Orme, C.D.L. Betapart: An R package for the study of beta diversity. *Methods Ecol. Evol.* **2012**, *3*, 808–812. [CrossRef]
79. Bowman, A.W.; Azzalini, A. R Package "sm: Non-Parametric Smoothing Methods. 2018. Available online: https://cran.r-project.org (accessed on 4 July 2021).
80. Blomberg, S.P.; Garland, T.; Ives, A.R. Testing for phylogenetic signal in comparative data: Behavioral traits are more labile. *Evolution* **2003**, *57*, 717–745. [CrossRef] [PubMed]
81. Tomao, A.; Antonio Bonet, J.; Castaño, C.; de-Miguel, S. How does forest management affect fungal diversity and community composition? Current knowledge and future perspectives for the conservation of forest fungi. *For. Ecol. Manag.* **2020**, *457*, 117678. [CrossRef]
82. Winter, M.; Devictor, V.; Schweiger, O. Phylogenetic diversity and nature conservation: Where are we? *Trends Ecol. Evol.* **2013**, *28*, 199–204. [CrossRef]
83. Arraiano-Castilho, R.; Bidartondo, M.; Niskanen, T.; Zimmermann, S.; Frey, B.; Brunner, I.; Senn-Irlet, B.; Hörandl, E.; Gramlich, S.; Suz, L.M. Plant-fungal interactions in hybrid zones: Ectomycorrhizal communities of willows (Salix) in an alpine glacier forefield. *Fungal Ecol.* **2020**, *45*, 100936. [CrossRef]
84. Ning, C.; Mueller, G.M.; Egerton-Warburton, L.M.; Xiang, W.; Yan, W. Host phylogenetic relatedness and soil nutrients shape ectomycorrhizal community composition in native and exotic pine plantations. *Forests* **2019**, *10*, 263. [CrossRef]
85. Nagati, M.; Roy, M.; Manzi, S.; Richard, F.; Desrochers, A.; Gardes, M.; Bergeron, Y. Impact of local forest composition on soil fungal communities in a mixed boreal forest. *Plant Soil* **2018**, *432*, 345–357. [CrossRef]
86. Tedersoo, L.; Brundrett, M.C. Evolution of Ectomycorrhizal Symbiosis in Plants. In *Biogeography of Mycorrhizal Symbiosis*; Springer: Cham, Switzerland, 2017. [CrossRef]

87. Glassman, S.I.; Wang, I.J.; Bruns, T.D. Environmental filtering by pH and soil nutrients drives community assembly in fungi at fine spatial scales. *Mol. Ecol.* **2017**, *26*, 6960–6973. [CrossRef] [PubMed]
88. Pescador, D.S.; de Bello, F.; López-Angulo Je Valladares, F.; Escudero, A. Spatial Scale Dependence of Ecological Factors That Regulate Functional and Phylogenetic Assembly in a Mediterranean High Mountain Grassland. *Front. Ecol. Evol.* **2021**, *9*, 482. [CrossRef]
89. Lekberg, Y.; Koide, R.T.; Rohr, J.R.; Aldrich-Wolfe, L.; Morton, J.B. Role of niche restrictions and dispersal in the composition of arbuscular mycorrhizal fungal communities. *J. Ecol.* **2007**, *95*, 95–105. [CrossRef]
90. Lladó, S.; López-Mondéjar, R.; Baldrian, P. Forest Soil Bacteria: Diversity, Involvement in Ecosystem Processes, and Response to Global Change. *Microbiol. Mol. Biol. Rev.* **2017**, *81*, e0063-16. [CrossRef]
91. Kivlin, S.N.; Winston, G.C.; Goulden, M.; Treseder, K. Environmental filtering affects soil fungal community composition more than dispersal limitation at regional scales. *Fungal Ecol.* **2014**, *12*, 14–25. [CrossRef]
92. Rousk, J.; Bååth, E.; Brookes, P.C.; Lauber, C.L.; Lozupone, C.; Caporaso, J.G.; Knight, R.; Fierer, N. Soil bacterial and fungal communities across a pH gradient in an arable soil. *ISME J.* **2010**, *4*, 1340–1351. [CrossRef] [PubMed]
93. Carrino-Kyker, S.R.; Kluber, L.; Petersen, S.M.; Coyle, K.P.; Hewins, C.R.; De Forest, J.; Smemo, K.A.; Burke, D.J. Mycorrhizal fungal communities respond to experimental elevation of soil pH and P availability in temperate hardwood forests. *FEMS Microbiol. Ecol.* **2016**, *92*, fiw024. [CrossRef] [PubMed]
94. Lauber, C.L.; Hamady, L.; Knight, R.; Fierer, N. Pyrosequencing-based assesment of pH as a predictor of soil bacterial community structure at continental scale. *Appl. Environ. Microbiol.* **2009**, *75*, 5111–5120. [CrossRef] [PubMed]
95. Goldmann, K.; Schöning, I.; Buscot, F.; Wubet, T. Forest management type influences diversity and community composition of soil fungi across temperate forest ecosystems. *Front. Microbiol.* **2015**, *6*, 1300. [CrossRef]
96. Zhang, T.; Wang, N.-F.; Liu, H.-Y.; Zhang, Y.-Q.; Yu, L.-Y. Soil pH is a Key Determinant of Soil Fungal Community Composition in the Ny-Ålesund Region, Svalbard (High Arctic). *Front. Microbiol.* **2016**, *7*, 227. [CrossRef]
97. Read, D.J.; Perez-Moreno, J. Mycorrhizas and nutrient cycling in ecosystems—A journey towards relevance? *New Phytol.* **2003**, *157*, 475–492. [CrossRef]
98. Sundersan, N.; Kumar, S.A.; Ganeshan, E.J.; Pandi, M. Evaluation of ITS molecular morphometrics effectiveness in species delimitation of Ascomycota—A pilot study. *Fungal Biol.* **2019**, *123*, 517–527. [CrossRef]
99. Liu, J.; Liu, J.; Shan, Y.; Ge, X.; Burgess, K. The use of DNA barcodes to estimate phylogenetic diversity in forest communities of Southern China. *Ecol. Evol.* **2019**, *9*, 5372–5379. [CrossRef]
100. Badotti, F.; de Oliveira, F.S.; Garcia, C.F. Effectiveness of ITS and sub-regions as DNA barcode markers for the identification of Basidiomycota. *BMC Microbiol.* **2017**, *17*, 42. [CrossRef] [PubMed]

Journal of
Fungi

MDPI

Article

The Genus *Leccinum* (Boletaceae, Boletales) from China Based on Morphological and Molecular Data

Xin Meng [1,2,3], Geng-Shen Wang [1,2,3], Gang Wu [1,2], Pan-Meng Wang [1,2,3], Zhu L. Yang [1,2,*] and Yan-Chun Li [1,2,*]

[1] Key Laboratory for Plant Diversity and Biogeography of East Asia, Kunming Institute of Botany, Chinese Academy of Sciences, Kunming 650201, China; Mengxin@mail.kib.ac.cn (X.M.); wanggengshen@mail.kib.ac.cn (G.-S.W.); wugang@mail.kib.ac.cn (G.W.); wangpanmeng@mail.kib.ac.cn (P.-M.W.)

[2] Yunnan Key Laboratory for Fungal Diversity and Green Development, Kunming Institute of Botany, Chinese Academy of Sciences, Kunming 650201, China

[3] College of Life Sciences, University of Chinese Academy of Sciences, Beijing 100049, China

* Correspondence: fungi@mail.kib.ac.cn (Z.L.Y.); liyanch@mail.kib.ac.cn (Y.-C.L.)

Citation: Meng, X.; Wang, G.-S.; Wu, G.; Wang, P.-M.; Yang, Z.L.; Li, Y.-C. The Genus *Leccinum* (Boletaceae, Boletales) from China Based on Morphological and Molecular Data. *J. Fungi* **2021**, *7*, 732. https://doi.org/10.3390/jof7090732

Academic Editors: Anush Kosakyan, Rodica Catana and Alona Biketova

Received: 28 May 2021
Accepted: 31 August 2021
Published: 6 September 2021

Abstract: *Leccinum* is one of the most important groups of boletes. Most species in this genus are ectomycorrhizal symbionts of various plants, and some of them are well-known edible mushrooms, making it an exceptionally important group ecologically and economically. The scientific problems related to this genus include that the identification of species in this genus from China need to be verified, especially those referring to European or North American species, and knowledge of the phylogeny and diversity of the species from China is limited. In this study, we conducted multi-locus (nrLSU, *tef1-α*, *rpb2*) and single-locus (ITS) phylogenetic investigations and morphological observations of *Leccinum* from China, Europe and North America. Nine *Leccinum* species from China, including three new species, namely *L. album*, *L. parascabrum* and *L. pseudoborneense*, were revealed and described. *Leccinum album* is morphologically characterized by the white basidioma, the white hymenophore staining indistinct greenish blue when injured, and the white context not changing color in pileus but staining distinct greenish blue in the base of the stipe when injured. *Leccinum parascabrum* is characterized by the initially reddish brown to chestnut-brown and then pale brownish to brown pileus, the white to pallid and then light brown hymenophore lacking color change when injured, and the white context lacking color change in pileus but staining greenish blue in the base of the stipe when injured. *Leccinum pseudoborneense* is characterized by the pale brown to dark brown pileus, the initially white and then brown hymenophore lacking color change when injured, and the white context in pileus and stipe lacking color change in pileus but staining blue in stipe when bruised. Color photos of fresh basidiomata, line drawings of microscopic features and detailed descriptions of the new species are presented.

Keywords: boletes; taxonomy; morphology; phylogeny; new taxa

1. Introduction

The genus *Leccinum* Gray is a species-rich genus of Boletaceae and is characterized by a whitish or yellow hymenophore, a white to cream context unchanging or staining blue or red when injured, a brown to blackish scabrous to dotted squamules on the surface of the stipe, and comparatively long and smooth basidiospores. Generally, most species of the genus are widely spread in the subarctic, boreal, temperate and Mediterranean regions, with a few secondary expansions to the neotropics [1–12]. Species in *Leccinum* are both ecologically and economically important. Most species of this genus exhibit mycorrhizal host specificity. Species of *Leccinum* sect. *Scabra* Smith & Thiers are associated with plants of *Betula*, while species of *L.* sect. *Fumosa* (A.H. Smith, Thiers & Watling) Gelardi are associated with plants of *Populus*. In *L.* sect. *Leccinum*, species are found exclusively associated with plants of *Populus* (e.g., *L. albostipitatum* den Bakker & Noordel. and *L. insigne*

A.H. Sm., Thiers & Watling), *Betula* (e.g., *L. atrostipitatum* A.H. Sm., Thiers & Watling), Pinaceae (e.g., *L. vulpinum* Watling and *L. piceinum* Pilát & Dermek) and Ericaceae that form arbutoidmycorrhizas (e.g., *L. manzanitae* Thiers and *L. monticola* Halling & G.M. Muell.). However, there are species in section *Leccinum* that are not host specific, i.e., *L. aurantiacum* (Bull.) Gray. This species is associated with plants of *Betula*, *Populus*, *Quercus*, *Salix* and sometimes with *Tilia* [13,14]. Some species of this genus are well-known edible mushrooms, such as *L. quercinum* (Pilát) E.E. Green & Watling, *L. scabrum* (Bull.) Gray and *L. versipelle* (Fr. & Hök) Snell, which are collected in China during the mushroom season.

The genus *Leccinum* was established by Gray in 1821 [13], based on the type species *L. aurantiacum*. Subsequently, more and more mycologists noticed the morphological distinctness and described many new species of this genus. As currently circumscribed, the genus comprises roughly 150 species [1–3,6–56]. North America is the species diversity center of this genus, and in total 118 species have been recorded from this area [19]. Some of the most important works are the serial works of Smith and Thiers [1,15–17], in which three sections of this genus were proposed (*L.* sect. *Leccinum* Smith & Thiers, *L.* sect. *Luteoscabra* Smith & Thiers and *L.* sect. *Scabra*), with 68 species described from Michigan. Twelve species from Central America were described: one species from Belize, eight species from Costa Rica and three species from Colombia [20–24]. In Europe, Singer divided species of this genus into four sections, including two known sections, *L.* sect. *Luteoscabra* and *L.* sect. *Leccinum*, and two newly proposed sections, *L.* sect. *Roseoscabra* and *L.* sect. *Eximia* [3]. In Singer's infrageneric classification, *L.* sect. *Scabra*, established by Smith and Thiers, was merged to *L.* sect. *Leccinum*. Recent molecular phylogenetic evidence has revealed that species of *L.* sect. *Luteoscabra*, *L.* sect. *Roseoscabra* and *L.* sect. *Eximia* belong to divergent clades of Boletaceae and represent many new genera (32,52–54). Thus, the genus *Leccinum* is restricted to the section *Leccinum* (Singer's infrageneric classification) [3]. den Bakker and Noordelos revised the European *Leccinum* species based on morphology and nrLSU sequences and documented sixteen species [14]. In their subsequent study, they treated the three subclades revealed by den Bakker et al. in *L.* section *Leccinum* [33,57] as three subsections (viz. *L.* subsect. *Leccinum*, *L.* subsect. *Fumosa* A.H. Sm., Thiers & Watling and *L.* subsect. *Scabra* Pilat & Dermek) [14]. This infrageneric subdivision was followed in the treatment of the genus in this study. In the Southern Hemisphere, four species have been reported, including one from New Zealand and three from Australia [27–29].

In Asia, six species of *Leccinum* have been reported from Malaysia [6]; ten species from Japan [7–10]; and a total of 31 species have been reported from China based on an extensive literature review [34–36,38–52,56]. Among these Chinese species, twelve species, viz. *L. albellum* (Peck) Singer, *L. chromapes* (Frost) Singer, *L. crocipodium* (Letell.) Watling, *L. eximium* (Peck) Singer, *L. extremiorientale* (Lar. N. Vassiljeva) Singer, *L. griseum* (Quél.) Singer, *L. hortonii* (A.H. Sm. & Thiers) Hongo & Nagas., *L. nigrescens* (Richon & Roze) Singer, *L. rubropunctum* (Peck) Singer, *L. rubrum* M. Zang, *L. rugosiceps* (Peck) Singer and *L. subglabripes* (Peck) Singer have been transferred to other genera [5,11,35,52–55]; eight species, viz. *L. duriusculum* (Schulzer ex Fr.) Singer, *L. intusrubens* (Corner) Høil., *L. oxydabile* (Singer) Singer, *L. quercinum*, *L. rufum* (Schaeff.) Kreisel, *L. subleucophaeum* E.A. Dick & Snell, *L. subradicatum* Hongo and *L. variicolor* Watling were reported without specimen support [39–43,49,51]; and eleven species, viz. *L. ambiguum* A.H. Sm. & Thiers, *L. atrostipitatum* A.H. Sm., Thiers & Watling, *L. aurantiacum*, *L. holopus* (Rostk.) Watling, *L. olivaceopallidum* A.H. Sm., Thiers & Watling, *L. potteri* A.H. Sm., Thiers & Watling, *L. roseofractum* Watling, *L. scabrum*, *L. subgranulosum* A.H. Sm. & Thiers, *L. subleucophaeum* var. *minimum* C.S. Bi and *L. versipelle* were reported with specimen citations [34,38,44–48]. Among these eleven species reported with specimen citations, only *L. subleucophaeum* var. *minimum* was originally described from China, and the remaining species were identified as species originally described from Europe and North America based on general morphological similarities. Indeed, a few species described from Europe and North America do occur in China, especially in northeastern and northwestern China. However, most species found in China have evolved independently in the southern part of China. Thus, identification of the Chinese *Leccinum* species needs to be reconfirmed.

In this study, we used both morphological data and molecular sequences from the nuclear ribosomal internal transcribed spacer (ITS), the large subunit of the nuclear ribosomal RNA (nrLSU), the translation elongation factor 1-alpha (*tef1-α*) and the RNA polymerase II second largest subunit (*rpb2*), together with ecological data to (1) elucidate species diversity of *Leccinum* in China; (2) evaluate the phylogenetic relationships of species within *Leccinum*; (3) make morphological and ecological comparisons between closely related species.

2. Materials and Methods

2.1. Taxon Sampling

Nineteen specimens of the genus *Leccinum* from China were examined. For each collection, a part of the basidioma was dried with silica gel for DNA extraction. The remaining materials were then air-dried at 45–50 °C using an electric food dehydrator. Specimens studied in this work were deposited in the Herbarium of the Kunming Institute of Botany, Chinese Academy of Sciences (KUN). Genera are abbreviated as follows: *L.* for *Leccinum*, *Le.* for *Leccinellum*, *O.* for *Octaviania*, *R.* for *Rossbeevera*, *Ru.* for *Rugiboletus*, *T.* for *Turmalinea*, *Ca.* for *Castanopsis*, *Li.* for *Lithocarpus*, *P.* for *Pinus* and *Q.* for *Quercus*.

2.2. Morphological Observation

The macroscopic descriptions are based on the detailed field notes and photographs of fresh basidiomata. Color codes of the form "4B2" indicate the plate, row, and color block from Kornerup and Wanscher [58]. For microscopic studies, the microscopic features of each part of the basidioma were observed under microscope (Leica DM2000, Leica Microsystems, Wetzlar, Germany), including basidiospores, basidia, cheilocystidia, pleurocystidia and pileipellis, using 5% KOH as a mounting medium to revive the dried materials. Microscopic studies follow Zhou et al. [59]. In the description of Basidiospores, the abbreviation n/m/p means n basidiospores measured from m basidomata of p collections in 5% KOH solution. The notation of the form (a) b–c (d) stands for the dimensions of the basidiospores; the range b–c contains a minimum of 90% of the measured values, a or d given in parentheses stands for extreme values. Q is used to mean "length/width ratio" of a basidiospore in a side view; Q_m means average Q of all basidiospores $\pm$ sample standard deviation. Measurements of basidospores, cystidia, basidia and terminal cells in pileipellis are presented as length $\times$ width. All microscopic structures were drawn freehand from rehydrated material under the microscope with 10$\times$ eyepiece and 100$\times$ objective (the total magnification is 1000$\times$).

2.3. Molecular Procedures

Genomic DNA was extracted from silica gel dried materials or herbarium specimens using the CTAB (Cetyltrimethyl ammonium bromide) method [60]. Polymerase chain reactions (PCRs) were performed to amplify partial sequences of nrLSU, *tef1-α*, *rpb2* and ITS using the extracted DNA. The nrLSU region was amplified with primers LROR/LR5 and LROR/LR3 [61]; *tef1-α* was amplified with primer pair EF1-983F and EF1-1567R [62]; *rpb2* was amplified with primers bRPB2-6F and bRPB2-7.1R [63] and ITS was amplified with primer pair ITS1 and ITS4 [64]. Protocols for the polymerase chain reactions (PCRs) and sequencing followed those in Wu et al. [65] and the references therein.

2.4. Sequence Alignments and Phylogenetic Analyses

The newly generated sequences of each locus were blasted in GenBank, and the most closely related sequences (nucleotide identities >95%) were downloaded for further alignment. Sequences were aligned separately for each of the loci using MAFFT v7.130b with the E-INS-I strategy and manually optimized on BioEdit v7.0.9 [66,67]. Two datasets, the ITS dataset and the multi-locus (nrLSU + *tef1-α* + *rpb2*) dataset, were analyzed using RAxML and Bayesian methods, respectively. For the multi-locus dataset, single-gene analyses were conducted to assess incongruence among individual genes using the ML method (results not shown). Because no well-supported bootstrap value (BS > 70%) [55] conflict was detected among the topologies of the three genes, their sequences were then concatenated together for further multi-locus analyses.

For ML analyses, the multi-locus and ITS datasets were analyzed using RAxML (https://www.phylo.org/, accessed on 26 August 2021) under the model GTRGAMMA [68]. Statistical supports for the phylogenetic analyses were determined using nonparametric bootstrapping with 1000 replicates. For BI analyses, the parameter model was selected by the Akaike information criterion (AIC) as the best-fit likelihood model with Modeltest 3.7 (Free Software Foundation, Boston, MA, USA) [69]. The models employed for each of the four loci were GTR + I + G for ITS, nrLSU and *tef1-α*, and SYM + I + G for *rpb2*. Posterior probabilities (PP) were determined twice by running one cold and three heated chains in parallel mode, saving trees every 1000th generation. Other parameters were kept at their default settings. Runs were terminated once the average standard deviation of split frequencies went below 0.01 [70]. Chain convergence was determined using Tracer v1.5 (http://tree.bio.ed.ac.uk/software/tracer/, accessed on 26 August 2021) to confirm sufficiently large ESS values (>200). Subsequently, the sampled trees were summarized after omitting the first 25% of trees as burn-in using the 'sump' and 'sumt' commands implemented in MrBayes.

3. Results

3.1. Molecular Phylogenetic Analysis

A total of 57 sequences, including fifteen for nrLSU, fifteen for *tef1-α*, fourteen for *rpb2* and thirteen for ITS, were newly generated in this study and aligned with sequences downloaded from GenBank. Sequences retrieved from GenBank and obtained in this study for the multi-locus phylogenetic analyses are listed in Table 1. The multi-locus dataset (Supplementary File S1) contained 122 sequences (49 for nrLSU, 41 for *tef1-α*, 32 for *rpb2*), representing 51 samples, and the alignment contained 2195 nucleotide sites, of which 530 were parsimony informative. *Borofutus dhakanus* Hosen & Zhu L. Yang and *Spongiforma thailandica* Desjardin, Manfr. Binder, Roekring & Flegel were chosen as the outgroup [71,72]. ML and Bayesian analyses produced very similar estimates of tree topologies, and thus only the tree inferred from ML analysis is displayed (Figure 1). The monophyly of *Leccinum* was highly supported (BS = 100% and PP = 1) in our analyses. Four main clades were recovered, and three of them correspond to the three known subsections, viz. *L.* subsect. *Leccinum*, *L.* subsect. *Fumosa* and *L.* subsect. *Scabra* of *L.* sect. *Leccinum* [14]. Three new species, namely *L. album*, *L. parascabrum* and *L. pseudoborneense*, were revealed in our multi-locus phylogenetic analyses. *Leccinum parascabrum* formed the remaining clade with BS = 100% and PP = 1, while *L. pseudoborneense* and *L. album* nested in *L.* subsect. *Scabra* and clustered together with *L. flavostipitatum* E.A. Dick & Snell, *L. subradicatum* and *L. variicolor* with low supported lineage (BS = 54%).

Table 1. Information on specimens used in multi-locus phylogenetic analyses and their GenBank accession numbers. Sequences newly generated in this study are indicated in bold.

Species	Voucher	Locality	GenBank Number				Reference
			ITS	nrLSU	*tef1-α*	*rpb2*	
Leccinellum corsicum	Buf 4507	USA	-	KF030347	KF030435	-	[55]
Le. crocipodium	MICH:KUO-07050707	USA	-	MK601749	MK721103	MK766311	[5]
Le. aff. *griseum*	KPM-NC-0017381	Japan	-	JN378508	JN378449	-	[73]
Le. lepidum	K(M)-142974	Italy	-	MK601751	MK721105	MK766312	[5]
Le. pseudoscabrum	CFMR:DPL-11432	USA	-	MK601752	MK721106	MK766313	[5]
Le. rugosiceps	CFMR:BOS-866	USA	-	MK601770	MK721124	MK766329	[5]
Leccinum album	KUN-HKAS53417	China	MZ392872 MZ392873	MW413907 HQ326880	MW439267 HQ326861	MW439259 MW439260	**This study**
L. aurantiacum	L-0342207	France	-	MK601759	MK721113	MK766318	[5]
L. cerinum	MK11800	Finland	-	AF139692	-	-	[4]
L. duriusculum	KUN-HKAS101160	Uzbekistan	MZ485402	MZ675541	MZ707785	MZ707779	**This study**
L. duriusculum	GL4676	France	-	AF139699	-	-	[4]

Table 1. *Cont.*

Species	Voucher	Locality	ITS	nrLSU	tef1-α	rpb2	Reference
L. flavostipitatum	MENMB10801	USA	-	MH620342	-	-	GenBank
L. holopus	MICH:KUO09150707	USA	-	MK601763	MK721117	MK766322	[5]
L. holopus	9109303	France	-	AF139700			[4]
L. holopus	KUN-HKAS111906	Austria	-	MW413906	MW439266	MW439258	This study
L. manzanitae	NY-14041	USA	-	MK601765	MK721119	MK766324	[5]
L. melaneum	KUN-HKAS57220	China	MZ485409	MZ675542	MZ707786	MZ707780	This study
L. monticola	NY-00815448	Costa Rica	-	MK601767	MK721121	MK766326	[5]
L. monticola	NY-760388	Costa Rica	-	MK601766	MK721120	MK766325	[5]
L. palustre	MK11107	Germany	-	AF139701	-	-	[4]
"*L. palustre*"	hdb030	Netherlands	AF454586	-	-	-	[57]
L. parascabrum	KUN-HKAS99903	China	MZ392874	MW413911	MW439271	MW439264	This study
L. parascabrum	KUN-HKAS59447	China	MZ392875	MW413912	MW439272	MW439265	This study
L. pseudoborneense	KUN-HKAS110156	China	MZ412902	MW413908	MW439268	MW439261	This study
L. pseudoborneense	KUN-HKAS110157	China	MZ412903	MW413909	MW439269	MW439262	This study
L. pseudoborneense	KUN-HKAS110158	China	MZ412904	MW413910	MW439270	MW439263	This study
L. pseudoborneense	KUN-HKAS89139	China	-	MZ536631	MZ543306	MZ543308	This study
L. pseudoborneense	KUN-HKAS92401	China	-	MZ536632	MZ543307	MZ543309	This study
L. quercinum	KUN-HKAS63502	China	-	KF112724	KF112250	KF112724	[65]
L. quercinum	Lq1	Germany	-	DQ534612	-	-	[74]
"*L. scabrum*"	KUN-HKAS56371	China	-	KT990587	KT990782	KT990423	[11]
L. scabrum	KUN-HKAS57266	China	-	KF112442	KF112248	KF112722	[65]
L. scabrum	KUN-HKAS98029	China	MZ485407	MZ675543	MZ707787	-	This study
L. scabrum	KPM-NC-0017840	UK	-	JN378515	JN378455	-	[73]
L. schistophilum	KUN-HKAS98024	China	MZ503508	MZ675544	MZ707788	-	This study
L. schistophilum	VDKO1128	Belgium	-	-	KT824055	KT824022	[75]
L. subradicatum	KPM-NC-24518	Japan	MT934814	MT812736	MT874822	-	[76]
L. sp.	KPM-NC-0017830	Japan	KC552009				[77]
L. variicolor	Lvar1	Germany	-	AF139706	-	-	[4]
L. variicolor	Hdb327	Canada	AY538843	-	-	-	[33]
L. variicolor	Watling6753	UK	AY853538	-	-	-	[78]
L. versipelle	KUN-HKAS76669	China	-	KF112443	KF112249	KF112723	[65]
L. versipelle	CFMR DLC2002-122	USA	-	MK601778	-	-	[5]
L. versipelle	-	Sweden	AF454573	-	-	-	[57]
L. versipelle	KUN-HKAS97997	China	MZ485404	MZ675545	MZ707789	MZ707781	This study
L. versipelle	KUN-HKAS99380	China	MZ485401	MZ675546	MZ707790	MZ707782	This study
L. violaceotinctum	CFMR:BZ-1676	Belize	-	MK601780	MK721133	MK766337	[5]
L. violaceotinctum	CFMR:BZ-3169	Belize	-	-	MK721134	MK766338	[5]
Borofutus dhakanus	KUN-HKAS73789	Bengal	-	JQ928616	JQ928576	JQ928597	[72]
Chamonixia brevicolumna	DBG:F023359	USA	-	MK601728	MK721082	MK766290	[5]
Octaviania japonimontana	KPM-NC-0017812	Japan	-	JN378486	JN378428	-	[73]
O. tasmanica	NY-02449788	USA	-	MK601798	MK721152	MK766355	[5]
Rossbeevera griseobrunnea	GDGM45913	China	-	MH537793	-	-	[79]
R. eucyanea	KPM-NC-0023895	Japan	-	KP222896	KP222915	-	[76]
Rugiboletus andinus	NY-00796145	USA	-	MK601758	MK721112	MK766317	[5]
Ru. andinus	NY-181460	USA	-	MK601757	MK721111	MK766316	[5]
Spongiforma thailandica	DED7873	Thailand	-	NG_042464	KF030436	MG212648	[71]
Turmalinea mesomorpha	KPM-NC-0017743	Japan	-	KC552050	-	-	[77]
T. yuwanensis	KPM-NC0023377	Japan	-	KJ001098	KJ001083	-	[77]
Tylocinum griseolum	KUN-HKAS52612	China	-	KT990631	KT990825	-	[11]

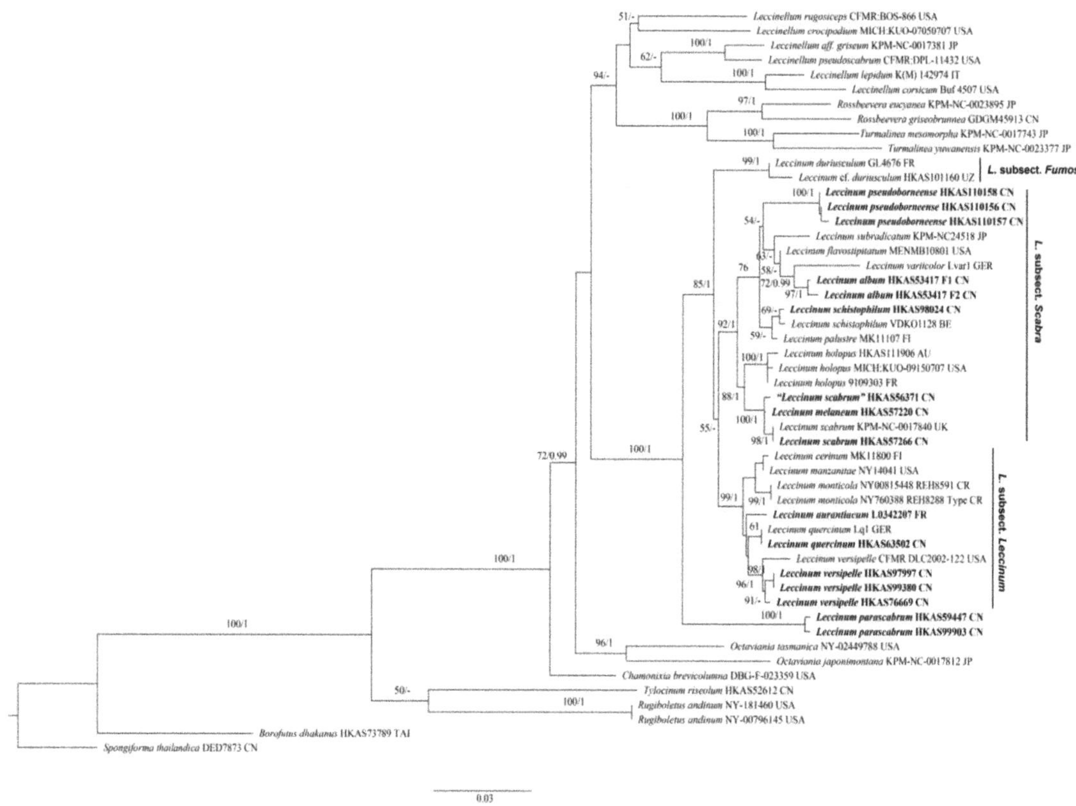

Figure 1. Maximum-likelihood phylogenetic tree generated from a three-locus (nrLSU + *tef1-α* + *rpb2*) dataset. BS > 50% in ML analysis and PP > 0.95 in Bayesian analysis are indicated as RAxML BS/PP above or below supported branches. Species of this genus from China and type species of this genus (*L. aurantiacum*) are indicated in bold. Voucher specimens and localities where the specimens were collected are provided behind the species names. AU = Austria, BE = Belgium, CN = China, CR = Costa Rica, FI = Finland, FR = France, GER = Germany, IT = Italy, JP = Japan, TAI = Thailand, UK =United Kingdom, USA = United States of America and UZ = Uzbekistan.

For the ITS dataset, as revealed by den Bakker et al. [57] and our primary analysis, the ITS1 region contains a minisatellite, which is characterized by the repeated presence of CTATTGAAAAG and CTAATAGAAAG core sequences and mutational derivatives. Moreover, some species contain a minisatellite in the ITS2 region, e.g., the newly described species *L. album* (GenBank Acc. No.: MZ392872 for clone 1 and MZ392873 for clone 2), with a region of 212 bp that consists of tandem repeats (see Supplementary Material for details). Though there is length variation in either the ITS1 or ITS2 spacers, it can also provide some phylogenetic signals. We performed phylogenetic analyses of the ITS dataset. In this dataset (Supplementary File S2), 51 samples were included. The length of the dataset was 1416 bp, of which 377 were parsimony informative. *Leccinellum albellum* (Peck) Bresinsky & Manfr. Binder was chosen as outgroup. ML and Bayesian analyses also produced very similar estimates of tree topologies, and only the tree inferred from ML analysis is displayed (Figure 2). The monophyly of *Leccinum* was also well supported (BS = 100% and PP = 1) in our analyses. Three new species viz. *L. album*, *L. parascabrum* and *L. pseudoborneense*) were revealed. *Leccinum album* is closely related to *L. pseudoborneense* yet without statistical support, while *L. parascabrum* forms an independent lineage. Species to which *L. parascabrum* is phylogenetically related remain as yet unknown.

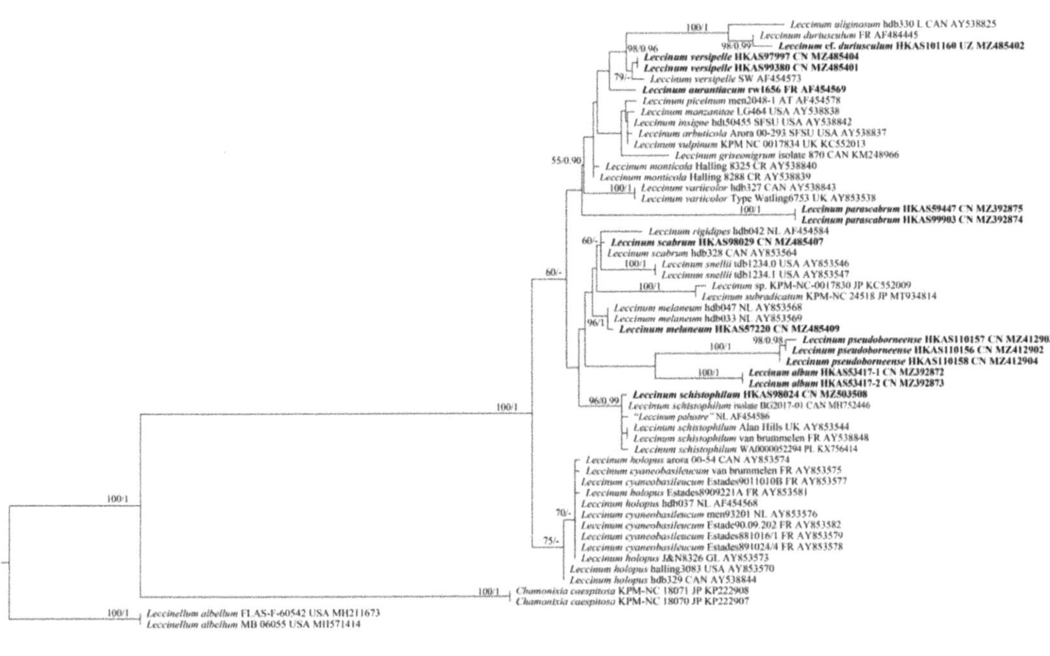

Figure 2. Maximum-likelihood phylogenetic tree generated from ITS dataset. BS > 50% in ML analysis is indicated above or below supported branches. Species of this genus from China and type species of this genus (*L. aurantiacum*) are indicated in bold. Voucher specimens, localities and GenBank numbers are provided behind the species names. AT = Austria, CAN = Canada, CN = China, CR = Costa Rica, FR = France, GL = Greenland, JP = Japan, NL = The Netherlands, PL = Poland, SW = Sweden, UK =United Kingdom, USA = United States of America and UZ = Uzbekistan.

Our ML and Bayesian analyses of ITS and multi-locus datasets revealed the existence of eight *Leccinum* species from China, including five known species viz. *L. melaneum* (Smotl.) Pilát & Dermek, *L. quercinum*, *L. scabrum*, *L. schistophilum* and *L. versipelle* and three new species viz. *L. album*, *L. parascabrum*, and *L. pseudoborneense*. The final alignments of both datasets were deposited in TreeBASE (S27490).

3.2. Taxonomy

Leccinum album X. Meng, Yan C. Li & Zhu L. Yang, sp. nov., (Figures 3g–h and 4). MycoBank: MB 838917.

Diagnosis: This species differs from other species in *Leccinum* in the combination of the entirely white pileus, the white pileal context not changing color when injured, the white hymenophore staining indistinct greenish blue when hurt, the white stipe coarsely covered with initially white and then darkened verrucose squamules, and the white stipe context always staining greenish blue at the base when injured.

Holotype: CHINA. Hunan Province: Chenzhou, Zhanghua County, Mangshan National Forest Park, E 112°92′, N 24°94′, alt. 850 m, associated with *Castanopsis fissa*, *Cyclobalanopsis glauca*, *Lithocarpus glabra* and *Pinus kwangtungensis*, 3 September 2007, Y.C. Li1072 (KUN-HKAS53417, GenBank Acc. No.: MZ392872 and MZ392873 for ITS, MW413907 and HQ326880 for nrLSU, MW439267 and HQ326861 for *tef1-α*, and MW439259 and MW439260 for *rpb2*).

Figure 3. Basidiomata of *Leccinum* species. (**a–c**) *Leccinum parascabrum* (KUN-HKAS99903, holotype); (**d–f**) *Leccinum pseudoborneense* ((**d**) from KUN-HKAS110157; (**e,f**) from KUN-HKAS110156, holotype); (**g,h**) *Leccinum album* (KUN-HKAS53417, holotype).

Etymology: Latin "*album*" means white, referring to the color of the basidiomata.

Basidiomata small to medium-sized. Pileus 3–5.5 cm in diam., hemispherical when young, subhemispherical to convex or plano-convex when mature, white (1A1) when young, white to cream (2B2–3) when mature; surface covered with concolorous farinose to pubescent squamules; context 5–10 mm thick in the center of pileus, taste mild, white (1A1) to pallid, not changing color when bruised; Hymenophore adnate when young, adnate to slightly depressed around apex of stipe; surface white (1A1), staining indistinct greenish blue (25B5–7) when injured; pores subangular to roundish, 0.3–1.5 mm wide; tubes up to 5 mm long, white to dirty pinkish (13A2), not changing color when bruised. Stipe 8–10 × 0.8–1.2 cm, clavate to subcylindrical, always enlarged downwards; surface white (1A1), densely covered with white (1A1) verrucose squamules, staining light greenish blue at base when injured; context whitish (1A1), staining blue at base when injured; basal mycelium white (1A1), lacking color change when injured.

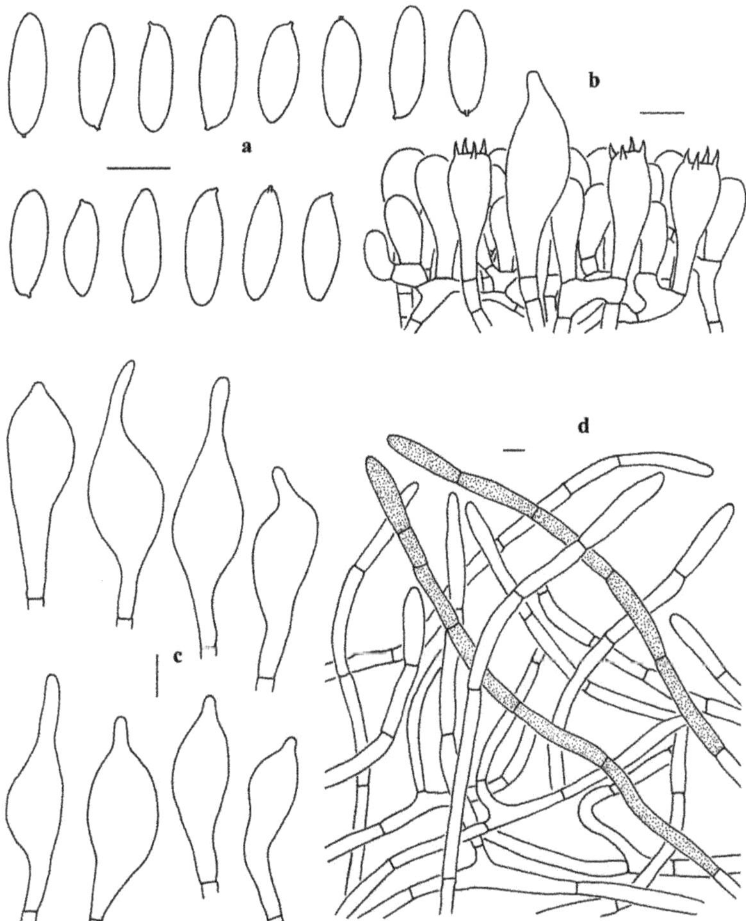

Figure 4. Microscopic features of *Leccinum album* (KUN-HKAS53417, holotype). (**a**) Basidiospores. (**b**) Basidia and pleurocystidium. (**c**) Cheilocystidia and pleurocystidia. (**d**) Pileipellis. Bars = 10 μm. Drawings by Y.-C. Li.

Basidiospores (40/2/1) 15–19 × 5–7 μm, Q = 2.5–3, Q_m = 2.75 ± 0.15, subfusiform to narrowly ellipsoid in side view with slight suprahilar depression, subcylindrical to fusiform in ventral view, smooth, somewhat slightly thick-walled (up to 0.5 μm thick), hyaline to yellowish in KOH, brownish yellow to olivaceous brown in Melzer's Reagent. Basidia 23–33 × 10–13 μm, clavate, 4-spored, hyaline to yellowish in KOH, yellowish to brownish yellow in Melzer's Reagent. Hymenophoral trama boletoid, hyphae subcylindrical, 4–10 μm wide, hyaline to yellowish in KOH, yellowish to yellow in Melzer's Reagent. Cheilo- and pleurocystidia 42–60 × 11–17.5 μm, abundant, subfusiform to fusiform, thin-walled, yellowish in KOH, yellowish to brownish yellow in Melzer's Reagent. Pileipellis a trichoderm, composed of more or less vertically arranged 5–10 μm wide hyphae, hyaline to yellowish in KOH, yellowish to yellow in Melzer's Reagent. Pileal trama made up of 6–12 μm wide filamentous hyphae, thin-walled, yellowish in KOH, yellowish to brownish yellow in Melzer's Reagent. Clamp connections absent in all tissues.

Habitat and distribution: Solitary or scattered in tropical forests dominated by plants of the families Fagaceae (*Castanopsis fissa*, *Cyclobalanopsis glauca* and *Lithocarpus glabra*) and Pinaceae (*Pinus kwangtungensis* or *P. armandii*); on acidic, humid and loamy soils;

distribution insufficiently known, rather rare in China and currently found in central and southeastern China (Hunan and Fujian Provinces).

Additional Specimen examined: CHINA. Fujian Province: Jianning County, E 116°84′, N 26°83′, alt. 900 m, associated with *Castanopsis fissa*, *Cyclobalanopsis glauca* and *Pinus armandii*, 16 July 1971, N.L. Huang 716 (KUN-HKAS39522).

Commentary: *Leccinum album* is characterized by the white pileus, the white hymenophore staining indistinct greenish blue when hurt, the white stipe densely covered with initially white and then darkened scabrous squamules, the white context in pileus not changing color when injured, and the white context in stipe unchanging or only staining distinct greenish blue at base when injured. Morphologically, *L. album* is close to *L. holopus*, *L. cyaneobasileucum* Lannoy & Estadès and *Le. albellum* (Peck) Bresinsky & Manfr. Binder in similar pileus colors. However, *L. holopus*, originally described from Europe (Germany), differs from *L. album* in its medium to large basidiomata (pileus 4–10 cm wide), becoming more viscid pileus with age, pure white or dirty white to pale buff or pale pallid pileus always with a glaucous green tinge, long hymenophoral tubes measuring 9–15 mm long, narrow and subcylindrical hymenial cystidia measuring 30–50 × 7.5–12.5 μm, narrow pileipellis hyphae measuring 3.5–5 μm wide, and association with trees of the genus *Betula* (Betulaceae) [80–82]. *Leccinum cyaneobasileucum*, originally described from France, is different from *L. album* in its white or greyish brown to light brown pielus, woolly stipe surface, slender basidiospores with $Q_m \geq 3$, relatively narrow hymenial cystidia measuring 32–44 × 5.5–7.5 μm, narrow pileipellis hyphae measuring 2–6.5 μm wide, and association with trees of the genus *Betula* [83]. *Leccinellum albellum*, originally described from New York, is characterized by its basidiomata not changing color when bruised and narrow basidiospores measuring 13–20 × 4–6 μm [16,17,30].

Phylogenetically, *L. album* is related to *L. variicolor* and *L. pseudoborneense* in the analyses of the multi-locus and ITS datasets, respectively (Figures 1 and 2). However, *L. variicolor* differs from *L. album* in its white to grey or cream pileal context staining vinaceous to brown when bruised, white stipe context staining pink to coral red in the upper part and green-blue in the lower part when bruised and association with plants of *Betula* [81]. *Leccinum pseudoborneense* is different from *L. album* in its pale brown to dark brown pileus, white context in pileus and stipe staining blue when bruised, narrow basidiospores measuring (11) 12–19 (20) × 4–5 (6) μm, narrow hymenial cystidia measuring 28–40 × 4–10 μm, and distribution in southwestern China.

Leccinum parascabrum X. Meng & Yan C. Li & Zhu L. Yang, sp. nov., (Figures 3a–c and 5). MycoBank: MB 838916.

Diagnosis: This species differs from other species in *Leccinum* by its initially reddish brown to chestnut-brown and then brown to pale brownish or even dirty white pileus, white pileal context lacking color change when injured, white to pallid and then light brown hymenophore lacking color change when injured, and the white stipe context staining greenish blue at the base when injured.

Holotype: CHINA. Hunan Province: Chenzhou, Zhanghua County, Mangshan National Forest Park, E 112°92′, N 24°94′, alt. 1100 m, associated with *Castanopsis fissa*, *Lithocarpus glabra* and *Pinus kwangtudgensis*, 12 September 2016, G. Wu 1784 (KUN-HKAS99903, GenBank Acc. No.: MZ392874 for ITS, MW413911 for nrLSU, MW439271 for *tef1-α*, and MW439264 for *rpb2*).

Etymology: Latin "*parascabrum*" refers to its similarity to *L. scabrum*.

Basidiomata small to medium-sized. Pileus 2.5–12.5 cm in diam., hemispherical when young, subhemispherical to convex or applanate when mature, reddish brown (12E8) to chestnut-brown (8C7–8) when young, brown (6C6) to pale brownish (7D7–8) or even dirty white (6A2) when mature; surface tomentose; context 6–13 mm thick in the center, white (1A1), not changing color when bruised; Hymenophore adnate when young, adnate to slightly depressed around apex of stipe; surface white to pallid (1A1) when young, and becoming light brown (6B4) when mature, not changing color when injured; tubes 6–14 mm long, 0.5–1.5 mm wide, creamy white (1A1), not changing color when bruised.

Stipe 12–14 × 1.1–2.2 cm, clavate, swollen downwards, always staining greenish blue at base when injured; surface white (1A1), covered with initially white (1A1) to light beige (5A4) and then brownish (7D8) squamules; context white (1A1), staining greenish blue (25B6–7) at base when injured; basal mycelium white (1A1).

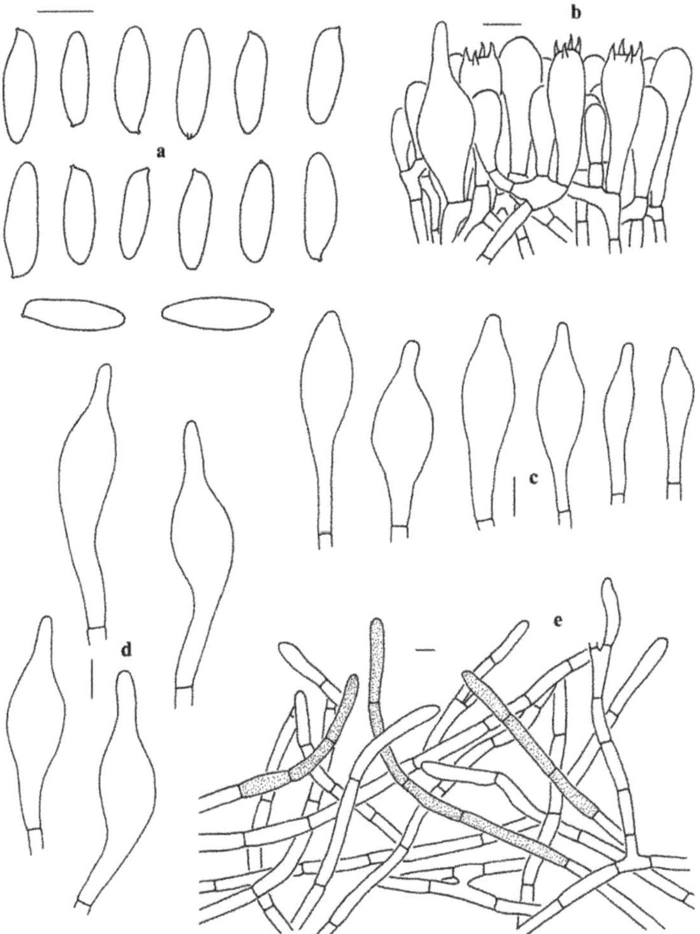

Figure 5. Microscopic features of *Leccinum parascabrum* (KUN-HKAS99903, holotype). (**a**) Basidiospores. (**b**) Basidia and pleurocystidium. (**c**) Cheilocystidia. (**d**) Pleurocystidia. (**e**) Pileipellis. Bars = 10 µm. Drawings by Y.-C. Li.

Basidiospores (80/2/2) 16–20 (–21) × 5–6 µm, Q = 3.2–3.8, Q_m = 3.43 ± 0.18, sub-fusiform to fusiform, slightly thick-walled (up to 0.5 µm thick), yellowish to yellowish brown in KOH, yellow to yellow-brown in Melzer's Reagent. Basidia 24–33 × 8–12 µm, clavate, 4-spored, hyaline to yellowish in KOH, yellowish to yellow in Melzer's Reagent. Hymenophoral trama boletoid, hyphae cylindrical, 3–7 µm wide, hyaline to yellowish in KOH, yellowish to yellow in Melzer's Reagent. Cheilo- and pleurocystidia 34–68 × 7.5–16 µm, abundant, subfusiform to fusiform, thin-walled, yellowish to pale yellowish brown in KOH, yellowish brown to brown in Melzer's Reagent. Pileipellis a trichoderm, composed of 5–9 µm wide filamentous hyphae, yellowish to pale brownish in KOH. Pileal trama made up of 5–10 µm wide filamentous hyphae, thin-walled, hyaline to yellowish in KOH, yellowish to brownish yellow in Melzer's Reagent. Clamp connections absent in all tissues.

Habitat and distribution: Solitary or scattered in tropical forests dominated by plants of the families Fagaceae (*Lithocarpus glabra*, *Castanopsis fissa* and Ca. *hystrix*) and Pinaceae (*Pinus kwangtudgensis* or *P. yunnanensis.*); on acidic or slightly alkaline, loamy soils; distribution insufficiently known, rather rare in China, currently known from central and southwestern China (Hunan and Yunnan Provinces).

Additional Specimen examined: CHINA. Yunnan Province: on the way from Tengchong County to Longling County, E 98°59′, N 24°81′, alt. 2010 m, associated with *Lithocarpus glabra*, *Castanopsis hystrix* and *Pinus yunnanensis*, 19 July 2009, Y.C. Li 1700 (KUN-HKAS59447, GenBank Acc. No.: MZ392875 for ITS, MW413912 for nrLSU, MW439272 for *tef1-α*, and MW439265 for *rpb2*).

Commentary: Leccinum parascabrum is characterized by the initially reddish brown to chestnut-brown and later brown to pale brownish or even dirty white pileus, the white pileal context not changing color when injured, the white to pallid and then light brown hymenophore not changing color when injured, the white stipe context with greenish blue color change at the base when injured, and the relatively large basidiospores measuring 16–20 (–21) × 5–6 µm, Q = 3.2–3.8. *Leccinum parascabrum* generally shares the similar colors of pileus and hymenophore, and the similar slender stems with *L. duriusculum*, *L. griseonigrum* A.H. Sm., Thiers & Watling, *L. scabrum*, *L. uliginosum* A.H. Sm. & Thiers and *Le. pseudoscabrum* (Kallenb.) Mikšík. However, *L. duriusculum*, originally described from Europe, can be distinguished from *L. parascabrum* by its pale grey-brown to dark greyish or reddish brown pileus, white context staining violaceous pink when bruised but yellow-green to blue-green in the base of stipe, relatively small basidiospores measuring 11.5–15.5 × 4.5–6 µm [84]. *Leccinum griseonigrum*, originally described from North America, differs from *L. parascabrum* in its avellaneous to dingy cinnamon-buff pileus, white pileal context staining blue when bruised, relatively small basidiospores measuring 13–16 × 4–5.5 µm, and association with trees of the genus *Populus* [16]. *Leccinum scabrum* differs from *L. parascabrum* in its wrinkled pileus, pale white hymenophore, pinkish discoloration when injured, and never bluish color change at the base of stipe [13,14,81]. *Leccinum uliginosum*, originally described from North America, is different from *L. parascabrum* in its dark fuscous to drab-grey pileus, white context in pileus becoming reddish and then fuscous when bruised, relatively small basidiospores measuring 14–18 × 3.5–5 µm, and small and inconspicuous hymenial cystidia [17]. *Leccinellum pseudoscabrum* differs from *L. parascabrum* in its initially red to purplish brown and then blackish brown context color change when injured, and the palisadoderm pileipellis composed of subglobose cells [14]. *Leccinum parascabrum* also shares the similar colors of pileus and hymenophore and the bluish color change at the base of stipe with *L. variicolor*. However, *L. variicolor* is different from *L. parascabrum* in its white to grey or cream pileal context staining vinaceous to brown when bruised, white stipe context staining pink to coral red in the upper part and green-blue in the lower part when bruised, relatively small basidiospores measuring (10) 13.5–17.5 (–20.0) × 5.0–6.5 (7.0) µm with Q = 2.4–3.1, and association with plants of *Betula* sp. [81]. In our phylogenetic analysis of the multi-locus and ITS datasets (Figures 1 and 2), *L. parascabrum* formed independent clades within *Leccinum*. It might represent a distinct section or subsection. However, formal change of the infrageneric division of this clade should await more molecular and morphological data from additional taxa. Species to which it is phylogenetically related remain as yet unknown.

Leccinum pseudoborneense X. Meng & Yan C. Li & Zhu L. Yang, sp. nov., (Figures 3d–f and 6).

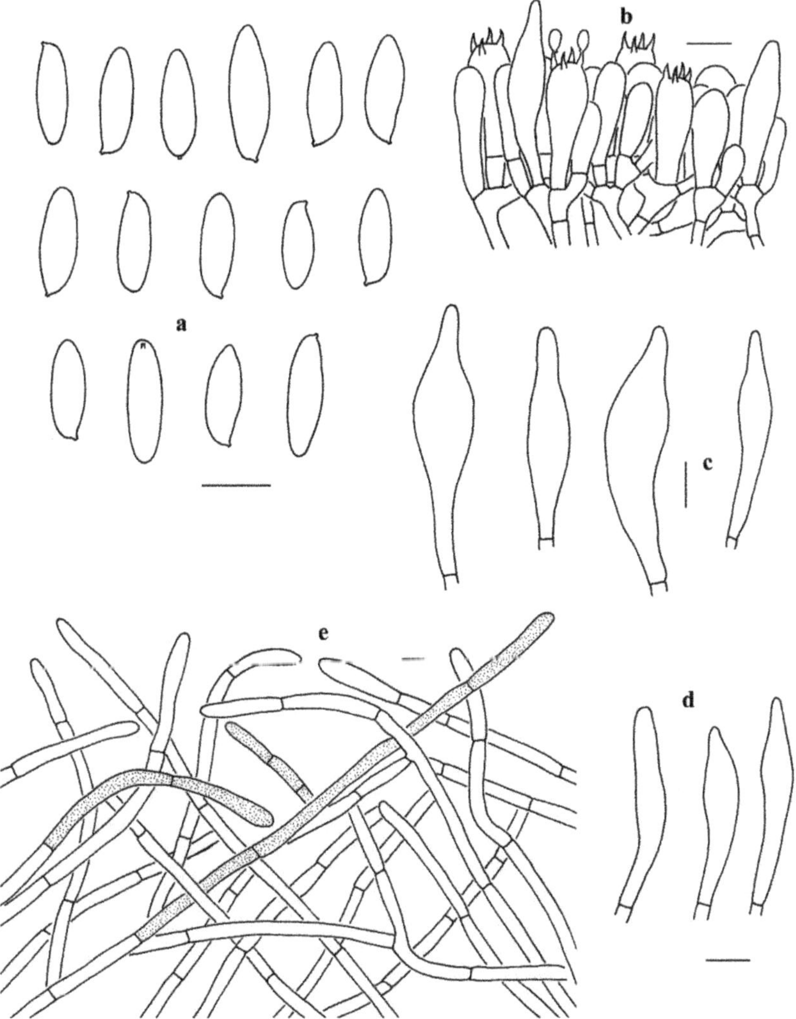

Figure 6. Microscopic features of *Leccinum pseudoborneense* (KUN-HKAS110156, holotype). (**a**) Basidiospores. (**b**) Basidia and pleurocystidia. (**c**) Pleurocystidia. (**d**) Cheilocystidia. (**e**) Pileipellis. Bars = 10 μm. Drawings by Y.-C. Li.

MycoBank: MB 838915.

Diagnosis: This species differs from other species in *Leccinum* in its nearly glabrous and pale brown to dark brown pileus, white context in pileus lacking color change when injured, white context in stipe staining blue when bruised, initially white and then brown hymenophore not changing color when injured, white stipe covered with ochraceous to dark brown squamules, and trichodermal pileipellis composed of 3–6 μm wide interwoven hyphae.

Holotype: CHINA. Yunnan Province: Xishuangbanna, Menghai County, Bada Town, E 100°12′, N 21°83′, alt. 1900 m, associated with *Castanopsis calathiformis*, *Castanopsis indica* and *Lithocarpus truncatus*, 22 June 2020, G.S. Wang 947 (KUN-HKAS110156, GenBank Acc. No.: MZ412902 for ITS, MW413908 for nrLSU, MW439268 for *tef1-α*, and MW439261 for *rpb2*)

Etymology: Latin *"pseudo"* = false, *"borneense"* = *L. borneense*, *"pseudoborneense"* is proposed because this species is similar to the species *L. borneense* originally described from Malaysia.

Basidiomata small to medium-sized. Pileus 4–10 cm diam, subhemispherical to convex or plano-convex; surface nearly glabrous, viscid when wet, pale brown (6D6–C5) to dark brown (6F6–E5); context 5–10 mm thick in the center, white (1A1), not changing color when bruised; Hymenophore adnate to depressed around apex of stipe; white (1A1) to pallid when young and becoming brown (6B5) when mature, not changing color when injured. Tubes 4–10 mm long, creamy white (1A1) when young, and becoming brownish yellow (5C7–8) when mature, not changing color when bruised; pores fine, no more than 1 mm wide. Stipe 10–15 × 2.1–2.9 cm, clavate, always swollen downwards; surface white (1A1), covered with ochraceous (2B3–5) to dark brown (6E7) squamules, staining asymmetric blue (23E7) when injured; context white (1A1), staining blue (23E7) when injured; basal mycelium white (1A1).

Basidiospores (100/5/5) (11–) 12–19 (–20) × 4–5 (–6) μm, Q = (2.75–) 3–3.58 (–3.6), Q_m = 3.31 ± 0.16, subfusiform to ellipsoid, slightly thick-walled (up to 0.5 μm thick), yellowish brown to olive brown in KOH, yellow-brown to dark olive-brown in Melzer's Reagent. Basidia 18–30 × 8–9 μm, clavate, 4-spored, hyaline to yellowish in KOH, yellowish to brownish yellow in Melzer's Reagent. Hymenophoral trama boletoid, hyphae cylindrical, 3–6 μm wide, hyaline to yellowish in KOH, yellowish to brownish yellow in Melzer's Reagent. Cheilo- and pleurocystidia 28–40 × 4–10 μm, abundant, subfusiform to fusiform, thin-walled, yellowish to brownish yellow in KOH, brownish to yellow-brown in Melzer's Reagent. Pileipellis a trichoderm, composed of more or less vertically arranged 5–12 μm wide filamentous hyphae, yellowish brown to brownish in KOH, brown to dark brown in Melzer's Reagent. Pileal trama made up of 6–12 μm wide filamentous hyphae, thin-walled, hyaline to yellowish in KOH, yellowish to yellow in Melzer's Reagent. Clamp connections absent in all tissues.

Habitat and Distribution: Scattered in tropical forests dominated by plants of the families Fagaceae (*Castanopsis calathiformis*, Ca. *orthacantha*, Ca. *indica*, *Lithocarpus truncatus*, *Li. mairei* and *Quercus griffithii*); on acidic, loamy or mossy, humid soils; moderately common in southwestern China (Yunnan Province).

Additional specimens examined: CHINA. Yunnan Province: Xishuangbanna, Menghai County, Bada Township, E 100°13′, N 21°84′, alt. 1900 m, associated with *Castanopsis calathiformis*, Ca. *indica* and *Lithocarpus truncatus*, 22 June 2020, G.S. Wang 960 (KUN-HKAS110157, GenBank Acc. No.: MZ412903 for ITS, MW413909 for nrLSU, MW439269 for *tef1-α*, and MW439262 for *rpb2*), the same location, 22 June 2020, G.S. Wang 965 (KUN-HKAS110158, GenBank Acc. No.: MZ412904 for ITS, MW413910 for nrLSU, MW439270 for *tef1-α*, and MW439263 for *rpb2*); Nanjian County, Gonglang Town, Huangcaoping, E 100°30′, N 24°54′, alt. 1200 m, associated with *Castanopsis orthacantha*, *Lithocarpus mairei* and *Quercus griffithii*, 30 June 2015, K. Zhao 773 (KUN-HKAS92401, GenBank Acc. No.: MZ536632 for nrLSU, MZ543307 for *tef1-α*, and MZ543309 for *rpb2*); Jinghong County, Dadugang Town, E 100°25′, N 21°26′, alt. 600 m, associated with *Castanopsis indica* and *Lithocarpus truncatus*, 30 June 2014, K. Zhao 476 (KUN-HKAS89139, GenBank Acc. No.: MZ536631 for nrLSU, MZ543306 for *tef1-α*, and MZ543308 for *rpb2*).

Commentary: Leccinum pseudoborneense is characterized by the nearly glabrous and pale brown to dark brown pileus, the white context in pileus not changing color when injured, the white context in stipe staining blue when bruised, the initially white and then brown hymenophore not changing color when injured, the white stipe covered with ochraceous to dark brown squamules, and the trichodermal pileipellis composed of 3–6 μm wide interwoven hyphae. *Leccinum pseudoborneense* is similar to *L. borneense* (Corner) E. Horak, originally described from Malaysia, in that they share a brown pileus, bluish color change of the context in stipe when bruised, and similar size of basidiospores. However, *L. borneense* differs from *L. pseudoborneense* in its yellow to olive yellow hymenophore staining blue when hurt, pale yellow to yellow pileal context staining blue when hurt, and deep yellow context in stipe staining blue but sometimes with reddish tint at base when injured [6,84]. *Leccinum pseudoborneense* is phylogenetically close to *L. album* in our phylogenetica analyses (Figures 1 and 2). However, *L. album* has a white basidioma, white hymenophore staining indistinctly greenish blue when hurt, white context in pileus not changing color when injured, white context in stipe unchanging or only staining distinctly greenish blue at base when injured, and relatively broad basidiospores measuring 15–19 × 5–7 μm.

4. Discussion

The genus *Leccinum* was defined and recognized variously by different mycologists. In an early molecular study, *Leccinum* was shown to be polyphyletic and proposed to be restricted to the sections *Leccinum* and *Scabra* by Binder and Besl [4]. Subsequently, Bresinsky and Besl [32] erected a genus *Leccinellum* Bresinsky & Manfr. Binder, to accommodate *L.* section *Luteoscabra*, including species with yellow hymenophores and/or context. In this study, the phylogenetic inferences based on the multi-locus dataset of nrLSU, *tef1-α* and *rpb2* largely coincide with those of Binder and Besl [4], Bresinsky and Besl [32] and den Bakker et al. [33]. Thus, we adopt the treatment of Bakker et al. [33] and treat *Leccinum* in a strict circumscription, which only includes species of *L.* sect. *Leccinum* (Singer's infrageneric classification with *L.* sect. *Scabra* merged to this section). Species in *Leccinum* are characterized by the white context lacking color changes or staining blue, gray or reddish tints when injured and the cutis-like pileipellis composed of interwoven filamentous hyphae

Eleven *Leccinum* species with specimen citations have been reported from China before this study, of which five species (*L. ambiguum*, *L. atrostipitatum*, *L. olivaceopallidum*, *L. potteri* and *L. subgranulosum*) were originally described from North America, five species (*L. aurantiacum*, *L. holopus*, *L. roseofractum*, *L. scabrum* and *L. versipelle*) were originally described from Europe, and only one taxon (*L. subleucophaeum* var. *minimum*) was originally described from China. Our molecular phylogenetic analyses along with morphological studies identified the existence of *L. quercinum*, *L. scabrum*, *L. subleucophaeum* var. *minimum* and *L. versipelle* in China. The distribution of other reported species have not yet been found, based on morphological and/or molecular data. In addition, three species new to science (*L. album*, *L. parascabrum* and *L. pseudoborneense*) and two species new to China (*L. melaneum* and *L. schistophilum*) were revealed in our study, based on molecular and morphology evidence. In conclusion, there are nine species of *Leccinum* in China.

Most species of *Leccinum* exhibit strong mycorrhizal host specificity. The host specificity along with climate type and edaphic factors appear to be important factors determining the distribution of different species. In China, *L. melaneum*, *L. scabrum*, *L. schistophilum* and *L. versipelle* are found in temperate forests and associated with plants of *Betula platyphylla* on acidic soils. *Leccinum album*, *L. parascabrum*, *L. pseudoborneense* and *L. subleucophaeum* var. *minimum* are found in tropical forests and associated with plants of Fagaceae (*Castanopsis calathiformis*, *Ca. hystrix*, *Ca. indica*, *Ca. orthacantha*, *Cyclobalanopsis glauca*, *Lithocarpus mairei*, *Li. truncatus* and *Quercus griffithii*) and/or Pinaceae (*Pinus kwangtudgensis* and *P. yunnanensis*) on acidic soils. It is noteworthy that *L. parascabrum* can be found in acidic or slightly alkaline habitats. *Leccinum versipelle* is found in subtropical forests and is associated with plants of *Populus yunnanensis* on acidic soils. The combination of the color of basidioma, the morphology of pileal surface, the size of basidiospores, the morphology of stipe, the color changes when injured, the climate type, the edaphic factors and the host preferences is very important in distinguishing species in this genus.

Supplementary Materials: The following are available online at https://www.mdpi.com/article/10.3390/jof7090732/s1, Alignment S1: alignment of multi-locus dataset; Alignment S2: alignment of ITS dataset.

Author Contributions: Conceptualization: Z.L.Y., Y.-C.L. and X.M.; field sampling: Z.L.Y., Y.-C.L., G.W. and G.-S.W.; molecular experiments and data analysis: X.M. and P.-M.W.; original draft—

writing: X.M.; original draft—review and editing: Y.-C.L. All authors have read and agreed to the published version of the manuscript.

Funding: This research was funded by the National Natural Science Foundation of China (Nos. 31872618, 31750001, 32070024, 31970015), the Natural Science Foundation of Yunnan Province (2018FB027), the Ten Thousand Talents Program of Yunnan (YNWR-QNBJ-2018-125), and the Key Research Program of Frontier Sciences, CAS (QYZDY-SSW-SMC029).

Institutional Review Board Statement: Not applicable.

Informed Consent Statement: Not applicable.

Data Availability Statement: Publicly available datasets were analyzed in this study. This data can be found here: https://www.ncbi.nlm.nih.gov/; http://www.mycobank.org/; https://www.treebase.org/treebase-web/home.html, accessed on 26 August 2021.

Acknowledgments: The authors thank X. H. Wang, Z. W. Ge, B. Feng and Q. Cai, Kunming Institute of Botany, CAS, for providing samples and/or related literatures.

Conflicts of Interest: The authors declare no conflict of interest.

References

1. Smith, A.H.; Thiers, H.D.; Watling, R. Notes on species of *Leccinum*. I. Additions to section *Leccinum*. *Lloydia* **1968**, *31*, 252–267.
2. Engel, H.; Krieglsteiner, G.J. *Rauhstielröhrlinge–die Gattung Leccinum in Europa*; Hilmar Schneider: Coburg, Germany, 1978; p. 76.
3. Singer, R. *The Agaricales in Modern Taxonomy*, 4th ed.; Koeltz Scientific Books: Koenigstein, Germany, 1986; pp. 1–981.
4. Binder, M.; Besl, H. 28S rDNA sequence data and chemotaxonomical analyses on the generic concept of *Leccinum* (Boletales). In *Micologia*; Associazone Micologica Bresadola, Ed.; Grafica Sette: Brescia, Italy, 2000; pp. 75–86. [CrossRef]
5. Kuo, M.; Ortiz-Santana, B. Revision of leccinoid fungi, with emphasis on North American taxa, based on molecular and morphological data. *Mycologia* **2020**, *112*, 1–15. [CrossRef]
6. Corner, E.J.H. *Boletus in Malaysia*; Government Printer: Singapore, 1972; p. 263.
7. Nagasawa, E. A preliminary checklist of the Japanese agaricales, 1: The boletineae. *Rep. Tottori. Mycol. Inst.* **1997**, *35*, 39–78.
8. Takahashi, H. Five new species of the Boletaceae from Japan. *Mycoscience* **2007**, *48*, 90–99. [CrossRef]
9. Terashima, Y.; Takahashi, H.; Taneyama, Y. *The Fungal Flora in Southwestern Japan: Agarics and Boletes*; Tokai University Press: Tokyo, Japan, 2016; pp. 1–303.
10. Katumoto, K. *List of Fungi Recorded in Japan*; Kanto Branch of the Mycological Society of Japan: Kyoto, Japan, 2010; pp. 1–954.
11. Wu, G.; Li, Y.C.; Zhu, X.T.; Zhao, K.; Han, L.H.; Cui, Y.Y.; Li, F.; Xu, J.; Yang, Z.L. One hundred noteworthy boletes from China. *Fungal Divers.* **2016**, *81*, 25–188. [CrossRef]
12. Li, T.H.; Song, B. Chinese boletes: A comparison of boreal and tropical elements. In Proceedings of the Tropical Mycology 2000, the Millennium Meeting on Tropical Mycology (Main Meeting 2000), British Mycological Society & Liverpool John Moores University, Liverpool, UK, 25–29 April 2000.
13. Gray, S.F.; Gray, J.E.; Shury, J. *A Natural Arrangement of British Plants*; Cradock and Joy: London, UK, 1821; pp. 1–647.
14. Den Bakker, H.C.; Noordeloos, M.E. A revision of European species of *Leccinum* Gray and notes on extralimital species. *Persoonia* **2005**, *18*, 511–587.
15. Smith, A.H.; Thiers, H.D.; Watling, R. A preliminary account of the North American species of *Leccinum*, Section *Leccinum*. *Mich. Bot.* **1966**, *5*, 131–178.
16. Smith, A.H.; Thiers, H.D.; Watling, R. A Preliminary account of the North American species of *Leccinum*, Sections *Luteoscabra* and *Scabra*. *Mich. Bot.* **1967**, *6*, 107–179.
17. Smith, A.H.; Thiers, H.D. *The Boletes of Michigan*; University of Michigan Press: Ann Arbor, MI, USA, 1971; p. 428.
18. Thiers, H.D. California *Boletes*. IV. The Genus *Leccinum*. *Mycologia* **1971**, *63*, 261–276. [CrossRef]
19. Both, E. *The Boletes of North America, a Compendium*; Buffalo Museum of Science: Buffalo, NY, USA, 1993; pp. 1–436.
20. Ortiz-Santana, B.; Halling, R.E. A new species of *Leccinum* (Basidiomycota, Boletales) from Belize. *Brittonia* **2009**, *61*, 172–174. [CrossRef]
21. Halling, R.E.; Mueller, G.M. *Leccinum* (Boletaceae) in Costa Rica. *Mycologia* **2003**, *95*, 488–499. [CrossRef]
22. Halling, R.E.; Mueller, G.M. *Common Mushrooms of the Talamanca Mountains, Costa Rica*; New York Botanical Garden Press: New York, NY, USA, 2005; pp. 1–197.
23. Halling, R.E. New *Leccinum* from Costa Rica. *Kew Bull.* **1999**, *54*, 747–753. [CrossRef]
24. Halling, R.E. A synopsis of Colombian boletes. *Mycotaxon* **1989**, *34*, 93–113.
25. Assyov, B.; Denchev, C.M. Preliminary checklist of Boletales s. str. in Bulgaria. *Mycol. Balc.* **2004**, *1*, 195–208.
26. Segedin, B.P.; Pennycook, S.R. A nomenclatural checklist of agarics, boletes, and related secotioid and gasteromycetous fungi recorded from New Zealand. *N. Z. J. Bot.* **2001**, *39*, 285–348. [CrossRef]
27. Watling, R. Australian boletes: Their diversity and possible origins. *Aus. Syst. Bot.* **2001**, *14*, 407–416. [CrossRef]
28. Segedin, P.B. An annotated checklist of Agarics and Boleti recorded from New Zealand. *N. Z. J. Bot.* **1987**, *25*, 185–215. [CrossRef]

29. Mcnabb, R.F.R. The Boletaceae of New Zealand. *N. Z. J. Bot.* **1968**, *6*, 137–176. [CrossRef]
30. Peck, C.H. Report of the botanist. *Ann. Rep. N. Y. St. Mus. Nat. Hist.* **1888**, *41*, 51–122.
31. Šutara, J. The delimitation of the genus *Leccinum*. *Czech. Mycol.* **1989**, *43*, 1–12. [CrossRef]
32. Bresinsky, A.; Besl, H. Beiträge zu einer Mykoflora Deutschlands: Schlüssel zur Gattungsbestimmung der Blätter-, Leisten- und Röhrenpilze mit Literaturhinweisen zur Artbestimmung. *Regensb. Mykol. Schr.* **2003**, *11*, 1–236.
33. Den Bakker, H.C.; Zuccarello, G.C.; Kuyper, T.W.; Noordeloos, M.E. Evolution and host specificity in the ectomycorrhizal genus *Leccinum*. *New Phytol.* **2004**, *163*, 201–215. [CrossRef]
34. Zang, M. Notes on the boletales from Eastern Himalayas and adjacent areas of China. *Acta Bot. Yunnanica* **1986**, *8*, 1–22.
35. Wu, G.; Zhao, K.; Li, Y.C.; Zeng, N.K.; Feng, B.; Halling, R.E.; Yang, Z.L. Four new genera of the fungal family Boletaceae. *Fungal Divers.* **2016**, *81*, 1–24. [CrossRef]
36. Wu, K.; Wu, G.; Yang, Z.L. A taxonomic revision of *Leccinum rubrum* in subalpine coniferous forests, Southwestern China. *Acta Edulis Fungi* **2020**, *72*, 92–100.
37. Halling, R.E.; Fechner, N.; Nuhn, M.; Osmundson, T.; Soytong, K.; Arora, D.; Binder, M.; Hibbett, D. Evolutionary relationships of *Heimioporus* and *Boletellus* (Boletales) with an emphasis on Australian taxa including new species and new combinations in *Aureoboletus*, *Hemileccinum*, and *Xerocomus*. *Aust. Syst. Bot.* **2015**, *28*, 1–22. [CrossRef]
38. Chiu, W.F. The boletes of Yunnan. *Mycologia* **1948**, *40*, 199–231. [CrossRef]
39. Mao, X.L.; Jiang, C.P.; Ouzhu, C.W. *Economic Macrofungi of Tibet*; Beijing Science and Technology Press: Beijing, China, 1993; pp. 1–651. (In Chinese)
40. Mao, X.L. *The Economic Fungi in China*; Science Press: Beijing, China, 1998; pp. 1–762. (In Chinese)
41. Mao, X.L. *The Macrofungi of China*; Henan Science and Technology Press: Zhengzhou, China, 2000; pp. 1–719. (In Chinese)
42. Wu, X.L.; Zang, M.; Xia, T.Y. *Coloured Illustrations of the Ganodermataceae and other Fungi*; Guizhou Science & Technology Press: Guiyang, China, 1997; pp. 1–358. (In Chinese)
43. Ying, J.Z.; Wen, H.A.; Zong, Y.C. *The Economic Macromycetes from Western Sichuan*; Science Press: Beijing, China, 1994; pp. 1–399. (In Chinese)
44. Bi, Z.S.; Zheng, G.Y.; Li, T.H.; Wang, Y.Z. *Macrofungus Flora of the Mountainous District of North. Guangdong*; Guangdong Science & Technology Press: Guangzhou, China, 1990; pp. 1–450. (In Chinese)
45. Yuan, M.S.; Sun, P.Q. *Sichuan Mushrooms*; Sichuan Science and Technology Press: Chengdu, China, 1995; pp. 1 737. (In Chinese)
46. Bi, Z.S.; Li, T.H.; Zhang, W.M.; Song, B. *A Preliminary Agaric Flora of Hainan Province*; Guangdong Higher Education Press: Guangzhou, China, 1997; pp. 1–388. (In Chinese)
47. Yeh, K.W.; Chen, Z.C. The boletes of Taiwan, I. *Taiwania* **1980**, *25*, 166–184.
48. Bi, Z.S.; Li, T.H.; Zheng, G.Y.; Li, C. Basidiomycetes from Dinghu Mountain of China. III. Some species of Boletaceae (2). *Acta Myco. Sinica* **1984**, *3*, 199–206. (In Chinese)
49. Song, B.; Li, T.H.; Wu, X.L.; Shen, Y.H. A preliminary review on Boletales resources in Yunnan, Guizhou and Guangxi, China. *Guizhou Sci.* **2004**, *22*, 91–97.
50. Fu, S.Z.; Wang, Q.B.; Yao, Y.J. An annotated checklist of *Leccinum* in china. *Mycotaxon* **2006**, *96*, 47–50.
51. Li, T.H.; Song, B. Bolete species known from China. *Guizhou Sci.* **2003**, *21*, 78–86.
52. Li, Y.C.; Feng, B.; Yang, Z.L. *Zangia*, a new genus of Boletaceae supported by molecular and morphological evidence. *Fungal Divers.* **2011**, *49*, 125–143. [CrossRef]
53. Halling, R.E.; Nuhn, M.E.; Fechner, N.; Osmundson, T.; Soytong, K.; Arora, D.; Hibbet, D.S.; Binder, M. *Sutorius*: A new genus for *Boletus eximius* (Boletaceae). *Mycologia* **2012**, *104*, 951–961.
54. Halling, R.E.; Nuhn, M.E.; Osmundson, T.; Fechner, N.; Trappe, J.M.; Soytong, K.; Arora, D.; Hibbett, D.S.; Binder, M. Affinities of the *Boletus chromapes* group to *Royoungia* and the description of two new genera, *Harrya* and *Australopilus*. *Austral. Syst. Bot.* **2012**, *25*, 418–431. [CrossRef]
55. Nuhn, M.E.; Binder, M.; Taylor, A.F.S.; Halling, R.E.; Hibbett, D.S. Phylogenetic overview of the boletineae. *Fungal Biol.* **2013**, *117*, 479–511. [CrossRef] [PubMed]
56. Wu, F.; Zhou, L.W.; Yang, Z.L.; Bau, T.; Li, T.H.; Dai, Y.C. Resource diversity of Chinese macrofungi: Edible, medicinal and poisonous species. *Fungal Divers.* **2019**, *98*, 1–76. [CrossRef]
57. Den Bakker, H.C.; Gravendeel, B.; Kuyper, T.W. An ITS Phylogeny of *Leccinum* and an Analysis of the Evolution of Minisatellite-like Sequences within ITS1. *Mycologia* **2004**, *96*, 102–118. [CrossRef]
58. Kornerup, A.; Wanscher, J.H. *Methuen Handbook of Colour*, 3rd ed.; Eyre Methuen: London, UK, 1983; pp. 1–252.
59. Zhou, M.; Dai, Y.C.; Vlasák, J.; Yuan, Y. Molecular Phylogeny and Global Diversity of the Genus *Haploporus* (Polyporales, Basidiomycota). *J. Fungi* **2021**, *7*, 96. [CrossRef]
60. Doyle, J.J.; Doyle, J.L. A rapid DNA isolation procedure for small quantities of fresh leaf tissue. *Phytochem. Bullet.* **1987**, *19*, 11–15.
61. Vilgalys, R.; Hester, M. Rapid genetic identification and mapping of enzymatically amplified ribosomal DNA from several *Cryptococcus* species. *J. Bact.* **1990**, *172*, 4238–4246. [CrossRef]
62. Rehner, S.A.; Buckley, E. A *Beauveria* phylogeny inferred from nuclear ITS and *EF1-α* sequences: Evidence for cryptic diversification and links to *Cordyceps teleomorphs*. *Mycologia* **2005**, *97*, 84–98. [CrossRef] [PubMed]
63. Matheny, P.B. Improving phylogenetic inference of mushrooms with *rpb1* and *rpb2* nucleotide sequences (*Inocybe*, Agaricales). *Mol. Phylogenet. Evol.* **2005**, *35*, 1–20. [CrossRef]

64. White, T.J.; Bruns, T.; Lee, S.; Taylor, J. Amplification and direct sequencing of fungal ribosomal RNA genes for phylogenetics. In *PCR Protocols. A Guide to Methods and Applications*; Innis, M.A., Gelfand, D.H., Sninsky, J.J., White, T.J., Eds.; Academic Press: San Diego, CA, USA, 1990; pp. 315–322. [CrossRef]
65. Wu, G.; Feng, B.; Xu, J.; Zhu, X.T.; Li, Y.C.; Zeng, N.K.; Hosen, M.I.; Yang, Z.L. Molecular phylogenetic analyses redefine seven major clades and reveal 22 new generic clades in the fungal family Boletaceae. *Fungal Divers.* **2014**, *69*, 93–115. [CrossRef]
66. Katoh, K.; Standley, D.M. MAFFT multiple sequence alignment software version 7: Improvements in performance and usability. *Mol. Biol. Evol.* **2013**, *30*, 772–780. [CrossRef] [PubMed]
67. Hall, T.A. BioEdit: A user-friendly biological sequence alignment editor and analyses program for Windows 95/98/NT. *Nucleic Acids Symp. Ser.* **1999**, *41*, 95–98.
68. Stamatakis, A.; Hoover, P.; Rougemont, J. A rapid bootstrap algorithm for the RAxML web-servers. *Syst. Biol.* **2008**, *75*, 758–771. [CrossRef]
69. Posada, D.; Buckley, T.R. Model selection and model averaging in phylogenetics: Advantages of the AIC and Bayesian approaches over likelihood ratio tests. *Syst. Biol.* **2004**, *53*, 793–808. [CrossRef] [PubMed]
70. Huelsenbeck, J.P.; Ronquist, F. Bayesian analysis of molecular evolution using MrBayes. In *Statistical Methods in Molecular Evolution*; Nielsen, R., Ed.; Springer: New York, NY, USA, 2005; pp. 183–232. [CrossRef]
71. Desjardin, D.E.; Binder, M.; Roekring, S.; Flegel, T. *Spongiforma*, a new genus of gasteroid boletes from Thailand. *Fungal Divers.* **2009**, *37*, 1–8. [CrossRef]
72. Hosen, M.I.; Feng, B.; Wu, G.; Zhu, X.T.; Li, Y.C.; Yang, Z.L. *Borofutus*, a new genus of Boletaceae from tropical Asia: Phylogeny, morphology and taxonomy. *Fungal Divers.* **2013**, *58*, 215–226. [CrossRef]
73. Orihara, T.; Smith, M.E.; Shimomura, N.; Iwase, K.; Maekawa, N. Diversity and systematics of the sequestrate genus *Octaviania* in Japan: Two new subgenera and eleven new species. *Persoonia* **2012**, *28*, 85–112. [CrossRef] [PubMed]
74. Binder, M.; Hibbett, D. Molecular systematics and biological diversification of Boletales. *Mycologia* **2006**, *98*, 71–81. [CrossRef]
75. Raspé, O.; Vadthanarat, S.; Kesel, A.D.; Degreef, J.; Hyde, K.D.; Lumyong, S. *Pulveroboletus fragrans*, a new Boletaceae species from Northern Thailand, with a remarkable aromatic odor. *Mycol. Progr.* **2016**, *15*, 38. [CrossRef]
76. Orihara, T.; Healy, R.; Corrales, A.; Smith, M.E. Multi-locus phylogenies reveal three new truffle-like taxa and the traces of interspecific hybridization in *Octaviania* (Boletales). *IMA Fungus* **2021**, *12*, 1–22. [CrossRef]
77. Orihara, T.; Lebel, T.; Ge, Z.W.; Smith, M.E.; Maekawa, N. Evolutionary history of the sequestrate genus *Rossbeevera* (Boletaceae) reveals a new genus *Turmalinea* and highlights the utility of ITS minisatellite-like insertions for molecular identification. *Persoonia* **2016**, *37*, 173–198. [CrossRef] [PubMed]
78. Den Bakker, H.C.; Zuccarello, G.C.; Kuyper, T.W.; Noordeloos, M.E. Phylogeographic patterns in *Leccinum* sect. *Scabra and the status of the arctic-alpine species L. rotundifoliae*. *Mycol. Res.* **2007**, *111*, 663–672. [CrossRef] [PubMed]
79. Hosen, M.I.; Zhong, X.J.; Gates, G.; Orihara, T.; Li, T.H. Type studies of *Rossbeevera bispora*, and a new species of *Rossbeevera* from south China. *MycoKeys* **2019**, *51*, 15–28. [CrossRef]
80. Watling, R. *British Fungus Flora—Agarics and Boleti. I. Boletaceae, Gomphidiaceae, Paxillaceae*; Her Majesty's Stationery Office: Edinburgh, UK, 1970.
81. Rostkovius, F.W.T. Die Pilze Deutschlands. In *Deutschlands Flora in Abbildungen nach der Natur mit Beschreibungen*; Sturm, J., Ed.; Gedruckt auf Kosten des Verfassers: Nürnberg, Germany, 1832; Part III; Volume 5, p. 132. [CrossRef]
82. Watling, R. British records. *Trans. Br. mycol. Soc.* **1960**, *43*, 691–694.
83. Noordeloos, M.E.; Kuyper, T.W.; Somhorst, I.; Vellinga, E.C. *Flora Agaricina Neerlandica*; Lidia Carla Candusso: Varese, Italia, 2018; Volume 7, pp. 178–181.
84. Horak, E. Revision of Malaysian species of Boletales s.l. (Basidiomycota) described by E.J.H. Corner (1972, 1974). *Malay. For. Rec.* **2011**, *51*, 1–283.

Journal of
Fungi

MDPI

Article

Beneficial Features of Biochar and Arbuscular Mycorrhiza for Improving Spinach Plant Growth, Root Morphological Traits, Physiological Properties, and Soil Enzymatic Activities

Dilfuza Jabborova [1,2,*], Kannepalli Annapurna [2,*], Sangeeta Paul [2], Sudhir Kumar [3], Hosam A. Saad [4], Said Desouky [5], Mohamed F. M. Ibrahim [6] and Amr Elkelish [7]

1   Institute of Genetics and Plant Experimental Biology, Uzbekistan Academy of Sciences, Tashkent Region, Kibray 111208, Uzbekistan
2   Division of Microbiology, ICAR-Indian Agricultural Research Institute, Pusa, New Delhi 110012, India; sangeeta_paul2003@yahoo.co.in
3   Division of Plant Physiology, ICAR-Indian Agricultural Research Institute, New Delhi 110012, India; sudhir_physiol@iari.res.in
4   Department of Chemistry, College of Science, Taif University, P.O. Box 11099, Taif 21944, Saudi Arabia; h.saad@tu.edu.sa
5   Department of Botany and Microbiology, Faculty of Science, Al-azhar University, Cairo 11884, Egypt; said.desouky@azhar.edu.eg
6   Department of Agricultural Botany, Faculty of Agriculture, Ain Shams University, Cairo 11566, Egypt; Ibrahim_mfm@agr.asu.edu.eg
7   Botany Department, Faculty of Science, Suez Canal University, Ismailia 41522, Egypt; amr.elkelish@science.suez.edu.eg
*   Correspondence: dilfuzajabborova@yahoo.com (D.J.); annapurna96@gmail.com (K.A.)

Citation: Jabborova, D.; Annapurna, K.; Paul, S.; Kumar, S.; Saad, H.A.; Desouky, S.; Ibrahim, M.F.M.; Elkelish, A. Beneficial Features of Biochar and Arbuscular Mycorrhiza for Improving Spinach Plant Growth, Root Morphological Traits, Physiological Properties, and Soil Enzymatic Activities. *J. Fungi* **2021**, 7, 571. https://doi.org/10.3390/jof7070571

Academic Editors: Anush Kosakyan, Rodica Catana and Alona Biketova

Received: 11 June 2021
Accepted: 14 July 2021
Published: 17 July 2021

Publisher's Note: MDPI stays neutral with regard to jurisdictional claims in published maps and institutional affiliations.

Abstract: Biochar and arbuscular mycorrhizal fungi (AMF) can promote plant growth, improve soil properties, and maintain microbial activity. The effects of biochar and AMF on plant growth, root morphological traits, physiological properties, and soil enzymatic activities were studied in spinach (*Spinacia oleracea* L.). A pot experiment was conducted to evaluate the effect of biochar and AMF on the growth of spinach. Four treatments, a T1 control (soil without biochar), T2 biochar alone, T3 AMF alone, and T4 biochar and AMF together, were arranged in a randomized complete block design with five replications. The biochar alone had a positive effect on the growth of spinach, root morphological traits, physiological properties, and soil enzymatic activities. It significantly increased the plant growth parameters, such as the shoot length, leaf number, leaf length, leaf width, shoot fresh weight, and shoot dry weight. The root morphological traits, plant physiological attributes, and soil enzymatic activities were significantly enhanced with the biochar alone compared with the control. However, the combination of biochar and AMF had a greater impact on the increase in plant growth, root morphological traits, physiological properties, and soil enzymatic activities compared with the other treatments. The results suggested that the combined biochar and AMF led to the highest levels of spinach plant growth, microbial biomass, and soil enzymatic activity.

Keywords: spinach; biochar; AMF; plant growth; root morphological traits; chlorophyll content; soil enzymes and microbial biomass

1. Introduction

Biochar is the carbon-rich material taken by pyrolysis using various biomasses. Biochar plays an important role in decreasing global warming in the world and reducing atmospheric CO_2 concentrations, as well as improving soil used in agriculture [1–6]. Biochar application has been noted to increase the activity of microbes in the soil and improve the physical and chemical properties of the soil [7–13]. Ścisłowska et al. [14] observed that biochar treatments improved the quality and productivity of soils. Laird et al. [15] reported that biochar treatments significantly increased the water-holding capacity, cation exchange capacity, and specific surface area of soils. Rice husk biochar significantly increased the organic matter in soil by 52.94% compared with the control [16]. Numerous reports have indicated that biochar alone increased the activities of dehydrogenase, alkaline phosphatase, acid phosphatase, acid phosphomonoesterase, alkaline phosphomonoesterase, protease, chymotrypsin, trypsin, phosphohydrolase, lipase-esterase, and esterase enzymes [17–19].

Biochar also plays an important role in soil nutrient availability and adsorption. Biochar promotes soil nutrients, such as nitrogen, potassium, calcium, magnesium, sodium, and total carbon [20,21]. Jabborova et al. [18] reported that the addition of biochar to soil significantly increased the content of nitrogen, phophorus, and potassium compared to the control.

Biochar had a positive effect on the growth, development, yield, and nutrient content of different plants. Several reports have indicated that biochar increased the seed germination, plant growth, and yield of soybeans [8,22–25]. The germination rate was highest when biochar was used compared to the control [20]. Concentrations of calcium and magnesium in maize leaves were significantly higher with a high biochar application rate than with the control [5]. Biochar application improved the root and shoot biomass of *Plantago lanceolata* [26]. Rice straw biochar significantly increased the plant height, number of bolls per plant, average boll weight, and seed cotton yield compared to the control [16]. Carter et al. [27] reported that rice husk biochar application increased the final biomass, root biomass, plant height, and number of leaves of lettuce (*Lactuca sativa*) and cabbage (*Brassica chinensis*) in comparison with plants that had not received biochar.

Biochar has a positive impact on plant physiological and biochemical properties. Several studies have shown that biochar application increased the plant photosynthesis, chlorophyll content, and transpiration rate [19,28,29]. Biochar amendment significantly increased the photosynthetic rate of okra (*Abelmoschus esculentus* L.) [25].

Arbuscular mycorrhizal fungi (AMF) is a major component of the rhizosphere microflora in natural ecosystems, and plays a significant role in ecosystems through nutrient cycling [30,31]. Mycorrhiza are microbes that promote plant growth and play an important role in enhancing plant nutrition [32–34]. AMF have a symbiotic relationship with plants. Inoculating crop roots with mycorrhiza improves the uptake of nutrients, such as nitrogen, potassium, phosphorus, calcium, and magnesium [35–38]. The plants inoculated with mycorrhiza had increased chlorophyll and carotenoid content and increased antioxidant enzymes, such as superoxide, dismutase, catalase, peroxidase, and ascorbate peroxidase [39]. The use of mycorrhiza helped to improve the branching of the plant root system and the growth and productivity of several field crops [40–42]. Several researchers have reported that inoculation with both mycorrhiza and plant growth-promoting rhizobacteria (PGPR) could be beneficial for agriculture [43,44]. Biochar and mycorrhiza have proved useful for enhancing plant growth and yield and reducing the intensity of disease. IIA [45] reported that dual inoculation with AMF and biochar significantly increased plant growth and the phosphorus content of maize.

Spinach (*Spinacia oleracea* L.) is high in nutrients that benefit humans, including bioactive compounds, vitamins, and minerals [46]. It is rich in bioactive compounds that contribute to anticancer, antiobesity, hypoglycemic, and hypolipidemic properties [47]. The production of spinach is largely affected by abiotic and biotic stress. In addition to abiotic stress, the combined application of biochar and AMF is also helpful for alleviating the negative impact of biotic stress [48]. However, little information is available on the interactive effect of biochar and AMF on spinach. In this study, we investigated the effect of the combined application of biochar and AMF on spinach growth, root morphological traits, physiological properties, and soil enzymatic activities. We hypothesized that the combined application of biochar and AMF would facilitate beneficial effects on plant growth, plant nutrients, physiological properties, and soil properties.

2. Materials and Methods

Soil collected from the Indian Agricultural Research Institute (IARI) was used for the experiment. The biochar used in the study was produced at 400 to 500 °C from a woody biomass (Amazon online shop, New Delhi, India), with a particle size of less than 2 mm. Seeds were obtained from the Division of Vegetable Science, IARI, New Delhi, India, and AMF were obtained from the Division of Microbiology, IARI, New Delhi, India.

2.1. Experimental Design

The effect of biochar and AMF on the growth of spinach was studied in pot experiments in a nethouse. All of the experiments were carried out in a randomized block design with five replications. Experimental treatments included a T1 control (soil without biochar), T2 = biochar alone, T3 = AMF alone, and T4 = biochar + AMF. Seeds were sown in plastic pots (20 cm in diameter and 20 cm in depth) containing 5.0 kg of soil. Each pot was watered every three days. At harvest, after 40 days, the shoot length, leaf length, leaf number, leaf width, fresh root weight, fresh shoot weight, dry root weight, and dry shoot weight were measured. Physiological parameters, such as relative water content, net photosynthetic rate, stomatal conductance, and transpiration rate, as well as photosynthetic pigments, were also determined after 40 days.

2.2. Measurement of Root Morphological Traits of Spinach

The roots were washed carefully with water. The whole root system was spread out and analyzed using a scanning system (Expression 4990, Epson, CA) with a blue board as a background. Digital images of the root system were analyzed using Win RHIZO software (Régent Instruments, Québec, QC, Canada). The total root length, root surface area, root volume, projected area, and root diameter were evaluated.

2.3. Physiological Parameters Measurement

Relative water content (RWC) was measured by the method of Abd El-Gawad, et al. [49]. One hundred mg of fully expanded fresh leaf sample (FW) were placed immediately after sampling in petri plates filled with double distilled water for 4 h at room temperature. The samples were then taken out and blotted dry, and the turgid weight (TW) was recorded. The samples were kept in an oven at 70 °C overnight, and the dry weight (DW) was recorded. Relative water content was calculated as:

$$RWC\ (\%) = [(FW - DW)/(TW - DW)] \times 100$$

Photosynthetic pigments were determined by the modified method of Hiscox and Israelstam [50]. Fresh leaves were collected in the morning. Fifty mg of fine pieces of fresh leaf sample 2 to 3 mm in size were cut and added to test tubes containing 5 mL of DMSO. Then the test tubes were incubated at 37 °C for 4 h in the dark. The incubation was continued until completely colorless tissue was obtained. The absorbance of the extract was taken at 470 nm, 645 nm, and 663 nm using a spectrophotometer against a DMSO blank. The chlorophyll a (Chl a), chlorophyll b (Chl b), total chlorophyll, and carotenoid contents were determined using the following equations:

$$Chl\ a\ (mg/g) = [12.7(A_{663}) - 2.69(A_{645})] \times V/W$$

$$Chl\ b\ (mg/g) = [22.9(A_{645}) - 4.68(A_{663})] \times V/W$$

$$Total\ Chl\ (mg/g) = [20.2\ (A_{645}) + 8.02\ (A_{663})] \times V/W$$

$$Carotenoids\ (mg/g) = [(1000 \times A_{470}) - (3.27 \times Chl\ a + 104 \times Chl\ b)] \times V/W$$

where, A = optical density; V = Volume of DMSO (in mL); W = sample weight.

The net photosynthetic rate, stomatal conductance, and transpiration rate were measured using a portable LI-6400XT photosynthesis measurement system between 9:00–11:00 a.m. The fully expanded youngest leaf was used for the measurement. The photosynthetic active radiation (PAR), temperature, and CO_2 concentration during the measurements were 300 mmol m^{-2} s^{-1}, 30 °C, and 400 mmol mol^{-1}, respectively.

2.4. Analysis of AMF Spores from Soil

The AMF spores were extracted from 10 g soil samples using wet sieving and decanting method. The soil sample was put over a series of soil sieves arranged in descending order of sieve sizes. The clean spores were mesh sieved and washed several times with distilled water before being transferred into water in a clean petri dish. The AMF spores were counted under a stereomicroscope [51].

2.5. Analysis of Soil Microbial Biomass Determination

The methods used to measure biomass C were based on those described by Vance et al. [52]. Three of six 17.5 g replicates of each soil sample were fumigated with purified $CHCl_3$ for 24 h. After removal of the $CHCl_3$, the C was extracted from fumigated and unfumigated samples with 0.5 M of K_2SO_4 for 1 h on an end-over-end shaker. Fumigated and unfumigated samples were filtered sequentially through filter paper (Whatman filter grade 42). The obtained supernatant liquid was measured at 280 nm using a spectrophotometer.

2.6. Analysis of Soil Enzymes

The alkaline phosphatase activities were assayed with the method by Tabatabai and Bremner [53]. For each soil, two sets of 1 g of soil were placed in conical flasks. One set was used as the control. Then, 0.2 mL of toluene and 4 mL of MUB (modified universal buffer) (pH = 11) were added, and 1 mL of *p*-nitrophenyl phosphate solution was added to the other set of samples. After swirling both flasks for a few seconds to mix the contents, these were placed in an incubator at 37 °C for 1 h. Calcium chloride (1 mL of 0.5 M) and 4 mL of 0.5 M of NaOH were added after incubation. Flasks were swirled for a few seconds, and 1 mL of *p*-nitrophenyl phosphate solution was added to the remaining set of samples. All suspensions were filtered through Whatman No. 1 filter paper quickly, and the yellow color intensity was measured at a 440 nm wavelength.

The fluorescein diacetate (FDA) hydrolytic activity was determined by the method of Green et al. [54]. A total of 0.5 mg of soil was added to 25 mL of sodium phosphate (0.06 M; pH = 7.6). To all assay vials, 0.25 mL of 4.9 mM of FDA substrate solution was added. All vials were vortexed and incubated in a water bath at 37 °C for 2 h. Then, soil suspension was centrifuged at 8000 rpm for 5 min. The clear supernatant was measured at 490 nm against a reagent blank solution using a spectrophotometer.

Dehydrogenase activity (DHA) was determined using the method described by Casida Jr. et al. [55]. Fresh homogenized soil samples (5 g) were placed in test tubes, and 5 mL of substrate 3% *v/w* 2,3,5-triphenyltetrazolium chloride (TTC) was added. The tubes were incubated at 25 °C for 24 h. A blank sample was similarly prepared, with 1 mL of a 3% TTC solution phosphate buffer being introduced. After incubation, the samples were centrifuged at 4500 rpm for 10 min. The supernatant liquid was discarded. The formed triphenyl-formazan (TPF) was extracted with methanol. To each tube, 5 mL of methanol were added, and then the tubes were vigorously shaken for a few min. The operation was repeated twice (10 mL of methanol was used for extraction). Again, the tubes were centrifuged. The obtained supernatant liquid was poured into a clean tube, and the absorbance of the solution was measured at 485 nm.

2.7. Statistical Analyses

Experimental data were analyzed with StatView Software using ANOVA. The significance of the effect of treatment was determined by the magnitude of the F value ($p < 0.05 < 0.001$).

3. Results

AMF application significantly increased the leaf number and leaf width (Figure 1). Biochar treatment significantly increased the shoot length by 77%, leaf length by 43%, leaf number by 50%, and leaf width by 45% compared to the control. The combined biochar and AMF significantly increased the shoot length and leaf length by 54% and 53%, respectively, over the control. Similarly, the combined biochar and AMF had a positive effect on the leaf number and leaf width, with a 30% and 36% increase, respectively, compared to the control.

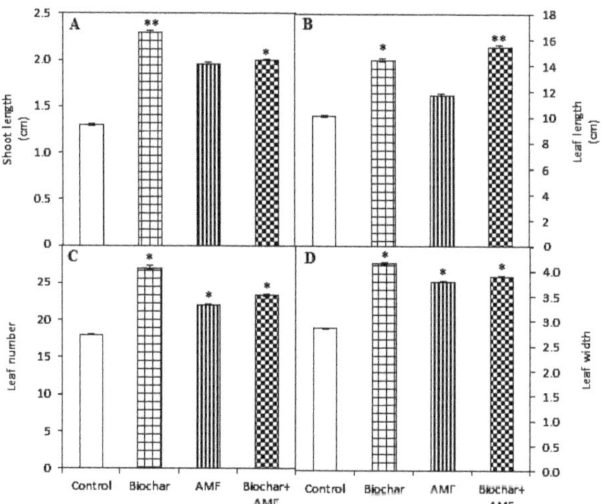

Figure 1. Biochar and AMF for the improvement of the shoot length (**A**), leaf length (**B**), leaf number (**C**), and leaf width (**D**) of spinach. Data are the means of three replicates ($n = 3$); * differed significantly at $p < 0.05$ *, $p < 0.01$ **.

The highest root and shoot fresh weight and dry weight were recorded with the biochar treatment and combined biochar and AMF (Figure 2). The root fresh weight (56%) and the root dry weight (52%) were significantly improved by the biochar treatment compared to the control (Figure 2). The biochar treatment significantly enhanced the shoot fresh weight by 38% and the shoot dry weight by 39% over the control. AMF treatment slowly increased the shoot fresh weight and shoot dry weight. AMF treatment significantly increased the root and shoot dry weight compared to the control. The combined application of biochar and AMF significantly enhanced the root fresh weight and dry weight by 45% and 45%, respectively, compared to the control. The shoot fresh weight (27%) and dry weight (28%) were also significantly enhanced by the combined application of biochar and AMF compared to the control.

AMF treatment enhanced the total root length, projected area, root diameter, and root volume by 43%, 35%, 50%, and 34%, respectively, compared to the control (Figure 3). Biochar treatment significantly increased the projected area and root volume by 63% and 59%, respectively, compared to the control. The total root length and root diameter were significantly enhanced by 76% and 80%, respectively, by the biochar treatment compared to the control. The combined biochar and AMF treatment significantly increased the total root length by 78% and root diameter by 90% over the control. The projected area and root volume were increased by 52% and 50%, respectively, by the combined biochar and AMF.

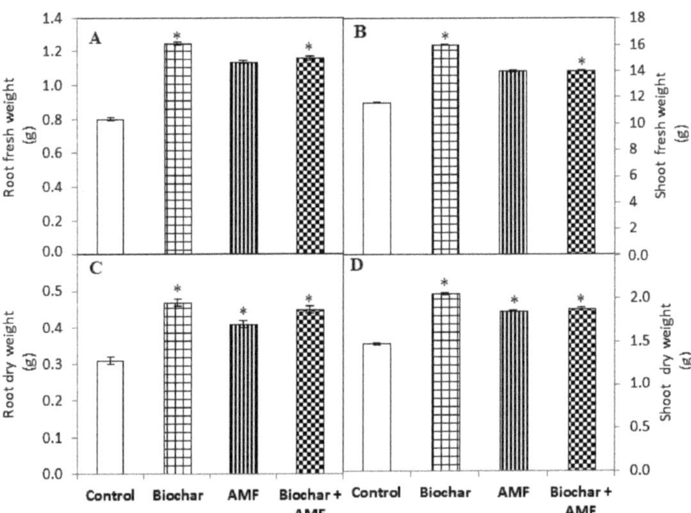

Figure 2. Biochar and AMF for the improvement of the root fresh weight (**A**), shoot fresh weight (**B**), root dry weight (**C**), and shoot dry weight (**D**) of spinach. Data are the means of three replicates (*n* = 3); * differed significantly at *p* < 0.05 *.

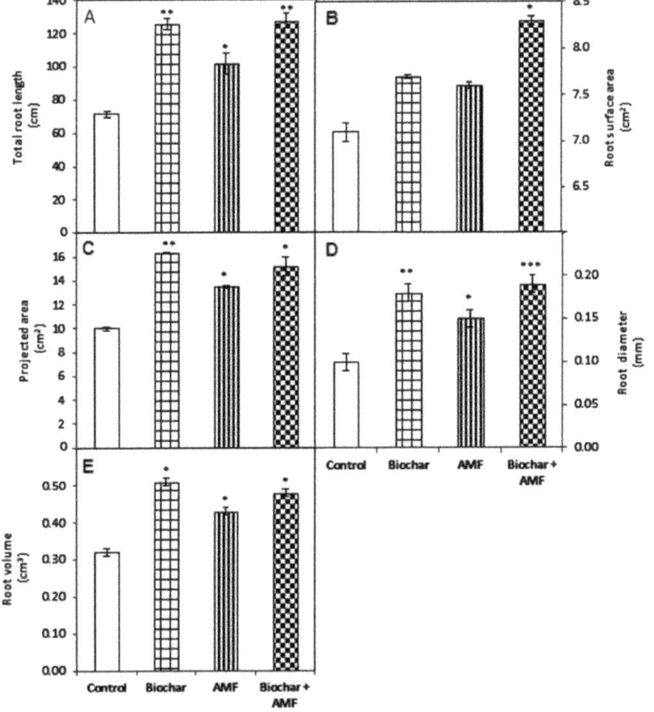

Figure 3. Biochar and AMF for the improvement of the root morphological traits of spinach. (**A**) Root length, (**B**) Root Surface are, (**C**) Root projected Area, (**D**) Root Diameter, (**E**) Root Volume. Data are the means of three replicates (*n* = 3); * differed significantly at *p* < 0.05 *, *p* < 0.01 **, *p* < 0.001 ***.

The biochar alone and the combined biochar and AMF significantly increased the photosynthetic rate by 50% and 74%, respectively, compared to the control (Figure 4). The stomatal conductance was significantly increased by the combined biochar and AMF compared to the control. The transpiration rate was significantly increased by all of the treatments compared with the control; the highest rate was recorded with the biochar treatment. Biochar treatment alone significantly enhanced the transpiration rate by 45% compared to the control.

All of the treatments improved the photosynthetic pigment content of the leaf compared to the control (Figure 5). The biochar significantly increased the content of total chlorophyll, chlorophyll a and b, and carotenoid of the leaf by 20%, 12%, 27%, and 46%, respectively, over the control (Figure 5). The AMF alone significantly increased the total chlorophyll content, chlorophyll a and b content, and carotenoid content of the leaf by 21%, 25%, 10%, and 33%, respectively. The combined treatment with biochar and AMF significantly increased the total chlorophyll content, chlorophyll a and b content, and carotenoid content of the leaf by 24%, 16%, 17%, and 38%, respectively, over the control.

All of the treatments, that is, the biochar alone, AMF alone, and biochar and AMF combined, increased the relative water content of the leaf compared to the control (Figure 6). The highest relative water content of the leaf was recorded for the combined biochar and AMF treatment; the content was 21% higher than with the control. In the treatment with biochar alone or AMF alone, there was an increase in the relative water content of 15% and 20%, respectively, compared to the control.

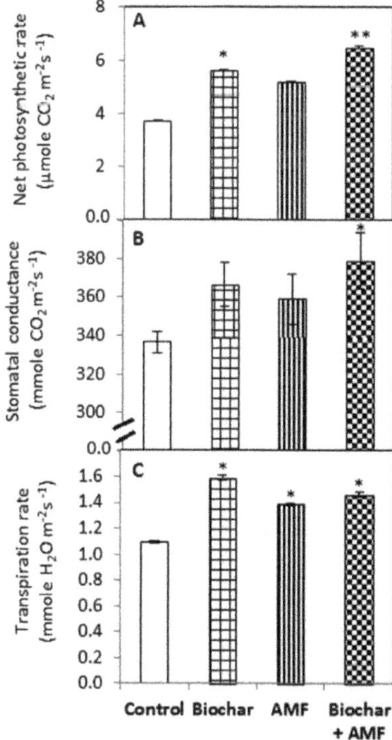

Figure 4. Biochar and AMF for the improvement of photosynthetic parameters. (**A**) Net photosynthesis, (**B**) Stomatal Conductance, (**C**) Transpiration rate. Data are the means of three replicates ($n = 3$); * differed significantly at $p < 0.05$ *, $p < 0.01$ **.

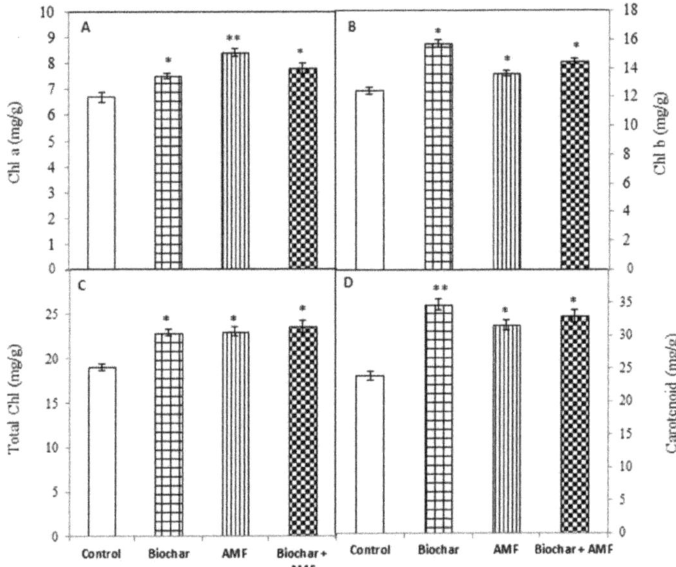

Figure 5. Biochar and AMF for the improvement of the photosynthetic pigment content of leaf (**A**) chlorophyll a (Chl a) content, (**B**) chlorophyll b (Chl b) content, (**C**) total chlorophyll content, and (**D**) carotenoid content. Data are the means of three replicates ($n = 3$); * differed significantly at $p < 0.05$ *, $p < 0.01$ **.

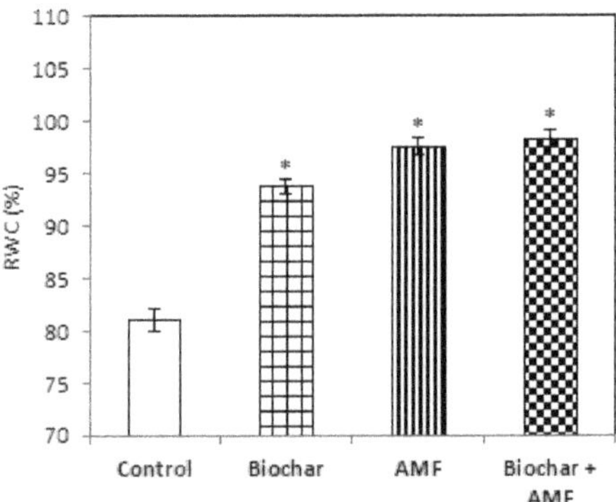

Figure 6. Biochar and AMF for improvement of the relative water content of the leaf. Data are the means of three replicates ($n = 3$); * differed significantly at $p < 0.05$ *.

Both the treatment with AMF alone and the treatment with AMF and biochar were more effective in increasing the AMF spores in soil than the control (Figure 7). The AMF spores in soil increased by 126% to 150% with the AMF alone and with the combined biochar and AMF compared to the control. There was an 82% increase in AMF spores in soil with the biochar treatment over the control.

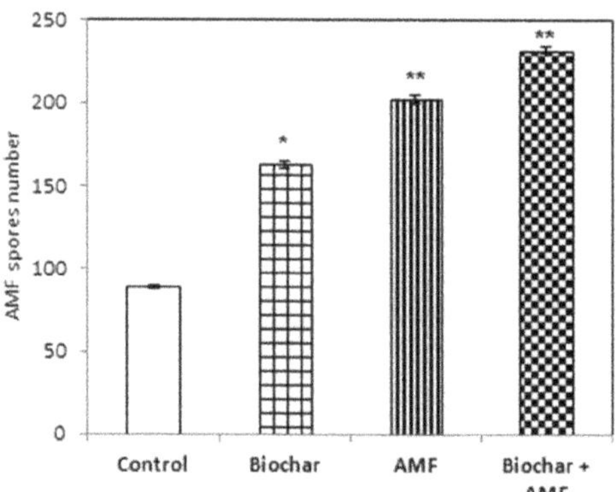

Figure 7. Biochar and AMF for the improvement of the AMF spores in soil. Data are the means of three replicates (*n* = 3); * differed significantly at $p < 0.05$ *, $p < 0.01$ **.

Both the treatment with biochar alone and the combined biochar and AMF treatment increased the carbon in the microbial biomass in soil compared to the control (Figure 8). The highest level of carbon was recorded with the combined biochar and AMF treatment compared with all other treatments; the level was 32% higher than with the control.

All of the treatments had a positive effect on the activity of alkaline phosphomonoes-trase and increased the enzymatic activity in different treatments; the range was 62% to 86% (Figure 9). The highest alkaline phosphomonoestrase acitivity in soil was recorded for the combined biochar and AMF treatment. The treatment, AMF alone (57%), and the treatment with biochar and AMF (55%) had a positive effect on the dehydrogenase activity of soil compared to the control. Furthermore, biochar and the treatment with combined biochar and AMF had a beneficial effect on the fluorescein diacetate activity of soil compared with the other treatments. The highest increase in the fluorescein diacetate activity was recorded for the combined biochar and AMF; it was 62% higher than with the control.

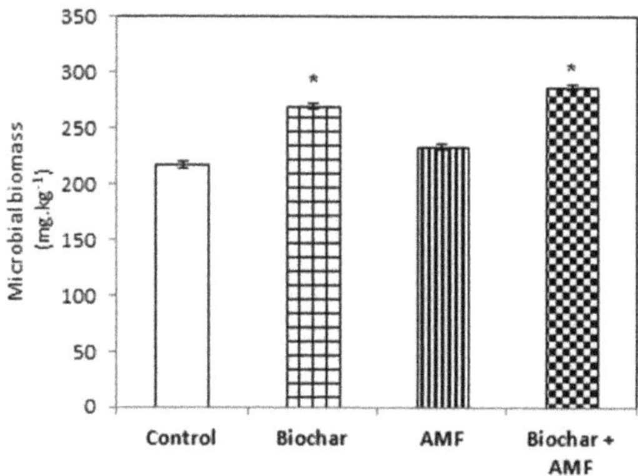

Figure 8. Biochar and AMF for the improvement of the microbial biomass in soil. Data are the means of three replicates (*n* = 3), * asterisk differed significantly at $p < 0.05$ *.

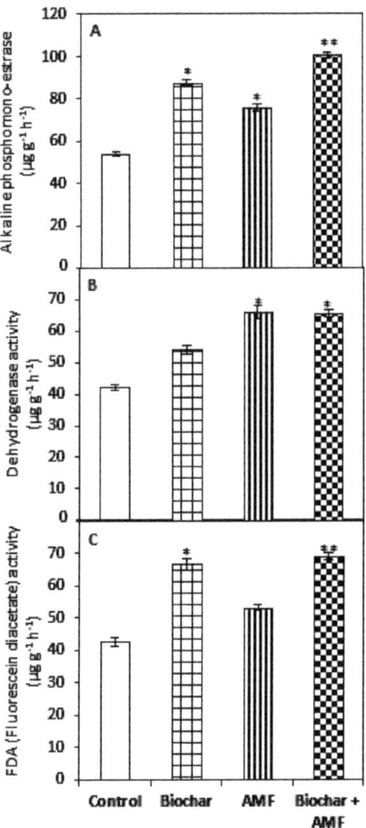

Figure 9. Biochar and AMF for the improvement of soil enzymes. (**A**) Alkaline phosphomonoester esterase, (**B**) Dehydrogenase, (**C**) Fluorescein diacetate activity. Data are the means of three replicates ($n = 3$); * differed significantly at $p < 0.05$ *, $p < 0.01$ **.

4. Discussion

4.1. Effect of Biochar and AMF on Growth of Spinach Plants

In general, biochar treatment promoted attributes of plant growth, namely, the shoot length, leaf length, leaf number, and leaf width, which were significantly higher compared to the control. Similarly, the biochar treatment significantly increased the root and shoot fresh weight and root and shoot dry weights compared to the control. This finding is consistent with the report of Bu et al. [56], who observed a significant enhancement in the growth of *Robinia pseudoacacia* L. with biochar. Similarly, Hilioti et al. [57] reported that castor stalk biochar enhanced the growth of the castor plant. Numerous researchers have reported that biochar increased the plant growth and yield in different crops [7,23–25,58]. The positive effect of biochar amendment on the root length, shoot length, root biomass, shoot biomass, and yield in French beans was noted by Saxena et al. [20]. Similarly, Carter et al. [27] observed that the application of rice husk biochar increased the final biomass, root biomass, plant height, and number of leaves of lettuce (*Lactuca sativa*) and cabbage (*Brassica chinensis*) compared to plants that did not receive the biochar treatment. Enhanced dry weight, leaf biomass, and root biomass with biochar applications were also reported by Trupiano et al. [19]. Similarly, rice straw biochar significantly promoted the plant height, number of bolls per plant, average boll weight, and seed cotton yield compared with the control treatment, as reported by Qayyum et al. [16].

There was a significant increase in the leaf number, leaf width, shoot fresh weight, and shoot and root dry weight with AMF treatment. Many studies have reported that the application of AMF increased plant growth parameters [59–62]. Sharma and Kayang [63] reported that the application of AMF noticeably increased plant growth parameters of tea (*Camellia sinensis* L.), such as the number of leaves, leaf area, plant height, shoot length, root length, and root and shoot weight. The combined application of biochar and AMF had a positive effect on the leaf number, leaf length, and shoot and root fresh and dry weights compared to the control. A study by Li and Cai [64] found that biochar and AMF both improved the growth performance of maize. Similar results were reported by Budi and Setyaningsih [65]; their study showed that biochar and AMF significantly increased the plant height, diameter, shoot dry weight, and root dry weight compared to the control plant.

4.2. The Effect of Biochar and AMF on Root Morphological Traits

Biochar treatment clearly resulted in improved root morphological parameters, such as the total root length, projected area, root diameter, and root volume, compared to the control. Several studies have reported that biochar application improved plant roots [18,22,66], thereby confirming our results. Bu et al. [56] reported a significant increase in the root length, root surface area, and root volume following the application of rice husk biochar and woodchip biochar. Similar results of significant improvement in root growth due to the addition of biochar were also reported by Trupiano et al. [19]. Zhang et al. [67] found that biochar addition increased the taproot length, root volume, and total root absorption area in tobacco. Li and Cai [64] indicated that the addition of biochar significantly altered the root morphology under both 40% FWC and 60% FWC.

There was a significant enhancement of the total root length, projected area, root diameter, and root volume with AMF treatment compared to the control. In tomato seedlings inoculated with AMF, there was an increase in the total root length and number of root tips [68]. Bi et al. [69] found that AMF could alleviate root damage stress by changing the root morphology. *Melia azedarach* inoculated with *Gigaspora margarita* had a significantly higher plant height, diameter, and shoot and root dry weight [65]. Data regarding the combined biochar and AMF treatment showed a significantly increased total root length and root diameter compared with all of the other treatments. This finding confirms an earlier report by Hashem et al. [48] that found that combined biochar and AMF treatment significantly increased the root length of the chickpea. Biochar in combination with AMF plays a key role in the plant utilization of underground water and nutrients, and this is also dependent on the root architecture [70].

4.3. Effect of Biochar and AMF on Plant Physiological Properties

The study showed that the addition of biochar had a positive effect on the physiological properties of spinach. The net photosynthesis rate and transpiration rate were significantly increased by biochar treatment alone. Biochar treatment also significantly increased the content of chlorophyll a, chlorophyll b, total chlorophyll, carotenoid, and relative water of the leaf over the control. Many researchers found that biochar application increased the photosynthesis, chlorophyll content, and transpiration rate in different plants [19,28,29,71]. He et al. [72] reported that the application of biochar significantly increased the photosynthesis rate and chlorophyll concentration in C_3 plants. Sarma et al. [25] found a strong positive effect of biochar amendment on the photosynthesis rate in okra. Hashem et al. [48] indicated that the application of biochar enhanced the amount of chlorophyll a, chlorophyll b, and total photosynthetic pigments.

The net photosynthetic rate, stomatal conductance, and transpiration rate were increased by treatment with AMF alone. AMF alone significantly increased the content of chlorophyll a, chlorophyll b, total chlorophyll, carotenoid, and relative water of the leaf. Similar results have been reported by Ren et al. [73] which showed that inoculation with AMF significantly improved the antioxidant enzymatic activity and net photosynthesis rate in Zea mays. AMF inoculation increased the chlorophyll content and photosynthesis rate of maize and chickpeas [48,64]. The combined application of biochar and AMF had a positive effect on the net photosynthesis rate, stomatal conductance, transpiration rate, relative water content, and photosynthetic pigments compared to the control (Figures 4–6). Similar results have been reported by Hashem et al. [48], showing that the combined application of AMF and biochar significantly increased the photosynthetic rate, relative water content, chlorophyll a, chlorophyll b and total chlorophylls in chickpea under normal condition. Similar findings confirming the significantly enhancement of the chlorophyll content and photosynthetic rate in maize were reported by Li and Cai [64]. As shown in Figure 10, they suggest that AMF and biochar increase nitrogen fixation and siderophore production, as well as enhance nutrient availability and absorption. Moreover, they trigger endogenous phytohormone synthesis and antioxidant production.

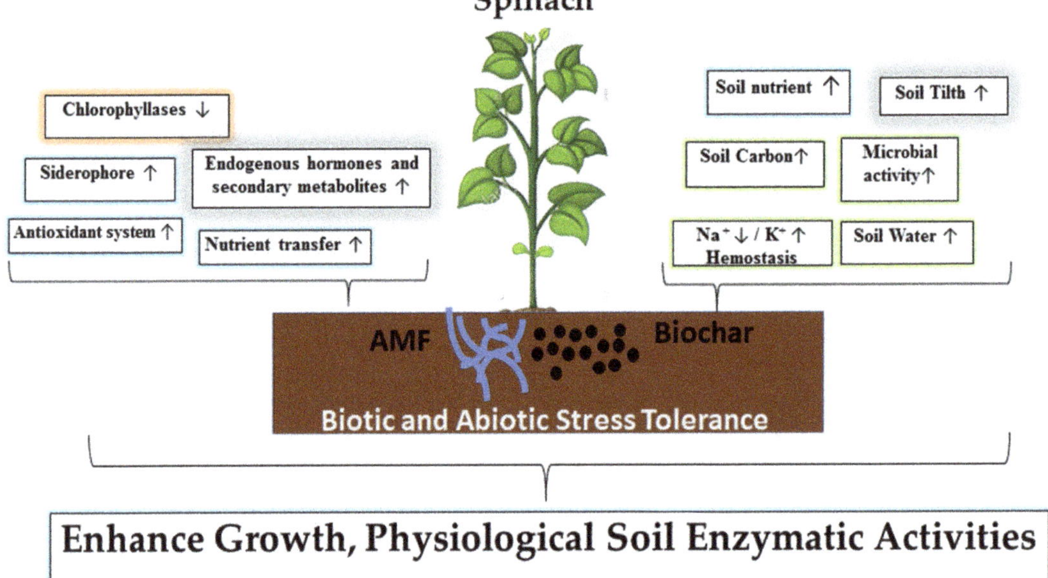

Figure 10. Summary for the action mechanism for the combined effect of biochar and AMF for the improvement of plant growth and soil enzymatic activity.

4.4. Effect of Biochar and AMF on AMF Spore Number, Microbial Biomass, and Soil Enzymatic Activity

Biochar treatment significantly increased the alkaline phosphomonoestrase activity and fluorescein diacetate activity compared with the control. Similar findings confirming enhanced soil enzymatic activity due to the addition of soybean biochar were reported by Jabborova et al. [18]. Bailey et al. [74] and Ma et al. [24] also observed increased enzymatic activity of soil due to biochar application. Several studies have reported an increase in the enzymes protease, chymotrypsin, trypsin, phosphohydrolase, lipase-esterase, and esterase with the application of biochar [17,19,75]. Oladele [76] reported significantly increased urease activity, invertase activity, and phosphatase activity with the application of biochar; the highest rate (12 t ha^{-1}) was at soil depths of 0 to 0.1 m. The microbial biomass and the AMF spores in soil also increased compared to the control. A similar increase in AMF spores due to biochar application was reported by Hashem et al. [48]. Numerous studies have shown that biochar application promoted AMF colonization rates [77–79].

AMF treatment alone increased alkaline phosphomonoestrase activity, dehydrogenase activity, microbial biomass, and AMF spores compared to the control. Similar findings confirming that biochar increased urease and phosphatase activity and the soil microbial biomass were reported by Zhaoxiang et al. [26] and Li and Cai [64]. The combined application of biochar and AMF significantly increased the activity of alkaline phosphomonoestrase, dehydrogenase, and fluorescein diacetate, as well as the microbial biomass and AMF spores. Similar results have been reported by Li and Cai [64], who found that the combined application of AMF and biochar significantly improved the soil microbial activity in the maize rhizosphere. The mechanism of the combined effect of AMF and biochar is summarized in Figure 10.

5. Conclusions

Biochar application has facilitated the improvement of root morphological traits and plant growth. It also positively influences soil enzymatic activity. The combined application of biochar and AMF had a significantly positive impact on spinach plant growth, root morphological traits, physiological properties, and soil enzymatic activities. We conclude that further work on specific interactions between biochars has the potential to decrease the application of mineral fertilizers. The combined application of biochar and AMF can be used as an efficient biofertilizer to promote the plant growth and yield of spinach in field conditions.

Author Contributions: Conceptualization, D.J., S.P., S.K. and K.A.; methodology, D.J., S.P., S.D., H.A.S., S.K. and K.A.; software, D.J., S.P., S.D., H.A.S., M.F.M.I. and A.E.; validation, D.J., S.P., M.F.M.I. and A.E.; formal analysis, D.J., S.P., S.K., K.A., M.F.M.I. and A.E.; investigation, D.J., S.P., S.K., K.A., M.F.M.I. and A.E.; resources, D.J., S.P., S.K., K.A., M.F.M.I. and A.E.; data curation, D.J., S.P., M.F.M.I. and A.E.; writing—original draft preparation, D.J., S.P., S.K. and K.A.; writing—review and editing, D.J., S.P., S.K., K.A., M.F.M.I. and A.E.; visualization, M.F.M.I., S.D., H.A.S. and A.E. All authors have read and agreed to the published version of the manuscript.

Funding: We appreciate and thank Taif University for financial support for the Taif University Researchers Supporting Project (TURSP-2020/07), Taif University, Taif, Saudi Arabia. This work has been financed by the Department of Biotechnology, the government of India (DBT), and TWAS.

Institutional Review Board Statement: Not applicable.

Informed Consent Statement: Not applicable.

Data Availability Statement: Not applicable.

Acknowledgments: We appreciate and thank Taif University for financial support for the Taif University Researchers Supporting Project (TURSP-2020/07), Taif University, Taif, Saudi Arabia. We thank our colleagues at the Division of Microbiology, ICAR-Indian Agricultural Research Institute, Pusa, New Delhi, India for providing the necessary support laboratory and net house facilities.

Conflicts of Interest: The authors declare no conflict of interest.

References

1. Lehmann, J.; Gaunt, J.; Rondon, M. Bio-char sequestration in terrestrial ecosystems—A review. *Mitig. Adapt. Strateg. Glob. Chang.* **2006**, *11*, 403–427. [CrossRef]
2. Soliman, M.H.; Abdulmajeed, A.M.; Alhaithloul, H.; Alharbi, B.M.; El-Esawi, M.A.; Hasanuzzaman, M.; Elkelish, A. Saponin Biopriming Positively Stimulates Antioxidants Defense, Osmolytes Metabolism and Ionic Status to Confer Salt Stress Tolerance in Soybean. *Acta Physiol. Plant.* **2020**, *42*, 114. [CrossRef]
3. Hashim, A.M.; Alharbi, B.M.; Abdulmajeed, A.M.; Elkelish, A.; Hozzein, W.N.; Hassan, H.M. Oxidative Stress Responses of Some Endemic Plants to High Altitudes by Intensifying Antioxidants and Secondary Metabolites Content. *Plants* **2020**, *9*, 869. [CrossRef]
4. Moustafa-Farag, M.; Mohamed, H.I.; Mahmoud, A.; Elkelish, A.; Misra, A.N.; Guy, K.M.; Kamran, M.; Ai, S.; Zhang, M. Salicylic Acid Stimulates Antioxidant Defense and Osmolyte Metabolism to Alleviate Oxidative Stress in Watermelons under Excess Boron. *Plants* **2020**, *9*, 724. [CrossRef] [PubMed]
5. Major, J.; Rondon, M.; Molina, D.; Riha, S.J.; Lehmann, J. Maize yield and nutrition during 4 years after biochar application to a Colombian savanna oxisol. *Plant Soil* **2010**, *333*, 117–128. [CrossRef]

6. Zamin, M.; Fahad, S.; Khattak, A.M.; Adnan, M.; Wahid, F.; Raza, A.; Wang, D.; Saud, S.; Noor, M.; Bakhat, H.F.; et al. Developing the First Halophytic Turfgrasses for the Urban Landscape from Native Arabian Desert Grass. *Environ. Sci. Pollut. Res. Int.* **2020**, *27*, 39702–39716. [CrossRef] [PubMed]

7. El Nahhas, N.; AlKahtani, M.D.F.; Abdelaal, K.A.A.; Al Husnain, L.; AlGwaiz, H.I.M.; Hafez, Y.M.; Attia, K.A.; El-Esawi, M.A.; Ibrahim, M.F.M.; Elkelish, A. Biochar and Jasmonic Acid Application Attenuates Antioxidative Systems and Improves Growth, Physiology, Nutrient Uptake and Productivity of Faba Bean (Vicia Faba L.) Irrigated with Saline Water. *Plant Physiol. Biochem.* **2021**, *166*, 807–817. [CrossRef] [PubMed]

8. Elsaeed, S.M.; Zaki, E.G.; Ibrahim, T.M.; Ibrahim Talha, N.; Saad, H.A.; Gobouri, A.A.; Elkelish, A.; Mohamed el-kousy, S. Biochar Grafted on CMC-Terpolymer by Green Microwave Route for Sustainable Agriculture. *Agriculture* **2021**, *11*, 350. [CrossRef]

9. Jiang, C.; Yu, G.; Li, Y.; Cao, G.; Yang, Z.; Sheng, W.; Yu, W. Nutrient resorption of coexistence species in alpine meadow of the Qinghai-Tibetan Plateau explains plant adaptation to nutrient-poor environment. *Ecol. Eng.* **2012**, *44*, 1–9. [CrossRef]

10. Lehmann, J.; Rillig, M.C.; Thies, J.; Masiello, C.A.; Hockaday, W.C.; Crowley, D. Biochar effects on soil biota–a review. *Soil Biol. Biochem.* **2011**, *43*, 1812–1836. [CrossRef]

11. Martinsen, V.; Alling, V.; Nurida, N.; Mulder, J.; Hale, S.; Ritz, C.; Rutherford, D.; Heikens, A.; Breedveld, G.D.; Cornelissen, G. pH effects of the addition of three biochars to acidic Indonesian mineral soils. *Soil Sci. Plant Nutr.* **2015**, *61*, 821–834. [CrossRef]

12. Oguntunde, P.G.; Fosu, M.; Ajayi, A.E.; Van De Giesen, N. Effects of charcoal production on maize yield, chemical properties and texture of soil. *Biol. Fertil. Soils* **2004**, *39*, 295–299. [CrossRef]

13. Yamato, M.; Okimori, Y.; Wibowo, I.F.; Anshori, S.; Ogawa, M. Effects of the application of charred bark of *Acacia mangium* on the yield of maize, cowpea and peanut, and soil chemical properties in South Sumatra, Indonesia. *Soil Sci. Plant Nutr.* **2006**, *52*, 489–495. [CrossRef]

14. Ścisłowska, M.; Włodarczyk, R.; Kobyłecki, R.; Bis, Z. Biochar to improve the quality and productivity of soils. *J. Ecol. Eng.* **2015**, *16*. [CrossRef]

15. Laird, D.A.; Fleming, P.; Davis, D.D.; Horton, R.; Wang, B.; Karlen, D.L. Impact of biochar amendments on the quality of a typical Midwestern agricultural soil. *Geoderma* **2010**, *158*, 443–449. [CrossRef]

16. Qayyum, M.F.; Haider, G.; Raza, M.A.; Mohamed, A.K.S.; Rizwan, M.; El-Sheikh, M.A.; Alyemeni, M.N.; Ali, S. Straw-based biochar mediated potassium availability and increased growth and yield of cotton (*Gossypium hirsutum* L.). *J. Saudi Chem. Soc.* **2020**, *24*, 963–973. [CrossRef]

17. Anderson, C.R.; Condron, L.M.; Clough, T.J.; Fiers, M.; Stewart, A.; Hill, R.A.; Sherlock, R.R. Biochar induced soil microbial community change: Implications for biogeochemical cycling of carbon, nitrogen and phosphorus. *Pedobiologia* **2011**, *54*, 309–320. [CrossRef]

18. Jabborova, D.; Wirth, S.; Kannepalli, A.; Narimanov, A.; Desouky, S.; Davranov, K.; Sayyed, R.; El Enshasy, H.; Malek, R.A.; Syed, A. Co-Inoculation of Rhizobacteria and Biochar Application Improves Growth and Nutrientsin Soybean and Enriches Soil Nutrients and Enzymes. *Agronomy* **2020**, *10*, 1142. [CrossRef]

19. Trupiano, D.; Cocozza, C.; Baronti, S.; Amendola, C.; Vaccari, F.P.; Lustrato, G.; Di Lonardo, S.; Fantasma, F.; Tognetti, R.; Scippa, G.S. The effects of biochar and its combination with compost on lettuce (*Lactuca sativa* L.) growth, soil properties, and soil microbial activity and abundance. *Int. J. Agron.* **2017**, *2017*. [CrossRef]

20. Saxena, J.; Rana, G.; Pandey, M. Impact of addition of biochar along with *Bacillus* sp. on growth and yield of French beans. *Sci. Hortic.* **2013**, *162*, 351–356. [CrossRef]

21. Wang, Y.; Yin, R.; Liu, R. Characterization of biochar from fast pyrolysis and its effect on chemical properties of the tea garden soil. *J. Anal. Appl. Pyrolysis* **2014**, *110*, 375–381. [CrossRef]

22. Głodowska, M.; Schwinghamer, T.; Husk, B.; Smith, D. Biochar based inoculants improve soybean growth and nodulation. *Agric. Sci.* **2017**, *8*, 1048–1064. [CrossRef]

23. Jeffery, S.; Abalos, D.; Prodana, M.; Bastos, A.C.; Van Groenigen, J.W.; Hungate, B.A.; Verheijen, F. Biochar boosts tropical but not temperate crop yields. *Environ. Res. Lett.* **2017**, *12*, 053001. [CrossRef]

24. Ma, H.; Egamberdieva, D.; Wirth, S.; Bellingrath-Kimura, S.D. Effect of biochar and irrigation on soybean-rhizobium symbiotic performance and soil enzymatic activity in field rhizosphere. *Agronomy* **2019**, *9*, 626. [CrossRef]

25. Sarma, B.; Borkotoki, B.; Narzari, R.; Kataki, R.; Gogoi, N. Organic amendments: Effect on carbon mineralization and crop productivity in acidic soil. *J. Clean. Prod.* **2017**, *152*, 157–166. [CrossRef]

26. Zhaoxiang, W.; Huihu, L.; Qiaoli, L.; Changyan, Y.; Faxin, Y. Application of bio-organic fertilizer, not biochar, in degraded red soil improves soil nutrients and plant growth. *Rhizosphere* **2020**, *16*, 100264. [CrossRef]

27. Carter, S.; Shackley, S.; Sohi, S.; Suy, T.B.; Haefele, S. The impact of biochar application on soil properties and plant growth of pot grown lettuce (*Lactuca sativa*) and cabbage (*Brassica chinensis*). *Agronomy* **2013**, *3*, 404–418. [CrossRef]

28. Agegnehu, G.; Bass, A.M.; Nelson, P.N.; Muirhead, B.; Wright, G.; Bird, M.I. Biochar and biochar-compost as soil amendments: Effects on peanut yield, soil properties and greenhouse gas emissions in tropical North Queensland, Australia. *Agric. Ecosyst. Environ.* **2015**, *213*, 72–85. [CrossRef]

29. Speratti, A.B.; Johnson, M.S.; Sousa, H.M.; Dalmagro, H.J.; Couto, E.G. Biochars from local agricultural waste residues contribute to soil quality and plant growth in a Cerrado region (Brazil) Arenosol. *GCB Bioenergy* **2018**, *10*, 272–286. [CrossRef]

30. Shokri, S.; Maadi, B. Effects of arbuscular mycorrhizal fungus on the mineral nutrition and yield of *Trifolium alexandrinum* plants under salinity stress. *J. Agron.* **2009**, *8*, 79–83. [CrossRef]

31. Tabassum, Y.; Tanvir, B.; Farrukh, H. Effect of arbuscular mycorrhizal inoculation on nutrient uptake, growth and productivity of chickpea (*Cicer arietinum*) varieties. *Int. J. Agron. Plant Prod.* **2012**, *3*, 334–345.

32. Behl, R.K.; Ruppel, S.; Kothe, E.; Narula, N. Wheat x Azotobacter x VA Mycorrhiza interactions towards plant nutrition and growth–a review. *J. Appl. Bot. Food Qual.* **2012**, *81*, 95–109.

33. Roesti, D.; Gaur, R.; Johri, B.; Imfeld, G.; Sharma, S.; Kawaljeet, K.; Aragno, M. Plant growth stage, fertiliser management and bio-inoculation of arbuscular mycorrhizal fungi and plant growth promoting rhizobacteria affect the rhizobacterial community structure in rain-fed wheat fields. *Soil Biol. Biochem.* **2006**, *38*, 1111–1120. [CrossRef]

34. Singh, S.; Kapoor, K. Inoculation with phosphate-solubilizing microorganisms and a vesicular-arbuscular mycorrhizal fungus improves dry matter yield and nutrient uptake by wheat grown in a sandy soil. *Biol. Fertil. Soils* **1999**, *28*, 139–144. [CrossRef]

35. Clark, R.; Zeto, S. Growth and root colonization of mycorrhizal maize grown on acid and alkaline soil. *Soil Biol. Biochem.* **1996**, *28*, 1505–1511. [CrossRef]

36. Javaid, A. Arbuscular mycorrhizal mediated nutrition in plants. *J. Plant Nutr.* **2009**, *32*, 1595–1618. [CrossRef]

37. Latef, A.A.H.A. RETRACTED ARTICLE: Influence of arbuscular mycorrhizal fungi and copper on growth, accumulation of osmolyte, mineral nutrition and antioxidant enzyme activity of pepper (*Capsicum annuum* L.). *Mycorrhiza* **2011**, *21*, 495–503. [CrossRef] [PubMed]

38. Meding, S.; Zasoski, R. Hyphal-mediated transfer of nitrate, arsenic, cesium, rubidium, and strontium between arbuscular mycorrhizal forbs and grasses from a California oak woodland. *Soil Biol. Biochem.* **2008**, *40*, 126–134. [CrossRef]

39. Latef, A.A.H.A.; Chaoxing, H. Arbuscular mycorrhizal influence on growth, photosynthetic pigments, osmotic adjustment and oxidative stress in tomato plants subjected to low temperature stress. *Acta Physiol. Plant.* **2011**, *33*, 1217–1225. [CrossRef]

40. Alizadeh, O.; Zare, M.; Nasr, A.H. Evaluation effect of mycorrhiza inoculate under drought stress condition on grain yield of sorghum.(*Sorghum bicolor*). In *Advances in Environmental Biology*; American-Eurasian Network for Scientific Information: Ma'an, Jordan, 2011; pp. 2361–2365.

41. Cavagnaro, T.R.; Jackson, L.; Six, J.; Ferris, H.; Goyal, S.; Asami, D.; Scow, K. Arbuscular mycorrhizas, microbial communities, nutrient availability, and soil aggregates in organic tomato production. *Plant Soil* **2006**, *282*, 209–225. [CrossRef]

42. Nunes, J.L.d.S.; Souza, P.V.D.d.; Marodin, G.A.B.; Fachinello, J.C. Effect of arbuscular mycorrhizal fungi and indolebutyric acid interaction on vegetative growth of 'Aldrighi' peach rootstock seedlings. *Ciência Agrotecnologia* **2010**, *34*, 80–86. [CrossRef]

43. Najafi, A.; Ardakani, M.R.; Rejali, F.; Sajedi, N. Response of winter barley to co-inoculation with Azotobacter and Mycorrhiza fungi influenced by plant growth promoting rhizobacteria. *Ann. Biol. Res.* **2012**, *3*, 4002–4006.

44. Ordookhani, K.; Khavazi, K.; Moezzi, A.; Rejali, F. Influence of PGPR and AMF on antioxidant activity, lycopene and potassium contents in tomato. *Afr. J. Agric. Res.* **2010**, *5*, 1108–1116.

45. IIA, A.E.S. Effect of biochar rates on A-mycorrhizal fungi performance and maize plant growth, Phosphorus uptake, and soil P availability under calcareous soil conditions. *Commun. Soil Sci. Plant Anal.* **2021**, *52*, 815–831.

46. Nemadodzi, L.E.; Araya, H.; Nkomo, M.; Ngezimana, W.; Mudau, N.F. Nitrogen, phosphorus, and potassium effects on the physiology and biomass yield of baby spinach (*Spinacia oleracea* L.). *J. Plant Nutr.* **2017**, *40*, 2033–2044. [CrossRef]

47. Roberts, J.L.; Moreau, R. Functional properties of spinach (*Spinacia oleracea* L.) phytochemicals and bioactives. *Food Funct.* **2016**, *7*, 3337–3353. [CrossRef]

48. Hashem, A.; Kumar, A.; Al-Dbass, A.M.; Alqarawi, A.A.; Al-Arjani, A.-B.F.; Singh, G.; Farooq, M.; Allah, E.F. Arbuscular mycorrhizal fungi and biochar improves drought tolerance in chickpea. *Saudi J. Biol. Sci.* **2019**, *26*, 614–624. [CrossRef]

49. Abd El-Gawad, H.G.; Mukherjee, S.; Farag, R.; Abd Elbar, O.H.; Hikal, M.; Abou El-Yazied, A.; Abd Elhady, S.A.; Helal, N.; ElKelish, A.; El Nahhas, N.; et al. Exogenous γ-aminobutyric acid (GABA)-induced signaling events and field performance associated with mitigation of drought stress in *Phaseolus vulgaris* L. *Plant Signal. Behav.* **2021**, *16*, 1853384. [CrossRef]

50. Hiscox, J.; Israelstam, G. A method for the extraction of chlorophyll from leaf tissue without maceration. *Can. J. Bot.* **1979**, *57*, 1332–1334. [CrossRef]

51. Yusif, S.; Dare, M.; Babalola, O.; Popoola, A.; Sharif, M.; Habib, M. The roles of biochar and arbuscular mycorrhizal inoculation on selected soil biological properties and tomato performance. *FUTY J. Environ.* **2018**, *12*, 1–8.

52. Vance, E.D.; Brookes, P.C.; Jenkinson, D.S. An extraction method for measuring soil microbial biomass C. *Soil Biol. Biochem.* **1987**, *19*, 703–707. [CrossRef]

53. Tabatabai, M.A.; Bremner, J.M. Use of p-nitrophenyl phosphate for assay of soil phosphatase activity. *Soil Biol. Biochem.* **1969**, *1*, 301–307. [CrossRef]

54. Green, V.S.; Stott, D.E.; Diack, M. Assay for fluorescein diacetate hydrolytic activity: Optimization for soil samples. *Soil Biol. Biochem.* **2006**, *38*, 693–701. [CrossRef]

55. Casida Jr, L.; Klein, D.A.; Santoro, T. Soil dehydrogenase activity. *Soil Sci.* **1964**, *98*, 371–376. [CrossRef]

56. Bu, X.; Xue, J.; Wu, Y.; Ma, W. Effect of Biochar on Seed Germination and Seedling Growth of *Robinia pseudoacacia* L. In Karst Calcareous Soils. *Commun. Soil Sci. Plant Anal.* **2020**, *51*, 352–363. [CrossRef]

57. Hilioti, Z.; Michailof, C.; Valasiadis, D.; Iliopoulou, E.; Koidou, V.; Lappas, A. Characterization of castor plant-derived biochars and their effects as soil amendments on seedlings. *Biomass Bioenergy* **2017**, *105*, 96–106. [CrossRef]

58. Agboola, K.; Moses, S. Effect of biochar and cowdung on nodulation, growth and yield of soybean (*Glycine max* L. Merrill). *Int. J. Agric. Biosci.* **2015**, *4*, 154–160.

59. Gogoi, P.; Singh, R. Differential effect of some arbuscular mycorrhizal fungi on growth of *Piper longum* L.(Piperaceae). *Indian J. Sci. Technol.* **2011**, *4*, 119–125. [CrossRef]

60. Ortas, I.; Ustuner, O. The effects of single species, dual species and indigenous mycorrhiza inoculation on citrus growth and nutrient uptake. *Eur. J. Soil Biol.* **2014**, *63*, 64–69. [CrossRef]

61. Sharma, D.; Kapoor, R.; Bhatnagar, A. Differential growth response of *Curculigo orchioides* to native arbuscular mycorrhizal fungal (AMF) communities varying in number and fungal components. *Eur. J. Soil Biol.* **2009**, *45*, 328–333. [CrossRef]

62. Singh, S.; Pandey, A.; Kumar, B.; Palni, L.M.S. Enhancement in growth and quality parameters of tea [*Camellia sinensis* (L.) O. Kuntze] through inoculation with arbuscular mycorrhizal fungi in an acid soil. *Biol. Fertil. Soils* **2010**, *46*, 427–433. [CrossRef]

63. Sharma, D.; Kayang, H. Effects of arbuscular mycorrhizal fungi (AMF) on *Camellia sinensis* (L.) O. Kuntze under greenhouse conditions. *J. Exp. Biol.* **2017**, *5*, 235–241. [CrossRef]

64. Li, M.; Cai, L. Biochar and Arbuscular Mycorrhizal Fungi Play Different Roles in Enabling Maize to Uptake Phosphorus. *Sustainability* **2021**, *13*, 3244. [CrossRef]

65. Budi, S.W.; Setyaningsih, L. Arbuscular mycorrhizal fungi and biochar improved early growth of neem (*Melia azedarach* Linn.) seedling under greenhouse conditions. *J. Manaj. Hutan Trop.* **2013**, *19*, 103–110.

66. Uzoma, K.; Inoue, M.; Andry, H.; Fujimaki, H.; Zahoor, A.; Nishihara, E. Effect of cow manure biochar on maize productivity under sandy soil condition. *Soil Use Manag.* **2011**, *27*, 205–212. [CrossRef]

67. Zhang, L.; Xu, M.; Liu, Y.; Zhang, F.; Hodge, A.; Feng, G. Carbon and phosphorus exchange may enable cooperation between an arbuscular mycorrhizal fungus and a phosphate-solubilizing bacterium. *New Phytol.* **2016**, *210*, 1022–1032. [CrossRef]

68. Berta, G.; Sampo, S.; Gamalero, E.; Massa, N.; Lemanceau, P. Suppression of Rhizoctonia root-rot of tomato by *Glomus mossae* BEG12 and *Pseudomonas fluorescens* A6RI is associated with their effect on the pathogen growth and on the root morphogenesis. *Eur. J. Plant Pathol.* **2005**, *111*, 279–288. [CrossRef]

69. Bi, Y.; Zhang, J.; Song, Z.; Wang, Z.; Qiu, L.; Hu, J.; Gong, Y. Arbuscular mycorrhizal fungi alleviate root damage stress induced by simulated coal mining subsidence ground fissures. *Sci. Total Environ.* **2019**, *652*, 398–405. [CrossRef] [PubMed]

70. Zhang, J.; Zhang, Z.; Shen, G.; Wang, R.; Gao, L.; Kong, F.; Zhang, J. Growth performance, nutrient absorption of tobacco and soil fertility after straw biochar application. *Int. J. Agric. Biol.* **2016**, 18. [CrossRef]

71. Petruccelli, R.; Bonetti, A.; Traversi, M.L.; Faraloni, C.; Valagussa, M.; Pozzi, A. Influence of biochar application on nutritional quality of tomato (*Lycopersicon esculentum*). *Crop Pasture Sci.* **2015**, *66*, 747–755. [CrossRef]

72. He, Y.; Yao, Y.; Ji, Y.; Deng, J.; Zhou, G.; Liu, R.; Shao, J.; Zhou, L.; Li, N.; Zhou, X. Biochar amendment boosts photosynthesis and biomass in C3 but not C4 plants: A global synthesis. *GCB Bioenergy* **2020**, *12*, 605–617. [CrossRef]

73. Ren, A.-T.; Zhu, Y.; Chen, Y.-L.; Ren, H.-X.; Li, J.-Y.; Abbott, L.K.; Xiong, Y.-C. Arbuscular mycorrhizal fungus alters root-sourced signal (abscisic acid) for better drought acclimation in *Zea mays* L. seedlings. *Environ. Exp. Botany* **2019**, *167*, 103824. [CrossRef]

74. Bailey, V.L.; Fansler, S.J.; Smith, J.L.; Bolton, H., Jr. Reconciling apparent variability in effects of biochar amendment on soil enzyme activities by assay optimization. *Soil Biol. Biochem.* **2011**, *43*, 296–301. [CrossRef]

75. Ouyang, L.; Tang, Q.; Yu, L.; Zhang, R. Effects of amendment of different biochars on soil enzyme activities related to carbon mineralisation. *Soil Res.* **2014**, *52*, 706–716. [CrossRef]

76. Oladele, S.O. Effect of biochar amendment on soil enzymatic activities, carboxylate secretions and upland rice performance in a sandy clay loam Alfisol of Southwest Nigeria. *Sci. Afr.* **2019**, *4*, e00107. [CrossRef]

77. Mickan, B.S.; Abbott, L.K.; Stefanova, K.; Solaiman, Z.M. Interactions between biochar and mycorrhizal fungi in a water-stressed agricultural soil. *Mycorrhiza* **2016**, *26*, 565–574. [CrossRef] [PubMed]

78. Solaiman, Z.M.; Blackwell, P.; Abbott, L.K.; Storer, P. Direct and residual effect of biochar application on mycorrhizal root colonisation, growth and nutrition of wheat. *Soil Res.* **2010**, *48*, 546–554. [CrossRef]

79. Vanek, S.J.; Lehmann, J. Phosphorus availability to beans via interactions between mycorrhizas and biochar. *Plant Soil* **2015**, *395*, 105–123. [CrossRef]

Journal of
Fungi

MDPI

Article

Inocybe brijunica sp. nov., a New Ectomycorrhizal Fungus from Mediterranean Croatia Revealed by Morphology and Multilocus Phylogenetic Analysis

Armin Mešić [1], Danny Haelewaters [2,3,4,*], Zdenko Tkalčec [1], Jingyu Liu [3], Ivana Kušan [1], M. Catherine Aime [3] and Ana Pošta [1]

[1]   Laboratory for Biological Diversity, Ruđer Bošković Institute, Bijenička cesta 54, HR-10000 Zagreb, Croatia; amesic@irb.hr (A.M.); ztkalcec@irb.hr (Z.T.); ikusan@irb.hr (I.K.); aposta@irb.hr (A.P.)

[2]   Faculty of Science, University of South Bohemia, Branišovská 31, 370 05 České Budějovice, Czech Republic

[3]   Department of Botany and Plant Pathology, Purdue University, 915 W. State Street, West Lafayette, IN 47907, USA; liu1643@purdue.edu (J.L.); maime@purdue.edu (M.C.A.)

[4]   Research Group Mycology, Department of Biology, Ghent University, K.L. Ledeganckstraat 35, 9000 Ghent, Belgium

*   Correspondence: danny.haelewaters@gmail.com

Citation: Mešić, A.; Haelewaters, D.; Tkalčec, Z.; Liu, J.; Kušan, I.; Aime, M.C.; Pošta, A. *Inocybe brijunica* sp. nov., a New Ectomycorrhizal Fungus from Mediterranean Croatia Revealed by Morphology and Multilocus Phylogenetic Analysis. *J. Fungi* **2021**, *7*, 199. https://doi.org/10.3390/jof7030199

Academic Editor: Anush Kosakyan

Received: 9 February 2021
Accepted: 8 March 2021
Published: 10 March 2021

Publisher's Note: MDPI stays neutral with regard to jurisdictional claims in published maps and institutional affiliations.

Abstract: A new ectomycorrhizal species was discovered during the first survey of fungal diversity at Brijuni National Park (Croatia), which consists of 14 islands and islets. The National Park is located in the Mediterranean Biogeographical Region, a prominent climate change hot-spot. *Inocybe brijunica* sp. nov., from sect. *Hysterices* (Agaricales, Inocybaceae), is described based on morphology and multilocus phylogenetic data. The holotype collection was found at the edge between grassland and *Quercus ilex* forest with a few planted *Pinus pinea* trees, on Veli Brijun Island, the largest island of the archipelago. It is easily recognized by a conspicuous orange to orange–red–brown membranaceous surface layer located at or just above the basal part of the stipe. Other distinctive features of *I. brijunica* are the medium brown, radially fibrillose to rimose pileus; pale to medium brown stipe with fugacious cortina; relatively small, amygdaliform to phaseoliform, and smooth basidiospores, measuring ca. 6.5–9 × 4–5.5 μm; thick-walled, utriform, lageniform or fusiform pleurocystidia (lamprocystidia) with crystals and mostly not yellowing in alkaline solutions; cheilocystidia of two types (lamprocystidia and leptocystidia); and the presence of abundant caulocystidia only in the upper 2–3 mm of the stipe. Phylogenetic reconstruction of a concatenated dataset of the internal transcribed spacer region (ITS), the nuclear 28S rRNA gene (nrLSU), and the second largest subunit of RNA polymerase II (*rpb2*) resolved *I. brijunica* and *I. glabripes* as sister species.

Keywords: 1 new taxon; Agaricomycetes; Basidiomycota; biodiversity; climate change; Inocybaceae; taxonomy

1. Introduction

The Brijuni archipelago consists of 14 islands and islets located in the Adriatic Sea (northern Mediterranean, Europe), near the southwestern coast of the Istrian peninsula. The archipelago is home to Brijuni National Park [1], which covers 33.9 km^2 of protected area, including the surrounding sea. The islands' surface area covers 7.4 km^2; Veli Brijun is the largest island with 5.7 km^2 and is devoid of permanent inhabitants. The National Park was established in 1983 to protect valuable marine and coastal (land) ecosystems and their biodiversity. The area is floristically rich, covered with evergreen vegetation and home to more than 400 native and exotic plant species mostly of Mediterranean origin. The Brijuni archipelago is characterized by a northern Mediterranean climate [1] with an average annual temperature of 13.9 °C, annual average precipitation of 817 mm, and a relatively high average air humidity of 76%.

During the 20th century, the air temperature increased globally by 0.74 °C, but the temperature rise in the Mediterranean area was higher—up to 1.5–4 °C depending on the region [2]. Following the Regional Climate Change Index (RCCI), the Mediterranean region is one of the most prominent climate change hot-spots in the world [3,4]. Models predicting the intensity of future climate change in this area are not optimistic. According to Mariotti et al. [5], land areas of the Mediterranean will gradually become drier; models predict 8% less precipitation in 2020–2049 compared to 1950–2000, a number that is projected to increase to 15% in 2070–2099. Drying in the northern Mediterranean [4,6] is projected to occur year-round, which will increase water stress for ecosystems if climate change continues at the current rate. Therefore, in the future, we can expect an increase of devastating climatic events (floods, storms, and droughts), more attacks of organisms that cause diseases, and a higher number of invasive species that will compete with indigenous species populations. These events could have a strong negative impact on the Mediterranean forest ecosystems [2] as well as on fruiting and existence of drought-sensitive fungal species in the area [7].

The ratio of plant species to macrofungal species is conservatively estimated as 1:6 [8]. Given the high number of plant species in the Brijuni Archipelago, an equally high diversity of fungal species is expected. Currently, however, there are no published data on fungi from this area. And even though fungal taxonomy has a long history in Europe, many species continue to be described from the continent. In 2019, 23% of newly described species of fungi were from Europe [9]. Given all this, it can be expected that undescribed species may be discovered at the Brijuni Islands. Initial field trips by Croatian mycologists aiming to document the fungal diversity of Brijuni National Park were carried out during the fall season in 2014, 2015, 2016, and 2020. In total, 546 records of basidiomycete fungi were made; 184 samples were collected and deposited in the Croatian National Fungarium (CNF) in Zagreb, Croatia. One of the most common genera found was *Inocybe* (Fr.) Fr. (Agaricomycetes, Agaricales, Inocybaceae), with 28 collections.

Inocybe sensu lato (s.l.) is a highly diverse genus of ectomycorrhizal mushrooms [10] currently with about 1000 accepted species [11]. It belongs to the family Inocybaceae Jülich. The species diversity within *Inocybe* s.l. is best known in Europe, especially in its central countries—Austria, France, Germany, the Netherlands, and Switzerland—with more than 450 species recorded [12]. Current ongoing studies are exploring the diversity of the genus in Europe and many new species have been recently described [13–22].

According to the taxonomic treatment by Matheny et al. [23] based on a six-locus phylogeny, the family Inocybaceae now consists of seven genera: *Auritella* Matheny & Bougher, *Inocybe sensu stricto* (s.s.), *Inosperma* (Kühner) Matheny & Esteve-Rav., *Mallocybe* (Kuyper) Matheny, Vizzini & Esteve-Rav., *Nothocybe* Matheny & K.P.D. Latha, *Pseudosperma* Matheny & Esteve-Rav., and *Tubariomyces* Esteve-Rav. & Matheny. The largest genus remains *Inocybe* s.s. with about 850 known species worldwide. Members of *Inocybe* s.s. can be distinguished from other genera in the family by the presence of pleurocystidia and basidiospores (with a distinct apiculus) that range from amygdaliform to ellipsoid, subcylindrical, angular, nodulose, or spinose in shape [23].

On 16 November 2016, during our fungal diversity research on the island of Veli Brijun, basidiomata of an interesting fungus belonging to *Inocybe* s.s. were found. Its basidiomata were macroscopically striking by the presence of an orange to orange–red–brown membranaceous surface layer (possibly a remnant of universal veil) in the basal part of the stipe—an unusual feature in the genus. Further detailed molecular phylogenetic and morphological analyses confirmed that the species was hitherto unknown to science. Therefore, it is here described as *I. brijunica* sp. nov.

2. Materials and Methods

2.1. Description of the Research Area

The holotype collection of *Inocybe brijunica* was collected on Veli Brijun Island. The biogeographical position and a long history of human interventions have shaped the landscape of Veli Brijun Island, merging natural and anthropogenic elements. The island is mostly covered by a thermophilous forest with holm oak (*Quercus ilex*) (including those in the maquis degradation stage), planted alleys or groves of pine trees (*Pinus halepensis* Mill. and *P. pinea* L.), cypresses (*Cupressus sempervirens* L.), cedars (*Cedrus* spp.), and parks and lawns often used as golf courses.

The *Inocybe* collection was found on the edge of the mature thermophilous *Q. ilex* forest and a lawn grazed by large herbivores (fallow deer [*Dama dama* L.], axis deer [*Axis axis* Erxleben], and European mouflon [*Ovis gmelini musimon* Pall.]) and occasionally machine-mowed by park staff. In addition, a few mature planted trees of *P. pinea* were present at the forest edge. Basidiomata of *I. brijunica* were found at ca. 70 m from the sea, epigeous on the soil covered with a shallow layer of oak and pine litter intermixed with scattered short grasses. The understory of the surrounding forest was devoid of herbaceous plants and shrubs due to the presence of large herbivores.

2.2. Morphological Study

The species description is based on a single but large collection consisting of 20 basidiomata. Macroscopic characters were documented with a Canon EOS 5D digital camera equipped with a Canon MR-14EX macro ring flash (Canon Europe, Uxbridge, UK). Microscopic features were observed by brightfield and phase contrast microscopy using a BX51 optical microscope (Olympus, Hamburg, Germany) under magnification up to 1500× and photographed with a Canon EOS M50 digital camera. Descriptions and images of microscopic characters were made from rehydrated specimens mounted in 2.5% potassium hydroxide (KOH), except for cystidia that were observed in 3% ammonium hydroxide (NH_4OH). Micromorphological terminology mostly follows Clémençon [24]. Line drawings were made by J.L. with PITT artist pens (Faber–Castell, Nürnberg, Germany) based on digital images.

Amyloid and dextrinoid reactions of basidiospores were tested in Melzer's reagent [25]. Randomly selected basidiospores from photographs of lamellae mounts were measured with Motic Images Plus 2.0 software (Motic Europe, Barcelona, Spain). The length/width ratio of basidiospores is given as the "Q" value (min–av.–max). Average basidiospore and pleurocystidia lengths, widths, and Q values are shown in italics. Numbers in square brackets [X/Y/Z] denote X elements measured in Y basidiomata of Z collections. Measurements of cystidia do not include crystals present at the apex. Type material was preserved by drying on a flow of hot air at maximum temperature of 50 °C. The holotype is deposited at CNF, and an isotype is deposited at PUL (Kriebel Herbarium, Purdue University, West Lafayette, IN, USA).

2.3. DNA Extraction, PCR Amplification, and Sequencing

Genomic DNA was extracted from parts of the lamellae using the QIAamp DNA Micro Kit (Qiagen, Valencia, CA, USA). The first ~1100 bp of the nuclear 18S nuclear ribosomal RNA gene (nrSSU), the internal transcribed spacer region of the rDNA (ITS, consisting of ITS1–5.8S–ITS2), the first ~1400 bp of the nuclear 28S rRNA gene (nrLSU), and the second largest subunit of RNA polymerase II gene (*rpb2*) were amplified [26]. The following primers were used: NS1, NS2, and NS4 [27] for nrSSU; ITS9mun [28] and ITS4 [27] for ITS; LR0R, LR5, and LR7 for nrLSU [29,30]; and RBP2-b6F, RPB2-b7R, and RPB2-b7.1R for *rpb2* [31]. Amplifications were done in 25 μL reactions, containing 12.5 μL of Promega 2× PCR Master Mix (Promega Co., Madison, WI, USA), 1.25 μL of each 10 μM primer, 9.0 μL of H_2O, and 1.0 μL of template DNA. PCR conditions for nrSSU, ITS, and nrLSU followed Haelewaters et al. [32]. For *rpb2*, PCR conditions were as follows: initial denaturation at 95 °C for 5:00 min; followed by 40 cycles of denaturation at 95 °C for 30 s, annealing at 55 °C for 45 s, and extension at 72 °C for 45 s; and final extension at 72 °C for 7:00 min. All amplifications were performed using the Eppendorf Mastercycler EP Thermal Cycler (Hauppauge, NY, USA). Purification of successful PCR products and sequencing in both directions using the amplification primers were outsourced to Genewiz (South

Plainfield, NJ, USA). Sequence reads were assembled and edited using Sequencher 5.4.6 for Windows software (Gene Codes Corporation, Ann Arbor, MI, USA). Assembled sequences were deposited at the National Center for Biotechnology Information (NCBI) GenBank database, under accession numbers MN749503–MN749504 (nrSSU), MN749370–MN749371 (ITS), MN749492–MN749493 (nrLSU), and MT878448–MT878449 (*rpb2*).

2.4. Sequence Alignment and Phylogenetic Analysis

Newly obtained ITS sequences were BLAST searched against NCBI GenBank's standard *nr/nt* nucleotide database (https://blast.ncbi.nlm.nih.gov/Blast.cgi, accessed on 9 August 2020), resulting in three isolates of *Inocybe glabripes* Ricken (sect. *Hysterices* Stangl & J. Veselský) as top results, which shared between 91.49% and 91.90% identity (GenBank accession numbers KX602255, MH216096, MN947389). Following this result, ITS, nrLSU, and *rpb2* sequences of *Inocybe* sect. *Hysterices* species [26,31,33–35] were downloaded for phylogenetic analysis.

Sequences were aligned by locus using MUSCLE version 3.7 [36], available through the Cipres Science Gateway [37]. Next, sequences in the ITS dataset were trimmed at the conserved motifs 5′–CATTA–3′ (3′ end of the nrSSU) and 5′–GACCT(CAAA …)–3′ (5′ end of the nrLSU) [38]. Because the different portions of the ITS spacer region (the two spacers and 5.8S) have different rates of evolution [39,40], the ITS1 and ITS2 spacers and the 5.8S conserved gene were extracted and treated as individual partitions in the multilocus analysis. Sequences in the nrLSU dataset were also trimmed to start with the conserved motif 5′–GACCT(CAAA …)–3′. Ambiguously aligned regions were removed using trimAl version 1.3 [41], with -gt = 0.6 and -cons = 0.5.

Evolutionary models for nucleotide substitution were selected for each partition (ITS1, 5.8S, ITS2, nrLSU, *rpb2*) using ModelFinder Plus [42], considering the Akaike Information Criterion. The data for each locus were combined in MEGA7 [43] to create a supermatrix of 2752 characters for 28 isolates representing ten species in *Inocybe* sect. *Hysterices* and two species in *Inocybe* sect. *Lactiferae* serving as outgroup taxa (details in Table 1). Maximum likelihood (ML) was inferred under partitioned models using IQ-TREE 1.6.7 [44,45]. Ultrafast bootstrapping was done with 1000 replicates [46].

Table 1. Overview of *Inocybe* isolates used in phylogenetic analyses. Newly generated sequences are in boldface. [T] stands for type specimens.

Species	Section	Isolate	Locality	ITS	nrLSU	*rpb2*
Inocybe aeruginascens	*Hysterices*	JG270502	Germany	GU949590	JN974970	
Inocybe aeruginascens	*Hysterices*	JG310508	Germany	GU949591	MH220256	MH249787
Inocybe aeruginascens	*Hysterices*	PC111007	South Africa	GU949592	MH220257	
Inocybe chondroderma	*Hysterices*	PBM1760	British Columbia	GU949586	MH220258	
Inocybe chondroderma	*Hysterices*	PBM1776	Washington	GU949579	JN974967	MH249789
Inocybe brijunica [T]	*Hysterices*	D. Haelew. F-1610a	Croatia	**MN749370**	**MN749492**	**MT878448**
Inocybe brijunica [T]	*Hysterices*	D. Haelew. F-1610b	Croatia	**MN749371**	**MN749493**	**MT878449**
Inocybe dulciolens [T]	*Lactiferae*	PBM2646	Tennessee	MH216088	MH220265	MH249796
Inocybe dulciolens	*Lactiferae*	PBM2450	New York	MH216087	MH220264	MH249795
Inocybe dulciolens	*Lactiferae*	LVK13340	New Jersey	MH216084	MH220261	MH249792
Inocybe erinaceomorpha	*Lactiferae*	EL128/05	Sweden	AM882735	AM882735	
Inocybe erinaceomorpha	*Lactiferae*	JV14756F	Sweden	MH216089	MH220266	MH249797
Inocybe glabripes	*Hysterices*	JV7318F	Finland	MH216096		MH249803
Inocybe hystrix	*Hysterices*	HRL11842	Quebec	KX897428		
Inocybe hystrix	*Hysterices*	PBM3300	North Carolina	GU949588	MH220275	
Inocybe hystrix	*Hysterices*	RS31493	Finland		AY380380	AY337381
Inocybe hystrix	*Hysterices*	SJ020824	Sweden	AM882810	AM882810	
Inocybe aff. *hystrix*	*Hysterices*	REH7405	Costa Rica	GU949589	JN974969	MH249806

Table 1. *Cont.*

Species	Section	Isolate	Locality	ITS	nrLSU	*rpb2*
Inocybe melanopus [T]	*Hysterices*	Stz3641	Washington		HQ201359	
Inocybe melanopus	*Hysterices*	BJ920904	Sweden	AM882725	AM882725	
Inocybe melanopus	*Hysterices*	JV4986	Finland	AM882727	AM882727	
Inocybe melanopus	*Hysterices*	PBM3975	Tennessee		MH220276	MH249807
Inocybe melanopus	*Hysterices*	TAA185135	Estonia	AM882726		
Inocybe aff. *pallidobrunnea*	*Hysterices*	PBM1957	Washington	MH216098	MH220277	MH249808
Inocybe aff. *pallidobrunnea*	*Hysterices*	PBM2242	Washington	MH216099	JN974968	MH249809
Inocybe sp.	*Hysterices*	PBM578	Washington	MH216104	JN974961	MH249813
Inocybe sp.	*Hysterices*	TR170-02	New Guinea		JN974964	MH249814
Inocybe sp.	*Hysterices*	TR180-02	New Guinea		JN974965	

3. Results

3.1. Phylogenetic Inference

The final multilocus dataset (Supplementary File S1) consists of 2752 characters, of which 425 are parsimony-informative and 2197 are constant. The number of total and parsimony-informative characters by locus as well as their selected evolutionary models as selected by ModelFinder Plus are presented in Table 2. The best-scoring ML tree (-lnL = 8722.045120) is shown in Figure 1. The topology is mostly congruent with Matheny and Kudzma [26], although support has improved for certain nodes. *Inocybe brijunica* sp. nov. is retrieved as a sister species of *I. glabripes* with maximum support. This set (*I. brijunica*, *I. glabripes*) is highly supported as sister to the clade holding *I. chondroderma* D.E. Stuntz ex Matheny, Norvell & E.C. Giles and *I.* aff. *pallidobrunnea* Kauffman.

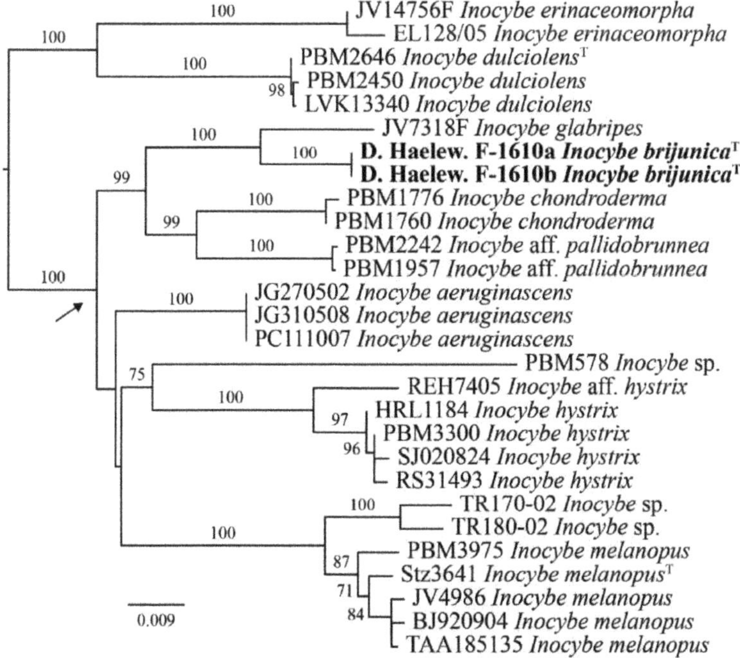

Figure 1. The best-scoring ML tree (-lnL = 8722.045120) of *Inocybe* sect. *Hysterices* (represented by the arrow) reconstructed from a concatenated ITS–nrLSU–*rpb2* dataset of 28 isolates. The tree topology is the result of ML inference performed in IQ-TREE. For each node, the ML bootstrap (≥70) is presented above or in front of the branch leading to that node. The new species is in boldface. [T] stands for type specimens.

Table 2. Overview of number of characters (total, informative, constant) and selected model of nucleotide substitution, by locus.

Locus	Sequences	Sites	Informative	Constant	Model	-lnL
ITS1	23	246	92	134	HKY + F + G4	1306.690
5.8S	23	158	4	153	TIM3e	249.420
ITS2	23	203	79	113	TPM3u + F + G4	1034.365
nrLSU	25	1379	94	1257	TN + F+I	3068.786
rpb2	16	766	156	540	TN + F+I	2935.101

3.2. Taxonomy

Inocybe brijunica Mešić, Tkalčec & Haelew., sp. nov.

Figures 2–4.

Mycobank MB838152

Typification: CROATIA. ISTRIA COUNTY: Brijuni National Park, Veli Brijun Island, 44°55′04″ N 13°46′33″ E, on the edge of grassland and forest of *Quercus ilex* L. with a few planted *Pinus pinea* L. trees along the forest edge, 16 November 2016, *A. Mešić & Z. Tkalčec* (holotype, CNF 1/7345; isotype, PUL F27673). GenBank (ex-isotype DNA isolate D. Haelew. F-1610a): nrSSU = MN749503, ITS = MN749370, nrLSU = MN749492, *rpb2* = MT878448; (ex-isotype DNA isolate D. Haelew. F-1610b): nrSSU = MN749504, ITS = MN749371, nrLSU = MN749493, *rpb2* = MT878449.

Etymology: Referring to the Brijuni archipelago, where the holotype was collected.

Description: Pileus 15–22 mm wide, obtusely (sub)conical with inflexed margin when young; convex to plano-convex, subumbonate and with deflexed margin at maturity; margin entire, occasionally with short radial splits; surface dry, finely radially fibrillose at first, then intensely fibrillose, rimulose to rimose, finally often partially cracked in small, shallow patches showing the paler flesh underneath; mostly medium brown, often with orange (fulvous brown) or reddish tones, less often light or dark brown where more deeply cracked; younger basidiomata often with rather inconspicuous, fibrillous, whitish veil remnants in marginal zone. Lamellae adnexed, subcrowded, L = ca. 40–50, l = 1–3, (sub)ventricose; whitish at first, then pale yellowish brown, finally light brown; edges fimbriate to slightly eroded, ± concolorous with sides. Stipe 17−30 × 2.5−4.5 mm, subcylindrical with slightly to moderately broadened base (up to 7 mm, sometimes submarginate); solid, surface dry, white flocculose at apex, becoming whitish longitudinally fibrillose toward base, fibrils more scattered with age, beneath the fibrils pale to medium brown; with more or less developed orange to dull orange–red or orange–red–brown, adhering, membranaceous surface layer (possibly a remnant of universal veil), at or just above the basal part of the stipe; basal tomentum scanty, whitish. Cortina (partial veil) present in young basidiomata, fibrillous, white, fugacious. Context cream, light brown when moist, not changing color on bruising, not darkening on drying. Odor spermatic. Taste not recorded.

Figure 2. *Inocybe brijunica* (CNF 1/7345, holotype). (**A**) Basidiomata in situ. (**B**) Basidiomata in lab. Bars: A, B = 10 mm.

Basidiospores [300/3/1] (6.2–)6.6–7.5–8.8(–9.8) × (4–)4.3–4.7–5.3(–5.7) μm, averages of different basidiomata 7.3–7.6 × 4.6–4.7 μm, Q = 1.35–1.6–1.96, av. Q = 1.58–1.62, a few very large spores occasionally present (up to ca. 12 × 7 μm); in frontal view ellipsoid, oblong or ovoid with rounded to subacute base and rounded to acute apex, in side view amygdaliform to phaseoliform, rarely subellipsoid, with rounded base and rounded to acute apex, sometimes subangulate in both views (especially in upper part); smooth, often with small, rather indistinct, apical germ-pore, moderately thick-walled (up to 0.8 μm), pale yellow–brown in KOH, pale brown in H$_2$O, non-amyloid and non-dextrinoid. Basidia 20–30 × 6.5–9 μm, clavate, predominantly 4-spored, occasionally 2-spored, thin-walled, hyaline to yellowish. Pleurocystidia of lamprocystidia-type, very abundant, [90/4/1] 34–50–65(–70) × 9–14–21 μm, Q = 2.39–3.65–5.45, predominantly utriform, lageniform, or fusiform, with obtuse apex of 6–9(–11) μm wide, sometimes (sub)clavate, conical with obtuse apex, narrowly ellipsoid or subcylindrical, in alkaline solutions mostly (sub)hyaline, less often with slightly yellowish wall, sometimes with dirty yellow cytoplasmic pigment, with strongly to poorly developed crystals at apex (soluble in KOH, rarely lacking), thick-walled, wall most often gradually thickened towards the apex (up to 1–5.5 μm). Lamellar edge heterogeneous. Cheilocystidia of two types: (a) lamprocystidia similar to pleurocystidia (although more often without crystals), scattered to abundant, and (b) leptocystidia (paracystidia) 9–30 × 5–14 μm, mostly clavate, less often (sub)fusiform or utriform, hyaline to subhyaline, thin to moderately thick-walled (up to ca. 0.6 μm), scattered to abundant. Pileipellis a cutis, composed of repent, thin-walled, smooth to minutely encrusted, hyaline to pale yellow–brown hyphae, 1–5(–7) μm wide. Cells of the upper part of pileal context with brown, intracellular and partially also encrusted extracellular pigment (brown pigmented layer ca. 80–150 μm wide). Stipitipellis a cutis, composed of repent, thin-walled, smooth, ca. 2–10 μm wide hyphae. Caulocystidia mostly abundant (often crowded) in upper 2–3 mm of stipe length, sparsely present toward middle of the stipe; many in the form of lamprocystidia, quite similar to pleurocystidia, others very variable, narrowly utriform, lageniform, (sub)cylindrical, clavate, urticoid or rather irregular, sometimes with subcapitate apex, some septate, hyaline, thin- to moderately thick-walled (up to ca. 1 μm); 12–100(–150) × 4–20 μm. Clamp connections present, conspicuous, rather abundant in all tissues.

Distribution and ecology: Thus far only known from the holotype collection. Ectomycorrhizal, found in the Mediterranean region of Croatia (Europe), on the island of Veli Brijun in Brijuni National Park, on the edge of *Quercus ilex* forest, with a few planted *Pinus pinea* trees, edging a neighboring grassland. An ITS sequence with accession number MH310748 [47], identified as *Inocybe* sp., from Italy shares 99% identity with *I. brijunica* (identities = 684/688 bp, gaps = 4/688) and may indicate a broader distribution in the Mediterranean basin.

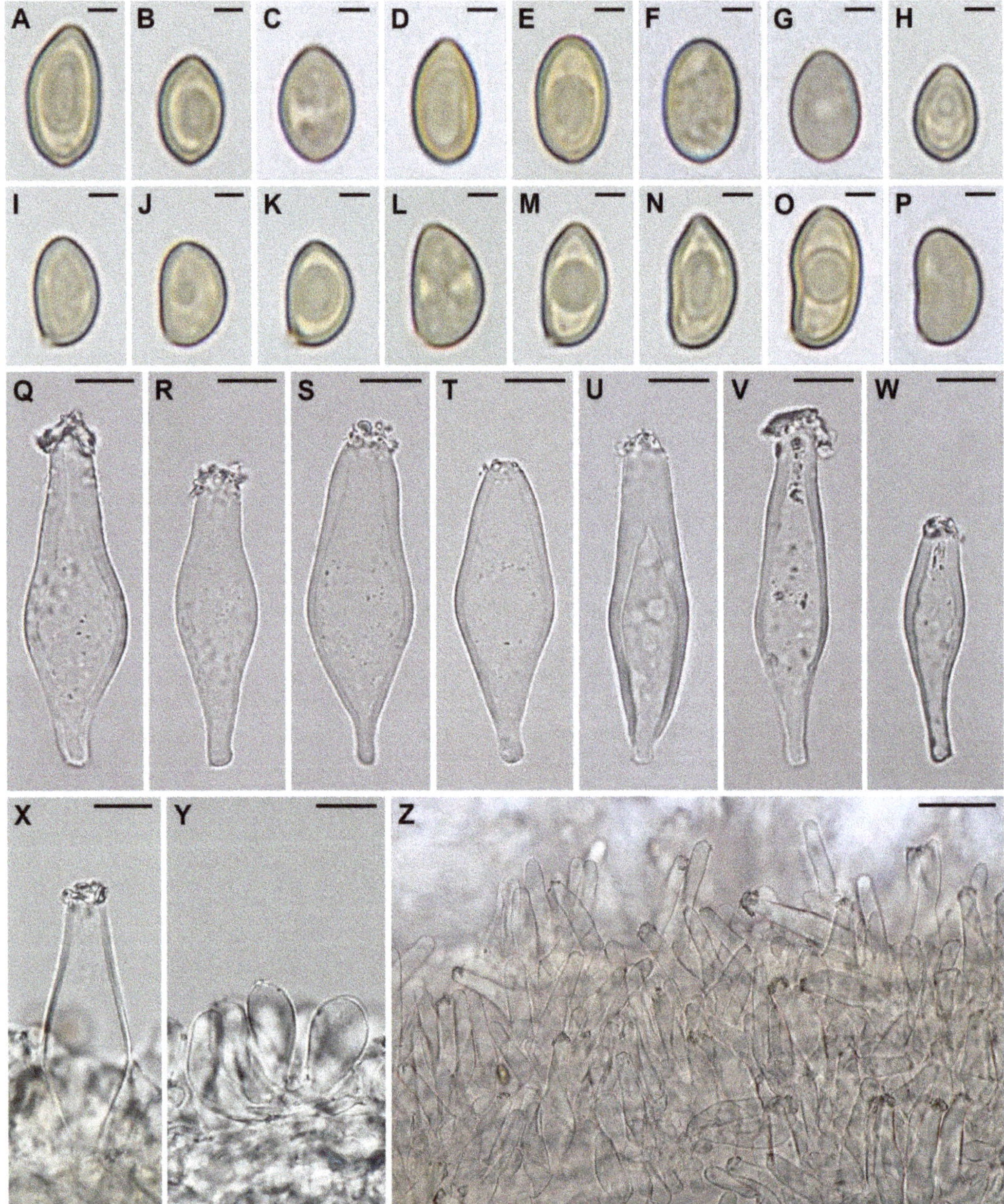

Figure 3. Inocybe brijunica (CNF 1/7345, holotype). (**A–H**) Basidiospores in frontal view. (**I–P**) Basidiospores in side view. (**Q–W**) Pleurocystidia. (**X**) Cheilolamprocystidium. (**Y**) Cheiloleptocystidia. (**Z**) Caulocystidia. Bars: (**A–P**) = 2 μm, (**Q–Y**) = 10 μm, Z = 30 μm.

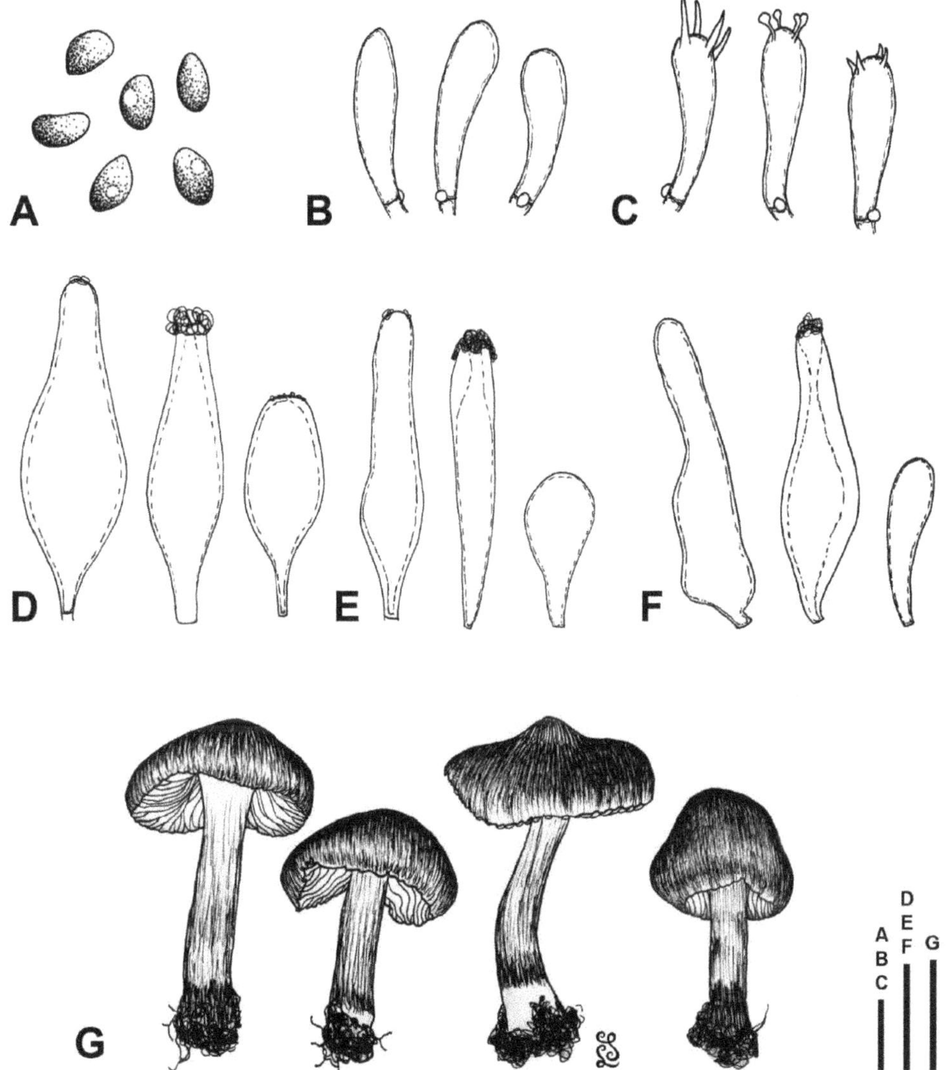

Figure 4. *Inocybe brijunica* (CNF 1/7345, holotype). (**A**) Basidiospores. (**B**) Basidioles. (**C**) Basidia. (**D**) Pleurocystidia. (**E**) Cheilocystidia). (**F**) Caulocystidia. (**G**) Basidiomata. Bars: (**A–C**) = 10 μm; (**D–F**) = 20 μm; (**G**) = 10 mm.

4. Discussion

The results of our multilocus phylogenetic analysis and morphological study place *I. brijunica* in sect. *Hysterices* [48]. Basidiomata produced by species belonging to this section (in the original sense) possess a squamulose pileus and stipe, lack violaceous tones, and have amygdaliform basidiospores. Matheny and Kudzma [26] emended the section to include taxa with non-squarrose basidiomata. Macromorphologically, *I. brijunica* can be easily recognized from all other *Inocybe* species by a conspicuous orange to orange–red–brown membranaceous surface layer present at or just above the basal part of the stipe. Other important morphological characters are: medium brown pileus with radially fibrillose to rimose surface; pale to medium brown stipe with slightly to moderately broadened (or sometimes submarginate) base; presence of fugacious cortina in young basidiomata; spermatic odor; color of context unchanged upon bruising; relatively small, amygdaliform to phaseoliform (and sometimes subangulate), smooth basidiospores (ca. 6.5–9 × 4–5.5 μm); pleurocystidia as utriform, lageniform, or fusiform, thick-walled (up to 1–5.5 μm) lamprocystidia, mostly with crystals and not yellowing in alkaline solutions; cheilocystidia of two types (lamprocystidia and leptocystidia); and presence of abundant caulocystidia only in the upper 2–3 mm of stipe length.

The basidiospores of the morphologically and phylogenetically closest species, *I. glabripes*, are very similar in size, measuring ca. 6–8 × 4–5 μm [49,50], but they are readily distinguished by being amygdaliform but not phaseoliform as in *I. brijunica*. In addition, the cystidial walls of *I. glabripes* are thinner (up to 2(–2.5) μm thick). *Inocybe glabripes* is a widespread species occurring in parks and open woodlands on predominantly alkaline soils from the Mediterranean region to the boreal zone of Europe [51], which forms ectomycorrhizae exclusively with broadleaved trees. So far, the species has been found in symbiotic relationship with trees in the genera *Betula* L., *Fagus* L., *Populus* L., *Quercus* L., *Tilia* L. [49–51], and with *Castanea sativa* Mill. [52]. It can be expected that its sister species *I. brijunica* also forms ectomycorrhizal relationships only with broadleaved trees, such as *Quercus ilex* at the holotype locality.

Inocybe pseudobrunnea Alessio, which grows under *Abies alba* Mill. (Pinales, Pinaceae), has a similar phaseoliform, but somewhat larger basidiospores, 8.5–10.5(11) × 4.5–6 μm, and its cystidia are rather bright yellow in ammonia solution [53], a characteristic that rarely occurs in *I. brijunica. Inocybe gracilenta* E. Ludw., only known from the type collection in Sweden, a damp locality under *Alnus* sp. (Fagales, Betulaceae), *Populus tremula* L. and *Salix* sp. (Malpighiales, Salicaceae), has amygdaliform but not phaseoliform basidiospores, which are otherwise similar in size compared to *I. brijunica*, 7–8.5(9.5) × 4.5–5.5 μm [51]. Additional differences are the papillate pileus and the slenderer (up to 2 mm wide) and white to faintly cream-colored stipe [51]. The Mediterranean species *I. barrasae* Esteve-Rav., described from Spain and fruiting in spring (April–May) in thermophilous Mediterranean *Quercus–Cistus* forests, has amygdaliform-shaped basidiospores that are larger and more elongated (8–11.5 × 4.5–5.5 μm, av. Q = 1.85), and bright yellow-colored cystidia in ammonia solution [54]. *Inocybe aerea* E. Ludw., another species only known from the holotype collection in Germany [51], has amygdaliform to ovoid and slightly larger basidiospores (7.5–10.5 × 5–6 μm), an ochraceous yellow and more slender (up to 2 mm wide) stipe, and thinner-walled cystidia (walls 0.2–2 (–3) μm thick). The North American species *I. pyrotricha* Stuntz [55] has orange to cinnamon or rusty–red fibrils on the stipe, like *I. brijunica*, but differs by slightly larger (7.5–10 × 4.5–6 μm) basidiospores, longer pleurocystidia (66–80 × 13.5–16.5 μm), and violaceous tinges in young lamellae and upper parts of the stipe.

Supplementary Materials: The following supplementary material is available online at https://www.mdpi.com/2309-608X/7/3/199/s1, File S1: Aligned, concatenated dataset of *Inocybe* sect. *Hysterices*, consisting of five partitions (ITS1, 5.8S, ITS2, nrLSU, *rpb2*).

Author Contributions: Conceptualization, A.M., D.H., and Z.T.; methodology, A.M., D.H., Z.T., and J.L.; phylogenetic analysis, D.H.; data curation, D.H., I.K., and A.P.; writing—original draft preparation, A.M., D.H., and Z.T.; writing—review and editing, A.M., D.H., Z.T., I.K., M.C.A., A.P.; visualization, D.H., Z.T., and J.L.; supervision, D.H. and A.P.; project administration, D.H.; funding acquisition, A.M., D.H., and M.C.A. All authors have read and agreed to the published version of the manuscript.

Funding: This work was supported in part by the Croatian Science Foundation (HRZZ-IP-2018-01-1736 to A.M., Z.T., I.K., A.P.; HRZZ-2018-09-7081 to A.P.), the National Science Foundation (DEB-2018098 to D.H.), and the USDA National Institute of Food and Agriculture (Hatch project 1010662 to M.C.A).

Acknowledgments: A.M. and Z.T. are grateful to Sandro Dujmović, former director of Brijuni National Park, for research support and to Martina Hervat for the help with the literature. P. Brandon Matheny (University of Tennessee–Knoxville, USA) is thanked for his generous help with North American *Inocybe* species.

Conflicts of Interest: The authors declare no conflict of interest. The funders had no role in the design of the study; in the collection, analyses, or interpretation of data; in the writing of the manuscript, or in the decision to publish the results.

References

1. Brijuni National Park Official Website. Available online: https://www.np-brijuni.hr/en/brijuni (accessed on 14 January 2021).
2. Blondel, J.; Aronson, J.; Bodiou, J.-Y.; Boeuf, G. *The Mediterranean Region—Biological Diversity in Space and Time*, 2nd ed.; Oxford University Press: New York, NY, USA, 2010.
3. Giorgi, F. Climate change hot-spots. *Geophys. Res. Lett.* **2006**, *33*, 1–4. [CrossRef]
4. Tuel, A.; Eltahir, E.A.B. Why is the Mediterranean a climate change hot spot? *J. Clim.* **2020**, *33*, 5829–5843. [CrossRef]
5. Mariotti, A.; Zeng, N.; Yoon, J.-H.; Artale, V.; Navarra, A.; Alpert, P.; Li, L.Z.X. Mediterranean water cycle changes: Transition to drier 21st century conditions in observations and CMIP3 simulations. *Environ. Res. Lett.* **2008**, *3*, 044001. [CrossRef]
6. Brogli, R.; Sørland, S.L.; Kröner, N.; Schär, C. Causes of future Mediterranean precipitation decline depend on the season. *Environ. Res. Lett.* **2019**, *14*, 114017. [CrossRef]
7. Antonelli, A.; Fry, C.; Smith, R.J.; Simmonds, M.S.J.; Kersey, P.J.; Pritchard, H.W.; Abbo, M.S.; Acedo, C.; Adams, J.; Ainsworth, A.M.; et al. *State of the World's Plants and Fungi 2020*; Royal Botanic Gardens: Kew, UK, 2020. [CrossRef]
8. Hawksworth, D.L. The fungal dimension of biodiversity: Magnitude, significance, and conservation. *Mycol. Res.* **1991**, *95*, 641–655. [CrossRef]
9. Cheek, M.; Nic Lughadha, E.; Kirk, P.; Lindon, H.; Carretero, J.; Looney, B.; Douglas, B.; Haelewaters, D.; Gaya, E.; Llewellyn, T.; et al. New scientific discoveries: Plants and fungi. *Plants People Planet* **2020**, *2*, 371–388. [CrossRef]
10. Saba, M.; Haelewaters, D.; Pfister, D.H.; Khalid, A.N. New species of *Pseudosperma* (Agaricales, Inocybaceae) from Pakistan revealed by morphology and multi-locus phylogenetic reconstruction. *MycoKeys* **2020**, *69*, 1–31. [CrossRef] [PubMed]
11. He, M.Q.; Zhao, R.L.; Hyde, K.D.; Begerow, D.; Kemler, M.; Yurkov, A.; McKenzie, E.H.C.; Raspé, O.; Kakishima, M.; Sánchez Ramírez, S.; et al. Notes, outline and divergence times of Basidiomycota. *Fungal Divers.* **2019**, *99*, 105–367. [CrossRef]
12. Bandini, D.; Oertel, B.; Ploch, S.; Ali, T.; Vauras, J.; Schneider, A.; Scholler, M.; Eberhardt, U.; Thines, M. Revision of some central European species of *Inocybe* (Fr.: Fr.) Fr. subgenus *Inocybe*, with the description of five new species. *Mycol. Prog.* **2018**, *18*, 247–294. [CrossRef]
13. Bandini, D.; Oertel, B.; Moreau, P.-A.; Thines, M.; Ploch, S. Three new hygrophilous species of *Inocybe*, subgenus *Inocybe*. *Mycol. Prog.* **2019**, *18*, 1101–1119. [CrossRef]
14. Bandini, D.; Oertel, B.; Ploch, S.; Thines, M. *Inocybe heidelbergensis*, eine neue Risspilz-Art der Untergattung *Inocybe*. *Z. Mykol.* **2019**, *85*, 195–213.
15. Bandini, D.; Oertel, B.; Schüssler, C.; Eberhardt, U. Noch mehr Risspilze: Fünzehn neue und zwei wenig bekannte Arten der Gattung *Inocybe*. *Mycol. Bavarica* **2020**, *20*, 13–101.
16. Bandini, D.; Sesli, E.; Oertel, B.; Krisai-Greilhuber, I. *Inocybe antoniniana*, a new species of *Inocybe* section *Marginatae* with nodulose spores. *Sydowia* **2020**, *72*, 95–106. [CrossRef]
17. Bandini, D.; Vauras, J.; Weholt, Ø.; Oertel, B.; Eberhardt, U. *Inocybe woglindeana*, a new species of the genus *Inocybe*, thriving in exposed habitats with calcareous sandy soil. *Karstenia* **2020**, *58*, 41–59. [CrossRef]
18. Cripps, C.L.; Larsson, E.; Vauras, J. Nodulose-spored *Inocybe* from the Rocky Mountain alpine zone molecularly linked to European and type specimens. *Mycologia* **2019**, *112*, 133–153. [CrossRef]
19. Crous, P.W.; Carnegie, A.J.; Wingfield, M.J.; Sharma, R.; Mughini, G.; Noordeloos, M.E.; Santini, A.; Shouche, Y.S.; Bezerra, J.D.P.; Dima, B.; et al. Fungal Planet description sheets: 868–950. *Persoonia* **2019**, *42*, 291–473. [CrossRef]
20. Crous, P.W.; Cowan, D.A.; Maggs-Kölling, G.; Yilmaz, N.; Larsson, E.; Angelini, C.; Brandrud, T.E.; Dearnaley, J.D.W.; Dima, B.; Dovana, F.; et al. Fungal Planet description sheets: 1112–1181. *Persoonia* **2020**, *45*, 251–409. [CrossRef]
21. Krieglsteiner, L.G. *Inocybe calosporoides*—Ein neuer Risspilz aus Portugal. *Südwestdeutsche Pilzrundsch.* **2019**, *55*, 68–72.
22. Krieglsteiner, L.G. Nomenclatural novelties. *Index Fungorum* **2019**, *411*, 1.
23. Matheny, P.B.; Hobbs, A.M.; Esteve-Raventós, F. Genera of Inocybaceae: New skin for the old ceremony. *Mycologia* **2019**, *112*, 83–120. [CrossRef] [PubMed]
24. Clémençon, H. *Cytology and Plectology of the Hymenomycetes*, 2nd ed.; Cramer: Stuttgart, Germany, 2012.
25. Erb, B.; Matheis, W. *Pilzmikroskopie*; Kosmos: Stuttgart, Germany, 1982.
26. Matheny, P.B.; Kudzma, L.V. New species of *Inocybe* (Inocybaceae) from eastern North America. *J. Torrey Bot. Soc.* **2019**, *146*, 213–235. [CrossRef]
27. White, T.J.; Bruns, T.; Lee, S.; Taylor, J. Amplification and direct sequencing of fungal ribosomal RNA genes for phylogenetics. In *PCR Protocols: A Guide to Methods and Applications*; Innis, M.A., Gelfand, D.H., Sninsky, J.J., White, T.J., Eds.; Academic Press: New York, NY, USA, 1990; pp. 315–322. [CrossRef]
28. Egger, K.N. Molecular analysis of ectomycorrhizal fungal communities. *Can. J. Bot.* **1995**, *73*, S1415–S1422. [CrossRef]

29. Vilgalys, R.; Hester, M. Rapid genetic identification and mapping of enzymatically amplified ribosomal DNA from several *Cryptococcus* species. *J. Bacteriol.* **1990**, *172*, 4238–4246. [CrossRef]
30. Hopple, J.S. Phylogenetic Investigations in the Genus Coprinus Based on Morphological and Molecular Characters. Ph.D. Thesis, Duke University, Durham, NC, USA, 1994.
31. Matheny, P.B. Improving phylogenetic inference of mushrooms with RPB1 and RPB2 nucleotide sequences (*Inocybe*, Agaricales). *Mol. Phylogenet. Evol.* **2005**, *35*, 1–20. [CrossRef]
32. Haelewaters, D.; Toome-Heller, M.; Albu, S.; Aime, M.C. Red yeasts from leaf surfaces and other habitats: Three new species and a new combination of *Symmetrospora* (Pucciniomycotina, Cystobasidiomycetes). *Fungal Syst. Evol.* **2020**, *5*, 187–196. [CrossRef] [PubMed]
33. Matheny, P.B.; Norvell, L.L.; Giles, E.C. A common new species of *Inocybe* in the Pacific Northwest with a diagnostic PDAB reaction. *Mycologia* **2013**, *105*, 436–446. [CrossRef] [PubMed]
34. Ryberg, M.; Matheny, P.B. Asynchronous origins of ectomycorrhizal clades of Agaricales. *Proc. R. Soc. B Biol. Sci.* **2012**, *279*, 2003–2011. [CrossRef] [PubMed]
35. Ryberg, M.; Nilsson, R.H.; Kristiansson, E.; Töpel, M.; Jacobsson, S.; Larsson, E. Mining metadata from unidentified ITS sequences in GenBank: A case study in *Inocybe* (Basidiomycota). *BMC Evol. Biol.* **2008**, *8*, 50. [CrossRef] [PubMed]
36. Edgar, R.C. MUSCLE: Multiple sequence alignment with high accuracy and high throughput. *Nucleic Acids Res.* **2004**, *32*, 1792–1797. [CrossRef]
37. Miller, M.A.; Pfeiffer, W.; Schwartz, T. Creating the CIPRES Science Gateway for inference of large phylogenetic trees. In Proceedings of the Gateway Computing Environments Workshop (GCE), New Orleans, Louisiana, 14 November 2010; Institute of Electrical and Electronics Engineers: Piscataway, NJ, USA, 2010; pp. 1–8. [CrossRef]
38. Dentinger, B.T.; Didukh, M.Y.; Moncalvo, J.M. Comparing COI and ITS as DNA barcode markers for mushrooms and allies (Agaricomycotina). *PLoS ONE* **2011**, *6*, e25081. [CrossRef]
39. Hillis, D.M.; Dixon, M.T. Ribosomal DNA: Molecular evolution and phylogenetic inference. *Q. Rev. Biol.* **1991**, *66*, 411–453. [CrossRef]
40. Haelewaters, D.; Dirks, A.C.; Kappler, L.A.; Mitchell, J.K.; Quijada, L.; Vandegrift, R.; Buyck, B.; Pfister, D.H. A preliminary checklist of fungi at the Boston Harbor islands. *Northeast. Nat.* **2018**, *25*, 45–76. [CrossRef]
41. Capella-Gutiérrez, S.; Silla-Martínez, J.M.; Gabaldón, T. TrimAl: A tool for automated alignment trimming in large-scale phylogenetic analyses. *Bioinformatics* **2009**, *25*, 1972–1973. [CrossRef] [PubMed]
42. Kalyaanamoorthy, S.; Minh, B.Q.; Wong, T.K.F.; von Haeseler, A.; Jermiin, L.S. ModelFinder: Fast model selection for accurate phylogenetic estimates. *Nat. Methods* **2017**, *14*, 587–589. [CrossRef]
43. Kumar, S.; Stecher, G.; Tamura, K. MEGA7: Molecular Evolutionary Genetics Analysis version 7.0 for bigger datasets. *Mol. Biol. Evol.* **2016**, *33*, 1870–1874. [CrossRef] [PubMed]
44. Nguyen, L.-T.; Schmidt, H.A.; von Haeseler, A.; Minh, B.Q. IQ-TREE: A fast and effective stochastic algorithm for estimating maximum likelihood phylogenies. *Mol. Biol. Evol.* **2015**, *32*, 268–274. [CrossRef] [PubMed]
45. Chernomor, O.; von Haeseler, A.; Minh, B.Q. Terrace aware data structure for phylogenomic inference from supermatrices. *Syst. Biol.* **2016**, *65*, 997–1008. [CrossRef]
46. Hoang, D.T.; Chernomor, O.; von Haeseler, A.; Minh, B.Q.; Vinh, L.S. UFBoot2: Improving the ultrafast bootstrap approximation. *Mol. Biol. Evol.* **2017**, *35*, 518–522. [CrossRef]
47. Wurzbacher, C.; Larsson, E.; Bengtsson-Palme, J.; Van den Wyngaert, S.; Svantesson, S.; Kristiansson, E.; Kagami, M.; Nilsson, R.H. Introducing ribosomal tandem repeat barcoding for fungi. *Mol. Ecol. Resour.* **2018**, *19*, 118–127. [CrossRef]
48. Stangl, J.; Veselský, J. Risspilze der Section *Lilacinae* Heim. *Česká Mykol.* **1982**, *36*, 85–99.
49. Kuyper, T.W. A revision of the genus *Inocybe* in Europe 1. Subgenus *Inosperma* and the smooth-spored species of subgenus *Inocybe*. *Persoonia* **1986**, *3*, 1–247.
50. Stangl, J. Die Gattung *Inocybe* in Bayern. *Hoppea* **1989**, *46*, 1–409.
51. Ludwig, E. *Pilzkompendium. Band 4*; Fungicon Verlag: Berlin, Germany, 2017.
52. Ferrari, E.; Bandini, D.; Boccardo, F. *Inocybe (Fr.) Fr., Terzo Contributo*; Fungi non delineati 73/74; Edizioni Candusso: Alassio, Italy, 2014.
53. Alessio, C.L. Complemento allo studio del Genere *Inocybe*: 8° contributo. *Riv. Micol. Assoc. Micol. Bresadola* **1987**, *30*, 79–89.
54. Esteve-Raventós, F. Two new species of *Inocybe* (Cortinariales) from Spain, with a comparative type study of some related taxa. *Mycol. Res.* **2001**, *105*, 1137–1143. [CrossRef]
55. Smith, A.H.; Stuntz, D.E. New or noteworthy fungi from Mount Rainier National Park. *Mycologia* **1950**, *42*, 80–134. [CrossRef]

MDPI

St. Alban-Anlage 66

4052 Basel

Switzerland

Tel. +41 61 683 77 34

Fax +41 61 302 89 18

www.mdpi.com

Journal of Fungi Editorial Office

E-mail: jof@mdpi.com

www.mdpi.com/journal/jof